Prison Systems

A Comparative Study Of Accountability in England, France, Germany, and The Netherlands

Jon Vagg

CLARENDON PRESS · OXFORD
1994

Oxford University Press, Walton Street, Oxford OX2 6DP
Oxford New York
Athens Auckland Bangkok Bombay
Calcutta Cape Town Dar es Salaam Delhi
Florence Hong Kong Istanbul Karachi
Kuala Lumpur Madras Madrid Melbourne
Mexico City Nairobi Paris Singapore
Taipei Tokyo Toronto
and associated companies in
Berlin Ibadan

Oxford is a trade mark of Oxford University Press

Published in the United States
by Oxford University Press Inc., New York

British Library Cataloguing in Publication Data
Data available

Library of Congress Cataloging in Publication Data
Vagg, Jon.
Prison systems: a comparative study of accountability in England,
France, Germany, and The Netherlands / Jon Vagg.
(Clarendon studies in criminology)
Includes bibliographical references and index.
1. Prison administration—European Economic Community countries—
Case studies. 2. Prisons—European Economic Community countries—
Finance—Case studies. 3. Prisoners—Legal status, laws, etc.—
European Economic Community countries—Case studies. 4. Prison
discipline—European Economic Community countries—Case studies.
I. Title. II. Series.
HV9640.5.V34 1994 365–dc20 94-7712
ISBN 0–19–825674–4

Set by Hope Services (Abingdon) Ltd.
Printed in Great Britain
on acid-free paper by
Biddles Ltd.

General Editor's Introduction

IT is not often that a new criminological series emerges. It is now over fifty years since Leon Radzinowicz and J.C.W. Turner began *English Studies in Criminal Science* under the Macmillan imprint, which, after ten volumes, became the internationally renowned *Cambridge Studies in Criminology*. Forty-one further volumes were published under the editorship of Sir Leon by Heinemann Educational Books, led by the distinguished publisher, the late Alan Hill. Ten more volumes, published by Gower and Avebury books, appeared under the general editorship of Anthony Bottoms before the prestigious Cambridge Series was unexpectedly and sadly wound-up.

The Cambridge Institute of Criminology then approached Oxford University Press in the hope that the series could be continued in another form at a time when, co-incidentally, the Press had begun to discuss the prospects for a criminological series with the Mannheim Centre for Criminology and Criminal Justice at the London School of Economics. With the energetic support of Richard Hart, these two institutions decided on a joint venture with the Oxford Centre for Criminological Research in which each would provide the members of the Editorial Board for a new series to be called *Clarendon Studies in Criminology*. I was honoured to have been asked by my colleagues (whose names are listed next to the title page) to be the General Editor for the first three years.

Clarendon Studies in Criminology aims to provide a forum for outstanding work in all aspects of criminology, criminal justice, penology, and the wider field of deviant behaviour. It will welcome works of theory and synthesis as well as reports of empirical enquiries and will be international in its scope. The first titles, Jon Vagg's *Prison Systems: A Comparative Study Of Accountability in England, France, Germany, and the Netherlands* and Philip Schlesinger and Howard Tumber's *The Media Politics of Criminal Justice* already indicate the potential range of the Series.

In this book, Jon Vagg provides the first comparative analysis of the way in which England and Wales, France, Germany and the

Netherlands attempt to render their prison systems accountable for the way in which they exercise their coercive power. It is based on fieldwork and an extensive knowledge of penal, legal, administrative and sociological sources in each of these countries. He describes the similarities and differences in the law and systems of administration, management, inspection, grievance and disciplinary procedures, as well as other measures for controlling prisoners. Dr Vagg deals not only with the structures of accountability but also with the way in which they work in practice within the realities of each penal system, how they impact upon and are perceived by both prisoners and prison staff and what happens when systems of control break down in riots and disturbances. The discussion is informed by a sophisticated conception of the role and meaning of accountability within the penal context. Of particular interest is his exploration of the relationship between 'macro-accountability'—control over prisons—and 'micro-accountability'—control within the prisons, to show how the notion of accountability is employed as a game or strategy by different parties with an interest in the system. The editors are delighted to introduce a book which not only fills a large gap in our knowledge but also brings a comparative perspective to bear on a subject which is fundamental to the legitimacy of penal systems.

Roger Hood

Contents

List of Tables

Acknowledgements

THE seeds of this book were sown in 1985, shortly after Mike Maguire, Rod Morgan, and I finished editing a book called *Accountability and Prisons*. That edited collection was an examination of various aspects of prisons' accountability in England, but for me it raised many more questions than it answered. In particular, it seemed to me that the English prison system was rather different—not so much in its organization as its world view and its social norms—from those in continental Europe or North America, and the continental and American experiences of imprisonment were anyway very far apart. Two related issues interested me. First, and bearing in mind that most European countries were signatories to various European conventions on prison matters, how similar or different were their prison systems in practice? Second, were there any general propositions that could be made about the nature or forms of accountability in relation to prisons?

The first step in this direction was a Council of Europe bursary which took me on brief trips to France and what was then West Germany. I was subsequently awarded a grant from the Economic and Social Research Council (No. E00232186). This paid for the bulk of the empirical research I did in England, France, West Germany, and the Netherlands in 1986–8. In 1988 I began to write up some of my material. I soon began to realize how slippery many of the concepts were, and how difficult it would be to provide any systematic description, let alone theoretical argument, on questions of accountability. During this process I moved from England to Hong Kong, had children, co-edited a couple of other books, and was frequently distracted from the work at hand. Every time I went back to my material I had to incorporate comments on new developments. In the time it has taken me to write this book, the English prison system alone has experienced two waves of major disturbances, introduced Fresh Start, opened several privatized prisons, and acquired a new 'executive agency' status. So this book is compiled partly from field research which is now some five years old and partly from subsequent letters, faxes, phone calls, visits, conferences, and general scurrying about. A large number of people have helped me at various times. In England, my mainstay was Rod Morgan, now at the University of Bristol. Brian Caffarey, in the Prisons Department, was very helpful at a number of stages, both in providing me with materials and in making factual corrections to some chapters I contributed to others' books (I have reused the material in a somewhat different form here). James Jacobs, at New York University Law School, attracted my attention to a number of useful items I would otherwise have missed. Richard Hart, at Oxford University Press, must have regretted ever accepting the book for publication but took every broken deadline in his stride. It was, however, David Downes, at the London School of Economics, and my colleague Carol Jones in Hong Kong, who finally goaded me into deciding that I should finish the book and have done with it.

In Germany, I relied heavily on Frieder Dünkel—then at the Max Planck Institute in Freiburg, now at the University of Griefswald—for advice, explanations, and practical help. Frieder was also the primary organizer of a series of seminars on prison affairs which helped me greatly in keeping up my comparative work on Europe while working in Hong Kong. These seminars took place in Germany in 1989, Poland in 1990, and Czechoslovakia in 1992. Professor Dr Günther Kaiser was also kind enough to provide a grant which enabled me to employ two research assistants (Uli Baumann and Ilse Klär) who conducted and translated most of the German interviews. On my return to England, Fee Hornig also spent a great deal of time translating interviews and documents for me.

In France, I owe most to Claude Faugeron at CESDIP—the Centre de Recherches Sociologiques sur le Droit et les Institutions Pénales, a CNRS research institute in Paris. She made many of the necessary practical arrangements for my visits to French prisons and Ministry of Justice officials. Monique Seyler, at CESDIP, and Dominique Bibal and Martine Barbarin, at the Ministry of Justice, all dealt efficiently with my many naïve queries and presented me with many kilos of reports and research materials. The Centre de Recherche Interdisciplinaire de Vaucresson, just outside Paris, was kind enough to provide me with accommodation and library facilities during most of my visits to France.

In the Netherlands, Max Kommer (at the WODC, the Ministry of Justice Research and Documentation Centre) was my main contact, though many others were helpful at different stages of the work. André Rook and Maria Brand-Koolen were particularly sympathetic at the beginning of my Dutch fieldwork. Most of the interviews in Dutch were done by Ralph Vossen, who also co-authored one conference paper with me. Constantijn Kelk, at the University of Utrecht, periodically updated me on Dutch developments.

The librarians at the Max Planck Institute, Freiburg, CESDIP in Paris, the Centre Interdisciplinaire de Vaucresson, and the WODC at the Ministry of Justice in The Hague were all extremely helpful not only in dealing with half-formed and badly-expressed requests but also in bringing my attention to material that I would otherwise have missed. The loose coalition of prisons researchers that has developed around Frieder Dünkel, and now meets more or less annually to discuss different aspects of imprisonment, has been a great help. Many individuals from that group have supplied me with material; and collectively, our discussions have stimulated some of the thoughts presented in this text. A great many other colleagues, friends, and institutions too numerous to list individually have also helped me at various points and in different ways.

Comparative research is fraught with difficulties at the best of times. I hope misunderstandings and factual errors have been removed from the text by repeated cross-checking of information with a large circle of people. It should go without saying, of course, that I bear responsibility for any that are left. And I should also say that many of those who provided me with data do not share the conclusions I have felt able to draw from such data.

J.V.

Hong Kong
March 1993

1

Introduction: Definitions and Problems

I HAD best start with answers to two questions: what do I mean by accountability, and why is it worth writing about in relation to prisons?

The idea of accountability is, above all else, a quality that defines certain kinds of relationship. It conveys the idea that certain individuals have a duty to report on, explain, and justify to others actions they have taken or decisions they have made. And it conveys the idea that this duty is imposed because those individuals have authority, powers, and perhaps budgets which are not theirs by right, but which form part of their occupational role.[1]

From this point on the idea rapidly becomes more complex, because the nature of those duties, reports, explanations, and justifications varies considerably across different areas of social life. The idea of accountability shifts radically as one moves between professional groups such as doctors and lawyers; state-sponsored provision such as education, social welfare, or housing; expert and scientific services such as factory or nuclear-installation inspectorates; and 'disciplined' services such as policing or prisons. It shifts in terms of the nature of the activities for which one is

---

[1] I remain agnostic about where these powers are delegated from. This is in itself an extraordinarily complex question. While, in a democratic state, one can argue that power ultimately rests with the electorate, it is clear that other power bases also exist. It is also fair to point out that in other relatively common state forms, such as military dictatorship, officials may be accountable to a head of state who is not himself accountable to anyone. And although most of the discussion in this book concerns powers delegated to government officials, it is clear that other forms of accountability also exist, as where—for example—one party to a commercial contract takes the other to court. In such a situation the machinery of the courts is used to bring one party to account to for its actions; and in such cases the account required must be cast not only in terms of the specifics of the contract, but also broader legal frameworks such as the law of contract.

accountable, to whom the 'service providers' can be required to report and explain, and the kinds of relationship that the 'service-providers' have with those to whom they render accounts.

Discussions of accountability must distinguish between these different kinds of relationship. One way of doing this is to use concepts such as 'stewardship', 'partnership', and 'directive' modes of accountability (terms which will be explained in greater detail below). These concepts reflect different emphases in relation to the accounting for decisions in individual cases as against the making of policy; the degree of professional or expert knowledge required to ask pertinent questions and understand the answers; the extent of democratic or lay control over service delivery; the relations between subordinates and superiors (for example, in policing and prisons) as opposed to the relations between professionals (as where doctors or lawyers appear before professional disciplinary bodies); and the respective rights, duties, and powers of the various participants in the 'accountability process'.

But even to take the one example of prisons, the number of different forms of accountability involved is large and any systematic discussion necessarily lengthy. Prison services are usually thought of as hierarchical; but at the topmost levels of the bureaucracy, although the director of the service 'fronts' for the service as a whole, many if not most decisions are made by committee, and almost everybody, therefore, has some say in others' spheres of interest. Relationships between prison systems and other bodies are equally complex. Prison system compliance with prison laws may be scrutinized by the courts (though precisely what sort of court varies across countries). Compliance with prison rules can be dealt with through several channels—organizational, legal, ministerial, and ultimately parliamentary—depending upon the rules' legal status. At the same time, prison service directors undoubtedly have a large hand in the drafting of those laws and rules. Accountability for policy is rather more diffuse, depending upon the nature of the policies, the role of various actors both within and outside prisons in formulating them, and the presence of a range of bodies (including, for example, the prison inspectorate in England) with the capacity to scrutinize both policies and their implementation. Moreover, in most countries prison authorities must comply with broader governmental policies and laws concerning matters including health and safety at work, freedom of information, and fiscal policies. And in addition to all this, one must remember the bottom line of parliamentary and ministerial control, the former often via committees with strong powers and wide terms of reference.

It is hardly possible in the space of one book to deal adequately with all these issues. I have concentrated on three broad areas in which I felt able to make substantial progress. One is broadly managerial, and is concerned with the linkages between different forms of accountability within the prison organization and the treatment of inmates within the prison system. A second is legal, and looks at the impact of structures of law, the role of the courts, and the uses of legal processes to influence the treatment of inmates and the development of penal policies. A third is political—not in the immediate sense of how ministers make policy or parliamentary committees sniff out problems, but in the broader matters of the linkages between accountability relationships, the setting of agendas about 'what the problems are', and the creation of solutions. This clearly leaves much untouched—in particular the questions of accountability for budgets and expenditure, and of patterns of accountability and decision-making behind the closed doors of ministerial and parliamentary committee rooms. But I feel no compunction in leaving others to tackle such areas.

While accountability is a complicated topic, complexity does not necessarily make it either interesting or important. None the less it is both, for the following reasons. Intellectually, we may characterize a study of accountability, to borrow Foucault's term, as a study in 'power/knowledge'. We know a great deal about the nature of control in prisons, and we know that flows of 'knowledge'—primarily about prisoners—are important in determining the ways in which controls are exercised. Accountability is thus the underside of control. None the less, research which explicitly links information and communication issues to the exercise of power is comparatively rare. A study of accountability is essentially a study of the information that the prison system gathers about its own operation, the ways that it uses that information, and the ways that that information is used to control prisons and, more generally, penal policy. It is therefore a study of the linkage between power and knowledge within the system, and that which is exercised over the system by political actors.

At the same time, a study of accountability also has a pragmatic appeal. The literature on prisons in England and elsewhere clearly indicates that major problems are endemic. Some have an immediately human dimension: overcrowding, suicides, inmate violence, riots, industrial action, and so on. Others are less dramatic, but none the less important, such as the rapid increase over the last 10 to 15 years in the prison budgets of many Western countries, despite the apparent lack of corresponding improvements in the position of either inmates or staff. These are issues which systems of accountability have addressed, which have often been the subject of new policy initiatives, but which are still with us and will be for the foreseeable future. A study of accountability is, therefore, a contribution to the blunt question: why is it so difficult to reform our prisons?

Accountability and History

Accountability is not a concept unique to contemporary forms of state organization. None the less, under those forms of organization it has assumed a very specific shape, and a brief historical excursion may help to set the scene for what follows.

The 'modern' nation-state, as Giddens (1985) notes, came into being relatively recently. It is typical of such a state that it is administered by a bureaucracy, operating according to fixed and written rules and maintaining files and records on individuals and on its internal operations. The concept of accountability pre-dates the modern nation-state by many centuries, as Normanton (1966) demonstrates in his discussion of mediaeval government. But the advent of the bureaucratically administered state with speedy systems of internal communication allowed a high degree of centralization of decision-making; a requirement to create documents detailing actions taken at subordinate levels; and the possibility for superiors to have recourse to those files as a means both of supervising subordinates and of generating data pertinent to policy-making. All this became possible in the first half of the 1800s, though in the four countries I am concerned with, complete centralized authority over the prison system had to wait for some decades, and took different forms in each country, influenced by a range of internal political factors including arguments about appropriate specifications for prison regimes; the ending of transportation in England and France; state budgets; central and local-government relations; and the formation of modern Germany as a federal republic.

Changes in the English penal climate—for example, the ending of transportation as a punishment, and the debates on the silent and separate systems—led to the creation of a prison system under centralized state control, and set in train the events leading to central control over all prisons.[2] That central control over a 'penal archipelago' was administered, of course, with the aid of those new bureaucratic technologies. In many areas—from new building specifications to prison regimes—attempts at standardization were made. The use of standard forms and styles of report, instructions on what to report, and inspectors to monitor the system, all came together into the 'underside' of control, the obligation to produce accounts of one's actions over an increasingly wide span of issues. Older forms of prison oversight survived, but changed their character significantly. Perhaps the best known English example is that of the Boards of Visitors. The old Visiting Magistrates, 'boards of governors' for local prisons, were the model for the Boards of Visitors of the centrally administered convict prisons (though they had fewer executive powers). Progressively undermined by the centralization of the local prison system, they were transformed into committees of local volunteers acting as the 'eyes and ears of the Home Secretary' at the level of the individual establishment, and ultimately themselves became administered and bureaucratized from the centre (Maguire and Vagg 1984).

In the Netherlands, the process of centralization appears to have been completed even more quickly. Franke's (1990) history of Dutch prisons suggests that the state was strong enough by the 1850s for parliamentary debates about the use of the Pennsylvania system of separate confinement to have been reflected in systematic changes in penal practice. More recently, experiences of large-scale incarceration during both World Wars kept parliamentary and social debates about imprisonment alive, while by the 1970s, steps in the direction of lay supervision of prisons had been taken. The structure of Dutch society, however, has tended to restrict the role

---

[2] See McConville (1981) for a detailed history of these events. Descriptions of the claims and counter-claims made over the silent (Pennsylvania) and separate (Auburn) systems of confinement can be found in Stefani et al. (1982). Garland (1985: 12) notes that after 1877, the Prison Commissioners in England began not only to standardize employment contracts for staff and designs for new prisons, but also to reorganize prison regimes broadly along the lines of the 'separate' system— that is, work in association but in silence. Certain prisons—in particular Pentonville—had earlier been designed with the 'silent' system in mind.

of the courts as such, for reasons explained in the following chapters.

In France, the process of centralization started earlier but took much longer. The criminal code of 1808 and the penal code of 1810 envisaged a joint system of centralized control over long-term prisons, and local, municipal control over the remainder. In fact the short-term prisons from 1811 were the responsibility of départements. The whole penal system came under centralized budgetary control in 1855, apparently as a result of departmental reluctance or inability to fund the short-term prisons. This, of course, gave the state a greater say in the management of prisons, and an 1875 law required the introduction of the 'separate system'—though this appears to have been largely ignored despite subsequent legislative attempts to enforce it. It was only in the process of reconstruction after World War II that the prison system as a whole become a single, centralized state system (Stefani *et al.* 1982: 416-18).[3] None the less, local influence over prisons continued at least in a symbolic form, by means of supervisory committees drawn from the local government structure, while the Napoleonic Code has been influential in giving the courts control over some aspects of prison policies which can be described as having to do with rehabilitation and reintegration.

Kaiser *et al.* (1983: 46-9) indicate that state control over prisons in Prussia was progressively imposed between 1804, with the publication of a general plan for the introduction of improved criminal law and the improvement of prison and punishment institutions, and 1870, the date of a new state-wide law introducing the Pennsylvania system. The German system is exceptional, however, in that the concept of the 'nation-state', post-World War II, has been diluted by the strong powers of the individual constituent Länder of the Federation and the constitutional restrictions on the ability of the federal government to make both law and policy. None the less, all the prison systems are governed by federal prison law and the Land governments collectively agreed a common set of prison rules in 1976-7 (which replaced earlier and also commonly-agreed rules). Though each Land maintains its own prison system, and there is no federal system, there is a high degree of formal and legal comparability across them. In this sense the rise of the 'mod-

---

[3] See also the accounts by Mestre (1987) and Pierre (1991).

ern' era of accountability is both similar and different to that of the other three countries I consider; similar in terms of the 'administrative technology' and comparability of formal and legal aspects of prison operation, but different in terms of how that uniformity has been achieved and, importantly for the later discussions in this book, different in the emphasis given to law as the dominant mode of accountability.

To be sure, the idea of a nation-state does not revolve simply around administrative centralization and the 'new technology of supervision'. But those forms of centralization, uniformity, and supervision, combined with the idea of the rule of law, and perhaps more importantly, the expansion of parliamentary democracy and the retention (and sometimes re-invention) of local or lay supervision, have together shaped the very ideas and forms of accountability that I have summarized above and will be discussing in later chapters.

One other point is also relevant to the development of systems of accountability: the rise of the 'expert' and the institutionalization of a variety of expert professions. Groups such as social workers and psychologists are now employed alongside prison officers in most penal institutions. Groups such as management consultants and academic criminologists are used in a variety of capacities; despite occasional publicly-expressed cynicism about the expertise of the latter, they are appointed to advisory committees, called upon to undertake research identified by the prison system as important, and sometimes become—formally or informally—part of a ministerial 'cabinet' or 'think tank'. Some pressure groups, capable of activating influential support networks, have substantial expertise in areas such as political lobbying. And the penal system itself has built up an internal expertise which has great significance, as we shall see, in the establishment of groups such as prison inspectorates. The growth of 'expert' scrutiny of prison systems has had two broad effects. First, much of what experts such as psychologists do in treating inmates can only, it is often alleged, be understood and evaluated by other similarly-trained professionals. And at the same time, much of that which goes on in prisons generally can only be properly analysed and interpreted by means of technical procedures used by persons such as academic researchers, or evaluated by persons with lengthy prison-professional experience, such as prisons inspectors. Despite the idea of lay participation built into, for example, the directive model, the idea that persons without relevant expert training have something to contribute is being increasingly eroded. Second, the proliferation not simply of experts but also of interest groups has placed the civil service, as Collins (1980) implies, increasingly in the position of being the arbiters of which forms of expertise should be given access to prisons, and which interest groups should be recognized as parties to serious debate.

Accountability and Theory

Although this book contains no new theory of accountability, a study of this sort must necessarily rest on some theoretical foundations. I shall in due course devote a complete chapter to this subject. In brief, however, the following are the major positions I shall be drawing upon.

First, I referred earlier to a linkage between power and knowledge, and specifically to accountability as the 'underside' of control. These references clearly resonate with Foucault's (1979) formulation of power/knowledge. Foucault argues that forms of knowledge and forms of power can be mapped precisely onto one another. The differentiation of groups of inmates from each other, the creation of privilege systems and systematic variation of regimes, and the keeping of records of individual behaviour and performance are structures of power, in that they enable institutional managers to manipulate inmates. They also, of course, specify the kinds of sanctions that can be imposed at every level up to and including the threat of force to ensure compliance, and the use of force to deal with non-compliance. At the same time, they are structures of knowledge. They systematically employ the notion of the 'official knowledge' of inmates, based on written records and supplemented by the observations of officials; they are a specific 'form' of knowledge, enabling both the tracking of individual cases over a period of time and the creation of statistics.

Yet such systems are not merely a form of power/knowledge in relation to inmates. They enable institutional managers to know about and control staff—which staff report inmate disciplinary infractions most frequently? which are most diligent? which encourage the best performance from inmates in their care? And, one level higher, they enable a central administration to lay down standards for the operation of establishments, and to make comparisons between their managers, while also creating the possibility for debates to take place about standards of treatment, such as those in the latter part of the 1800s concerning the imposition of the Pennsylvania system. As Franke (1990) reports, in the Netherlands at any rate, such debates were conducted using a variety of statistics including, for example, rates of suicide and insanity among prisoners as arguments for the abolition of the Pennsylvania model. The power/knowledge form is clearly a crucial part of any theoretical approach to accountability, albeit that in this context I shall apply it not to the 'control' of inmates, but to the relationships existing between prison governors and headquarters, and between the prison system as a whole and those outside it who have the capacity to influence its shape.

Second, since my conception of accountability is primarily one of the quality of a relationship, the question arises of how relationships might be described in order to distinguish pertinent qualities. Morgan and Maggs (1985) propose a threefold typology of relationships: of direction, stewardship, and partnership.[4] The directive mode is one where A is 'subordinate and obedient' to B. In practical prison terms, elected representatives would instruct the director of the service as to policies, and provide a framework of rules to govern the exercise of discretion by decision-makers. In principle the director of the service would have only 'executive' and not policy-making functions, however these two might be distinguished from each other. The stewardship concept suggests that the head of a service retains substantial discretion in policy as well as day-to-day matters, but is periodically 'audited' by the competent authority—whether an elected body or a professional organization appointed ultimately by that body—to ensure that policies have been made and implemented with regard to the prevailing standards of fiscal regularity, humanity, propriety, and so forth. The partner model is one in which the head of service retains executive powers but is expected to work closely with one or more representative

---

[4] Their model is based on the discussion of political forms of accountability in Marshall (1984). However Marshall concentrates on two forms. 'Executive' accountability, which corresponds to Morgan and Maggs' 'direct' form, implies direct control over the events for which accountability is required. 'Explanatory' accountability, which partly corresponds to Morgan and Maggs' 'stewardship', implies that a minister, for example, has no direct personal involvement in relevant matters but can call for (largely descriptive or justificatory) accounts which may then be presented to Parliament.

bodies in identifying and addressing issues in areas such as policy priorities, 'service delivery', and good practice. What interests might be represented on the 'partner' bodies, and how individuals are selected for membership, can be fairly variable.

Almost always, the relationships between prison systems and other bodies include elements of all three types. However, the distinction is theoretically important in that the major options, between them, cover every kind of relationship having to do with accountability: in crude terms, 'did you do what I told you?', 'how did you solve the problem?', and 'let's work on this together until we have a practical and honourable solution'. And in addition, they cover the three main problems that accountability relationships can encounter; directive models are open to self-interested manipulation by those who hold power in the relevant committees; stewardship is a *post hoc* exercise in which there is little room for intervention in the operational, day-to-day exercise of discretion; and partnership, if not buttressed by stronger powers on the side of the service's 'partner', leaves effective control in the hands of a service which may be obliged to consider, but not to follow, the partner's advice.

Third, it must also be said that these typologies are not enough to understand the way in which accountability is 'done'. Relationships are not static; if interactionist sociology has shown us anything, it is that the roles people possess must also be performed in order to accomplish anything. Being director or directed, steward or principal, executive or non-executive partner, are all roles that are performed in respect of particular sets of issues. The fundamental metaphors of interactionism are those of the performance, the play and the game. These terms can be used to describe the ways in which relationships are not simply connections between structural positions, but also ways of pursuing self-interest. I do not mean this cynically; and I do not mean to suggest that things are never what they seem, or that everyone has hidden agendas. It is simply that the scope of a relationship, the use of powers determined in law, and the provision of information, are all matters of negotiation in so far as any formal or legal provisions are both enabling and constraining on both parties.

These three broad approaches—power/knowledge, Morgan and Maggs's typologies, and interactionist perspectives—might not be expected to sit together very easily. I shall argue that they do, and, in brief, this is how. The typologies are a kind of 'ground zero'; they encompass almost the whole range of political relationships pertaining to accountability within the modern nation-state. They clearly leave to one side the question, raised earlier, of the role of expertise; this is incorporated only via the idea of the expert 'audit' conducted by groups recognized and acting under the general authority of the same principal to whom the 'steward' reports. The rise of the expert whose work can only be evaluated by other experts is most often discussed in situations such as complaints against doctors and the inspection and evaluation of highly technical facilities such as nuclear power plants. None the less, it is important enough in the prison context to be counted as a fourth type of relationship, since largely self-regulating professions, such as social work and psychology, have come to be included in the operation of prisons. Such professions are not easily amenable to detailed external scrutiny, and at the political level are often reduced to crude factors such as specialist-to-inmate ratios and the overall costs of service provision.

Yet whatever the general provisions governing the relationships between the various parties, they must still be implemented in the particular circumstances of individual prisons, where both (or all) parties can use the laws and rules as a resource to enable, as well as a constraint on, possible courses of action. The micro-level negotiations thus remain very important as the 'facework' of building and maintaining relationships in which accountability is claimed to inhere. And finally, the concept of power/knowledge is relevant as a tool to understand how both the micro- and macro-levels of negotiated relationships form structures in which the exercise of power leads to specific kinds of reporting, explanation, and justification, while those reports are the basis on which the exercise of power is made not only intelligible, but possible.

Languages of Accountability

Day and Klein (1987) make the point that, in many cases, there is little agreement as to the currency in which accountability must be paid. What the criteria of evaluation should be is frequently discussed as intently as the matter being evaluated. Many micro-sociological studies of prisons provide examples of this in face-to-face interaction between inmates, guards, and senior officers. Perhaps the best recent example is that provided by King and McDermott (1990), in which a potted plant became the subject of the creative use of prison rules by both an inmate and staff. In decisions on individual cases, often made on the basis of sets of written regulations, it remains true that no rule can legislate its own interpretation. There are always marginal cases, cases not specifically provided for, and cases where arguably the observance of the rules goes against common sense. At this level, there is likely to be controversy about whether the circumstances of a particular case really fall properly within the area defined implicitly or explicitly by the rules, and whether, in the light of available information, any other decision could or should have been taken. The debate will be about matters such as the factors that were or should have been taken into consideration when deciding that the rule applied, or in its application. And where the policies of an organization come to be scrutinized, particularly where there are few agreements as to the organizational aims or where aims may be contradictory (as with the police; see e.g. Chatterton 1987) then even while there may be a consensus about the issues that a policy raises, there may also be debate about the criteria in terms of which those issues can be decided.

But the idea of the 'languages of accountability' goes deeper than questions about which set of criteria actions, decisions, or policies should be judged against. Language can be used in a symbolic as well as a rational debate. Imagine that a member of parliament asks an apparently straightforward question, such as how many inmates had access to integral sanitation in a given prison on a given date. Theoretically at least, a figure should exist which would, having regard to all considerations, be considered 'satisfactory'. That figure might be satisfactory in that it is much higher than it was two or three years previously, thus indicating that 'progress is being made'. In real life, of course, the point of asking such a question is that the questioner knows the answer will highlight some defect or problem within the prison service. Such questions are often asked (at parliamentary level in the UK) not for the answer to be measured against an agreed level of satisfactory performance, but as a strategy to highlight a problem and to place on the political agenda a criterion—progress towards integral sanitation—by which service performance is likely to be judged in future. A second example would the use of the term 'independent' to describe boards of lay persons who monitor prison establishments, or the constitution of a prison inspectorate. The point of using such a term is persuasive rather than descriptive; it is intended to signal that such bodies are credible and impartial observers and reporters of prison affairs, and are thus to be treated as a means of making the system accountable rather than simply being a tool of management. Yet the very word hides a series of problems: independent of whom? independent in what ways? independent in their ability to appoint their own members? independent in that they have the expertise to interpret their findings without outside help?

Problems in Accountability

What can go wrong with the way in which supervisory and other processes operate? To some extent the answers can be derived from the points made above. First, it is possible for the total picture of accountability to contain sufficient gaps—areas where no supervisory procedures operate—for abuses of power to go unchecked. And this may also occur where there is duplication of powers and two supervisory bodies each assume that the other has the primary responsibility for a particular area. Second, there may be disagreements within bodies charged with oversight which stall the process of accountability; by the time that a consensus is reached about how a particular problem should have been handled, the problem itself may have become ancient history. Third, there is a substantial literature on the tendency for regulatory agencies to become 'colonized' by the bodies they scrutinize (see for example Lynxwiler *et al.* 1983). One criticism made of prison inspection processes is that they are staffed by persons who have worked in prisons and who may seek an accommodation with the prison system. Here, the problem of expertise raises its head. The professional prison governor working in an inspectorate, for example, will be able to find out much more than an amateur; but he may also share assumptions and values with his erstwhile colleagues which blunt his critical sensibilities, or be unwilling to criticize situations which he himself would feel are difficult for the person who must deal with them operationally. And finally, at critical times, chains of command and channels of information may break down. Many inquiries conducted after prison riots bear witness to this (e.g. Home Office 1977; HM Chief Inspector of Prisons 1987).

But at a routine level, matters can also become critical without ever becoming so obviously out of order. A lack of information—the lifeblood of accountability—or even overburdening with information can lead to matters being passed over without comment, or put aside until there is time to deal with them, by which time the issues have become stale. Accountability delayed is no accountability.

Some Concrete Issues

Many previous discussions of accountability in prisons have focused on specific procedures, institutions, and pragmatic considerations such as disciplinary systems, inspectorates, and the utility of minimum standards for prison conditions. Although these topics will be addressed in following chapters, a brief introduction to them at this stage may lend substance to what has been a rather abstract introduction.

Prisoners' rights probably constitute the single most important area of concern. The principle is accepted, at least in the Western world, that offenders go to prison as a punishment and not for punishment. On the whole they retain all those rights that are not explicitly taken away by the fact of imprisonment, or which follow necessarily from the removal of liberty. In principle, while their freedom of movement is curtailed, their rights in relation to matters such as unlawful violence, torture, freedom of thought and expression, correspondence and so on are broadly preserved. Other areas, such as the right to work (which is elaborated in the European Social Charter of 1961) are not. However, and notwithstanding documents such as the European Social Charter and the European Convention on Human Rights, the extent to which prisoners may be said to have rights in such areas depends on the extent to which national laws recognize rights as such. In the case of England, which has no bill of rights and no written constitution, the extent to which such rights can be claimed and enforced must largely be deduced from common law principles, precedents, and the obligations placed by law upon, for example, the Home Secretary.

In addition, it is often argued that prisoners are in an unusual legal position. They are subject to a range of provisions that do not apply to the general population: censorship, prison disciplinary proceedings, parole provisions and restrictions on personal belong-ings are all examples. In such circumstances, it can be argued that they should also have fixed and recognized rights in relation to such procedures and situations. This is the nub of the argument for 'special rights for prisoners'. Rights, however, come in various forms (for a more detailed discussion, see Richardson 1984; 1985). For my purposes the major sense of a right is that which is brought about because there is a duty placed upon a person to act in a certain way: if prison officials are obliged to allow inmates to write to their lawyer there is a right to correspondence with a lawyer. In prison, few rights are quite so clearly drawn as in this example. Prisoners may have the right to other correspondence too: but it may be qualified by provisions concerning the content of letters, the persons to whom they are addressed, the notepaper used, a clear indication of the identity of the sender and so forth. In as much as prisoners have rights, qualified rights of this kind are the norm. Further, the enforcement of rights against 'bad' decisions by prison staff or administrators may also be problematic. This is especially so where there are lengthy delays in legal processes. It is by no means rare, when a prisoner seeks to take his case to court, for him to have served his sentence and been released by the time of the hearing. This is perhaps the most striking example of the problems of obtaining a remedy against the prison authorities' decisions, but others will be discussed in due course. In short, the whole question of the extent to which prisoners do have rights, should have rights, and should be able to enforce them is still a major issue for prison reform, and it is closely linked to the procedures for *complaints* both to the prison administration and to the courts.

One area which has been discussed with particular vigour in England is that of *prison disciplinary hearings* (see e.g. Fitzgerald 1985; Home Office 1985). This is closely linked with debates about prisoners' rights and grievance procedures. Prisoners are subject to a disciplinary code which creates offences of doing or saying things that, outside prison, would attract no penalty; and, in addition, some matters which are the subject of criminal sanctions can also be dealt with by the internal disciplinary code. Typically, standards required of evidence are less stringent than in the courts, and the person who conducts the hearing—in most cases a member of the prison governor-grade staff—has extensive powers of punishment which may (for example, in England) include powers to alter the date of release, via the withdrawal of remission of sentence given automatically by the administration. The role of the disciplinary procedure varies across prison systems, since other control measures, including inducements to inmates to behave themselves, are also structured differently across the four prison systems I shall be discussing. Moreover, the opportunities for complaint against the conduct of hearings and the decisions made vary not only across the four countries, but also with the nature of the complaint itself. Thus in all four countries, the rights of prisoners in relation to such procedures, and the precise nature of the proceedings, the powers of governors and the extent of any oversight of the internal disciplinary system, and indeed other control measures which are also available, are matters for debate if not concern.

In recent years, arguments have been put forward in favour of the establishment of *prison regime standards* (see e.g. Casale 1985). Although the United Nations has promulgated Standard Minimum Rules for the Treatment of Prisoners, and these were adopted by the Council of Europe (and amended in 1987, becoming the 'European Prison Rules'), these rules are quite general, and a wide range of practices and conditions might be regarded as meeting any particular rule. Indeed, given even the climatic and cultural differences between member states of the Council of Europe, this would have to be the case in relation to matters such as ventilation, clothing, bedding, and food. But prison conditions, either in general or for particular groups of prisoners, are far from ideal in England and France, and can still be criticized, though perhaps to a lesser extent or in more specific cases, in the Netherlands and West Germany. The argument for regime standards is that they provide a 'baseline' against which regimes can be evaluated and priorities for improvement established. The need for such a baseline was touched upon earlier in this chapter. Across Europe as a whole, the resources devoted to prisons have grown dramatically over the last twenty or so years. But in England at least, while the numbers of prison staff have increased massively over that time, and the staff:prisoner ratio has decreased markedly, prison conditions have, on the whole, worsened (King and McDermott 1989). The situation is not quite so clear-cut in the other countries, but the degree of overcrowding in, for example, French prisons must have had a deleterious effect on daily life in the institutions. And in the Netherlands, where the prison population is kept within prison capacity by the expedient of creating a system of 'waiting lists' to enter prison, there is periodic pressure to allow overcrowding to take place despite a massive prison building programme. Prison conditions, however, include 'performance' aspects of the regime, such as staff decision-making, as well as actual physical conditions. Discussions of the quality of regimes must necessarily include assessments of the ways in which staff deal with inmates. One area worthy of particular consideration is the perception, in England and the Netherlands at least, that prison staff are becoming progressively less prepared to interact with inmates. This, if true, may be for a number of reasons, including staff worries about the extent to which inmates are obtaining rights and are prepared to enforce them through complaints procedures and the courts. Yet if staff are gradually withdrawing from interaction with inmates, the nature of prison control may be adversely affected—and, for example, the opportunities for inmates to prey on each other may be increasing.

Lastly, there has been much discussion in recent years about the *oversight and supervision* of prison systems. Much of this debate centres around the single issue of the extent to which the procedures are 'independent' of the the prison system. Independence, in this context, means separate staffing and published reports. One example will serve to demonstrate the issue. In the 1970s, the English prison system had long been criticized for its secretiveness. The prison inspectorate at that time was internal to the department and its reports were not published. It was sometimes claimed, though with what truth it is now impossible to tell, that the inspectorate was ineffective because it had no independent voice. In 1978 the proposal that the inspectorate be taken outside the department was first made. Subsequent events are too lengthy to relate here (they are discussed in Morgan 1985), but suffice it to say that in 1981 this was done. The inspectorate was now 'independent'. However, the goalposts moved. Early criticism of the new inspectorate revolved around the fact that most of its staff were drawn from the prison department; that its reports took too long to be published; that some sections of the reports were not included in the published versions; and that since the Chief Inspector reported to the Home Secretary, who also had charge of the prison system, there was no true public accountability. The practical and constitutional questions that these points raise will be discussed in a later chapter. The point here is that the meaning of 'independence' is not fixed. It is one of the terms that, when accountability is being discussed, is subject to different interpretations and evaluation according to different criteria.

Beyond all this, there are of course prison problems that are not, or at least not wholly, the animals of prison administrations. Prisons cannot be blamed for the courts' decisions to imprison ever-increasing numbers of offenders. But they may have some scope in the ways that they deal with the numbers. There are, for example, differences between the four countries I shall discuss as to the amount of control the prison authorities can exercise over the allocation of remand prisoners. In England, in the last few years, pressure on remand places has seen remands scattered far from the courts that will hear their cases. In France (and, incidentally, in Scotland) the regulations governing prisons give the authorities much less latitude as to where remand prisoners are held. There are similar differences at the other end of the system. While prison authorities do not themselves make decisions on parole, or on amnesties, ministers have powers in relation to such areas which could be, and in some cases quite openly have been, used to reduce prison populations. Such points raise interesting questions about who is accountable for prisons as well as the accountability of prisons themselves. In relation to remand prisoners one might well ask how, in countries with legal controls on where they are held, information about overcrowding is disseminated, to whom, and how these controls can be changed. In relation to parole, interesting questions emerge about the extent to which powers formally held at ministerial level are routinely delegated to the prison authorities.

Rights, grievances, discipline, standards, inspection; these are some of the key terms that give accountability its substance in discussions of prison life. The agencies and mechanisms by which this substance is given effect are described in the next chapter, which provides some short pen-pictures of the practical arrangements through which accountability is pursued in the four countries I studied.

Conclusions

This book offers no conclusions of the formal, hypothesis-related, 'QED' kind. My intention is to illustrate the nature of the problems one is up against in trying to render prison systems account-able. They are not problems which can be solved, but only man-aged, often with answers in one conceptual domain creating new problems and questions in others.

The view that mechanisms of accountability should form a 'net-work', crisscrossing the social (and political, and economic) space of prisons and subjecting every aspect of prisons to scrutiny is appealing. But that is not the reality. It would be nice to think of it even as an imperfect network, with holes and gaps; but again, such a view cannot be sustained. Prisons are scrutinized by and through the major political and legal institutions and procedures of our time, in the light of their own concerns. These have variously to do with state budgets and expenditure, with minimizing exposure to politically damaging situations and disturbances, with asserting (or refusing to assert) the power of the courts, and with officials, in the expressive American phrase, covering their asses.

Prime among the current 'answers' to prison problems are regime standards and inmate rights. Both may well be desirable, but not on the grounds that they will in and of themselves improve prison conditions. The chapter on rights indicates that most rights thus far granted have to do with procedures rather than substan-tive aspects of prison regimes, and that whatever rights are granted, sufficient operational discretion will almost inevitably be retained to render them unenforceable if this is deemed opera-tionally necessary. Their main utility lies in their creation of norms, and in the upward pressure they exert on privileges, since prison staff need to maintain a level of provision which it is at their discretion to give or withhold. Much the same argument can be advanced for regime standards.

Many of the problems in prison are at least partly the result of policies pursued outside it, in terms of the length and frequency of custodial sentences. Although much has been written on this sub-ject, and in both England and France there have been political ini-tiatives to expand prison capacity, in truth there is probably less co-ordination in policy implementation in this kind of area than in many others because of the difficulties, certainly in England, atten-dant upon politicians attempting to tell the judiciary how to sen-tence. There are no easy solutions here either, and no quick progress can be expected. Meanwhile, prisoners, who are not faced with such delicate problems, can take action into their own hands.

It is notable that while English prison riots up to the early 1980s often occurred in relation to conditions in long-term establishments, in 1986 and 1990 they primarily involved short-sentence and remand inmates. And it is no less true for being unpalatable that riots can result in improved prison conditions and relaxations in prison regimes, providing of course that there is some reservoir of sympathy for prisoners within the media and among key public and political figures—or alternatively, as was the case in France in the 1970s, providing that there is a certain lack of resolution within the administration as to how to handle the situation.

In the last analysis, accountability is a matter of practical politics. The issues are often seen as only loosely, if at all, interconnected. They are played out on a range of stages, some in the glare of the media, some behind closed doors, and some forming the basis of day-to-day interactions between prison staff and their charges. And the differing currencies in which accountability is paid, not being freely translatable, considerably complicate the situation. It is these cans of worms that I shall, in the following chapters, try to bring to some kind of academic order.

The next two chapters are largely descriptive. One introduces the various participants in the systems of accountability in the four countries—England, France, the Netherlands, and Germany—that I an concerned with. The other is a view of recent prison history in the four countries, through a lens especially tinted to highlight accountability questions. Chapters 4, 5, and 6 mark a change of direction, being concerned with the vantage point of inmates, staff, and senior management respectively, and the issues that emerge for them. Chapter 7 moves from a descriptive to a more analytical approach. It reviews the material presented up to that point and proposes a more ordered framework for dealing with questions of accountability, and the four subsequent chapters try to consider selected issues from the perspective it offers. They deal with riots, rights and complaints, inspections and standards, and discipline and control. The remaining chapters tackle some of the questions raised by the recent and current change of attitude towards the privatization of prisons, and end with a review and discussion of the viability of some theoretical approaches towards issues of accountability.

2

Prison Systems and Participants in Accountability

THERE is substantial variation across the prison systems I am concerned with as to how they are organized, who has what powers, and why. I should therefore begin with a descriptive account of who the major participants are in the areas that the last chapter identified as having to do with accountability. The following sections deal in turn with management structures and organization, prison populations, relevant legal factors, laws and administrative rules, and the various actors whose work will be discussed in the following chapters.

Management Structures and Prison Organization

In England, the administration of prisons falls to the Home Office, the 'ministry of home affairs'. In France and the Netherlands, the prison services are lodged within ministries of justice. In Germany, the federal nature of the republic means that each of the constituent states (Länder) has its own prison service, responsible to the ministry of justice for the Land. While the prison law, the Strafvollzugsgesetz, is applicable nationally, the national ministry of justice plays a rather restricted role in prison affairs, and there is no federal prison system.

The interface between prison services and their political masters is managed in a variety of different ways. In England, until 1993, a director-general, his deputy, and the regional directors comprised the key members of the Prisons Board, which dealt with important policy, planning, and operational matters. In 1993, the Prison Department became an 'executive agency', headed by a CEO; the Prisons Board became a board of management, and a new supervisory board was created to handle the interface between the department and ministers. The introduction of 'private prisons' further complicated the situation, although the agency remained ultimately responsible for their operation. Despite these complexities, political accountability for the prison system remains with the Home Secretary, assisted by two junior ministers with specific responsibility for prison affairs.[1] In France, the Direction de l'Administration Pénitentiaire (DAP) is headed by a director who also has responsibility for the probation service. The director reports to the minister of justice, who has his own cabinet of advisers who are political appointees. In Germany, each Land prison system is a distinct organization coming under the control of the Land Minister of Justice. The Dutch prison service is one of the responsibilities of the Minister of Justice, and it functions, headed by its own director, as a division within the ministry.

Prison systems vary in size and in the formal differentiations between kinds of establishments. In the late 1980s, the English service had some 130 establishments; the French, about 150; and the Dutch, just over 60. Among West German systems prior to the reunification of Germany, Bremen and Saarland were the smallest, with under 1,000 inmates each. The largest were Nordrhein-Westfalen, with around 15,000 prisoners, and Bayern and Baden-Württemberg, with about 9,000 and 7,000 inmates respectively, and each with around 40 prisons. All three of these Länder had prison populations larger than that of the Netherlands.

Total figures for the numbers of establishments are complicated by the way in which 'establishments' are defined. For example, large German prisons often have one or more 'satellite' institutions, usually small open prisons, under their wing. The main establishment and the satellites are directed by the same governor. Kaiser *et al.* (1983) cite a figure of 162 establishments in 1981, though without saying whether this included or excluded satellites. There is a similar indeterminacy in the Netherlands. The establishment at Den Haag comprises three distinct buildings—a prison, a young offender centre and a prison hospital—and the prison itself is divided into three wings, each officially counted as a separate

---

[1] In addition to these changes, a number of other policies were being implemented, of which the most important was the privatization of parts of the prison system. Action was also being taken on some recommendations of the Woolf Report, esp. in relation to inmate complaints procedures. These changes are discussed in later chapters.

establishment even though there is still a governor who has overall responsibility for all three.

The formal differentiations between establishments require brief commentary. In England, remand centres and local prisons both hold persons on remand from courts in their catchment area, and the latter also take convicted prisoners from the courts; short-term prisoners serve their whole sentence there, and long-termers are placed there pending transfer to a training prison, often a lengthy wait. Training prisons offer, in principle, a less crowded environment and more training-oriented regime.[2] Inmates may go into progressively less secure accommodation, ending in an open training prison. The dispersal prisons—whose future has been debated on a number of occasions—are closed training prisons with high levels of security; they were the outcome of two reports on security in the 1960s (Home Office 1966; Advisory Council on the Penal System 1968), the latter of which argued for the dispersal of security-risk prisoners among a number of establishments.[3] Some establishments are multi-purpose (for example, male establishments with a female wing), while dispersal prisons are by definition also closed training prisons. In addition, there has been a trend towards the use of special units intended for distinct and relatively small populations. In the 1980s, 'vulnerable-prisoner units' became common, and special units were commissioned around the country to deal

[2] However, figures for training prison populations in the late 1980s suggested that in each of the four regions the system was then divided into, one training prison was used to absorb a much larger share of overcrowding than the others. I do not know whether this was done by intention or accident. (Figures provided by South West Regional Office, Prison Dept.) On the subject of the allocation of inmates to prisons it is also worth noting that during almost the whole of the 1980s a considerable number of remand and sentenced inmates were held in police stations because of prison overcrowding and/or industrial action, while many others were transferred to prisons far away from their home areas largely because of regional variations in overcrowding and attempts to reduce pressure on the congested London establishments.

[3] Though my study only involved male prisons, a word on female and youth establishments is worthwhile. Female establishments are either separate institutions, or discrete sections within prisons which also take males. There are only 11 female units or establishments, so that the degree of specialization found on the male side is not replicated here. Youth Custody Centres were established by the Criminal Justice Act 1982, replacing the older 'Borstals'. They take offenders up to the age of 21 who have YC sentences—determinate sentences of a minimum 4 months. The Detention Centre regimes are officially described as 'brisk', meaning that they operate higher levels of discipline. There are no DCs for girls.

with relatively small groups, usually of long-termers, with particular security and control needs.

In France, allocation of inmates to maisons d'arrêt is based on catchment areas for courts, and allocation of longer-term inmates to maisons centrales and centres de détention is based, albeit more loosely, on catchment areas for maisons d'arrêt. There is also a national 'observation centre' at Fresnes prison, which makes recommendations for the allocation of certain categories of prisoner, including all inmates whose sentence will run for ten years or more from the point at which it is confirmed. In principle the centres de détention offer less restrictive regimes than the maisons centrales, though it is now recognized that there are few real differences between them. Several establishments or parts of establishments are reserved for females or young prisoners, and some specialized centres exist for physically handicapped, ill, or aged prisoners, and psychopaths. There are no longer high-security centres within certain establishments, these having been the subject of prison disturbances in the 1970s. Inmates thought to require high security are kept in the general population, although they are more closely watched, and in some prisons special security procedures are in place, such as passbooks which go with the inmate and must be signed off by staff to indicate who is responsible for them at any one time. One final noteworthy feature of the French system is that within many of the maisons d'arrêt, inmates are segregated on racial lines. Thus Fresnes, on the outskirts of Paris, has several wings primarily for persons of North African descent, and one for French inmates.

German states, for the most part, operate a system in which specific establishments take inmates sentenced from particular courts, without any distinction between long- and short-term inmates. However, certain establishments may be given specific functions; for example, ethnic minorities and foreigners may be allocated to one or two specific prisons within each Land. Special establishments or parts of establishments are reserved for women. One effect of the decentralized approach is that the systems vary considerably in, for example, the amount of open prison accommodation they offer. Kaiser *et al.* (1983) show that the proportion of prisoners in open conditions (in the old West German Länder) varied from 0.6 to 22.4 per cent in 1979 (the national average was 8.5 per cent).

In the Netherlands, a Huis van Bewaring acts both as a remand centre and a place of custody pending allocation to a prison. However, there is not only a waiting-list system for entry into prison from the courts (the 'walking sentences'); a waiting list also exists for allocation from remand accommodation to prison. Thus many short-termers are released before they are transferred to a designated prison.[4] Young prisoners, females and adult male inmates are all kept in separate accommodation and do not mix. Some institutions (for example Over-Amstel, in Amsterdam) have several units for different populations on the same site. Two establishments—Veenhuizen and Maastricht—have specially designated segregation units, while Den Haag prison has the 'bunker'—a high security cell block. There is a minimum/medium/maximum security distinction made between the prisons, though even in high security establishments, physical security tends to be less intrusive than in comparable institutions in other countries.

In England, France, and the Netherlands there is a third layer of administration interposed between institutions and headquarters. In West Germany, at the time of my fieldwork, only one Land has such a division, with two regions. The English system is divided into fifteen area offices, each headed by an area manager and covering about nine establishments.[5] The French prison system has nine regions, each headed by a regional director with wide supervisory powers over individual establishments—though the very largest prisons, such as Fleury-Mérogis, are for some purposes (such as budgets) outside the regional structure. Although Dutch

---

[4] This system is described briefly by Brand-Koolen (1987). However, the Netherlands, like England, has also begun to use police cells to house remand and sometimes short-term convicted inmates. Thus there is now one waiting list for the transfer of prisoners from police cells to HvBs; one for the transfer of sentenced inmates from HvBs to prisons; and a third for persons not held in pre-trial custody awaiting places to serve their sentence. Cf. Kelk (forthcoming).

[5] This has been the position since 25 Sept. 1990. Prior to that, and during the time of my fieldwork, the prison system was divided into 4 regions, each responsible for 30-40 establishments. The regional offices accommodated a Regional Director, 3 or 4 assistant directors with specific areas of responsibility, a psychologist and assistant psychologist, and support staff. While a wide range of functions were devolved to the regional offices—for example, many staffing arrangements and most prisoner petitions were dealt with at that level—there was always some ambiguity about how important the region really was in the total scheme of prison affairs. In most important respects the Area Managers have similar responsibilities to the previous Regional Directors, although they exercise them in relation to fewer establishments (which they can presumably attend to more effectively).

establishments are grouped into four regions, this structure was only introduced during my period of fieldwork, and the regional directors were still feeling their way into their posts at that time. They did not have the formal powers granted to them in France nor even the formal roles defined for them in England. Their staffing comprised little more than a regional director, a 'penitentiary adviser' and secretarial support. In consequence, they had a comparatively free-floating, advisory role.

Populations, Throughput, Consequences

Table 2.1 gives some 'snapshot' descriptive statistics on the prison populations in the four countries. The figures give rise to the following observations.

First, the prison population is, in proportional terms, larger in England than in the other countries. However, sentence lengths were typically shorter than in France or the old West Germany. Comparison with the Netherlands is complicated by the Dutch pol-

TABLE 2.1. *Prison Populations and Related Data, 1 September 1989*

	England and Wales	France	Germany	Netherlands
Prison population	48,481	45,102	51,729	6,461
% change on previous year	−0.7	−2.8	−0.7	+10.9
Detention rate per 100,000 inhabitants	96.2	78.5	83.8	44.6
% unconvicted	22.1	45.3	23.5	39.4
% female	3.7	4.5	4.2	3.6
% youths	21.6[b]	10.8[c]	10.8[d]	13.0[e]
% foreigners	1.4	27.8	14.5	24.2
Committals, 1988	147,093[f]	83,517	91,723	19,965
Mean period of detention (months), 1988	4.0	6.3[g]	6.8	3.5

[a] The old W. Germany; pre-dates reunification of E. and W. Germany.
[b] Up to age 21.
[c] Up to age 21.
[d] Up to age 25; based on convicted inmates only.
[e] Up to age 23.
[f] Figure includes substantial double-counting. Normally individuals are counted first as committed unconvicted, and again on committal under sentence. The mean period of detention may therefore be longer than the figure shown here, because it should be based on the smaller number of committals under sentence.
[g] Data for 1986.

Source: Council of Europe, *Prison Information Bulletin* No. 13/14, June/Dec. 1989.

icy on prison capacity, noted below, though it is worth mentioning that even though Dutch prison sentences were typically short, the incarceration rate—that is, the number of persons per 100,000 who entered prison each year—was not dissimilar to that in England. Second, for various structural reasons, the French prison population contains a very high percentage of unconvicted inmates. The implication is, quite clearly, that any discussion of prison conditions or inmate rights is missing almost half the population if it does not consider unconvicted as well as convicted prisoners. Third, the Dutch prison population contains a very large percentage of foreigners, most of whom are located in prisons in the Randstadt, the urban area along the coast between Amsterdam and Rotterdam. Many are sentenced for drug trafficking, Amsterdam being a major centre for drugs trans-shipment. And fourth, one must bear in mind that the Dutch prison population is to some extent artificially low, because the legal stricture of one person per cell means that, at any one time, many persons sentenced to imprisonment are actually at liberty, awaiting their call to serve their prison sentence.[6] The high figure for the increase in the number of Dutch inmates is more a reflection of increasing prison capacity than increasing use of imprisonment.

Some remarks on longer-term trends may also be useful. Though not reflected in Table 2.1, West Germany experienced a substantial fall in its prison population in the 1980s, brought about by a large (17 per cent from 1983–6) reduction in remand prisoners. This was at least partly the result of changes in remand practices. Dünkel (1987) notes that.[7]

at the end of the seventies and the beginning of the eighties, it was common practice to keep especially young offenders in pre-trial detention in order to further their education or to administer a short but sharp shock for deterrent purposes. The partial abolition of this practice was due to

[6] The Dutch Prisons Act requires, with few exceptions, a cell to contain one prisoner only. The prison population is managed via 'waiting lists', as outlined in n. 4 above. However there was an increase in the numbers of 'walking sentences'—people at liberty, awaiting prison places to serve their sentence—in the early 1980s, with the result that pressure was put on the service to use existing capacity more effectively while new accommodation was being prepared.

[7] This phenomenon is also discussed in Graham (1987), who notes that while there was a decline in the use of imprisonment for adults, the major drop was for juveniles and youths; his explanation for this supports Dünkel's claim, but also indicates that there was a rise in the number of projects to divert juveniles from custody.

criticisms in the mass media since 1983 (see e.g. the campaign of the German Bar Association).

The other major point that puts Table 2.1 into a longer-term context is that the size of the penal population has been a major issue in all of the countries except West Germany. In England and France, the issue has centred around the long-term rise in the prison population, the consequent high degree of overcrowding in short-term establishments, and shortfalls in regimes. The English solution was initially to embark on a major prison building programme, and more recently to establish privately managed prisons. In France, the 1986 capacity of about 37,000 places was, on the original plans, to have been expanded by some 15,000 privately built and maintained (though not privately staffed) places, though the plan was scaled down as it became increasingly problematic.[8]

Overcrowding has led to some crisis-management measures in both England and France. In France, a 'grâce collective' in July 1980 reduced the prison population by almost 2,000; a second in July 1981, together with an 'amnestie' in August, reduced it by around 9,000; another 'grâce collective' in July 1985 released a further 4,500. However, releasing inmates through amnesties does not solve the underlying problem. Projections of the French prison population (Barre 1987) suggested that the figure of about 45,000 at the beginning of 1985 would have risen to over 50,000 by the end of that year and to 55,000 by the end of 1986 had there not been the breathing space provided by the last 'grâce'. But that breathing space was very short. By the end of 1985 the population had risen to about the same level it had been at the beginning of the year.

In England, the idea of an amnesty has had little political appeal. But one result of the 1982 Criminal Justice Act was to release about 2,000 young offenders in 1983 as their indeterminate Borstal sentences were recalculated as determinate Youth Custody Orders, while a change in the rules affecting parole released about 1,000 adult prisoners in 1985. And in 1987, rules on remission were

---

[8] These trends are also discussed in Ch. 3, while English and French 'private prisons' are discussed further in Ch. 13. The question of overcrowding in English local prisons is partly, of course, to do with decisions about where to put the additional inmates; in France there seems to be less flexibility about such matters. Although there has also been some concern about the capacity of German prison systems, various measures have been taken to forestall an increase in the number of prisoners, as discussed in Heinz (1988). In practice, as noted above, the German prison population actually fell in the 1980s.

changed, which resulted in a marginally slower growth in the population than would otherwise have been expected.

Overcrowding has deleterious effects on regimes and staffing. Moreover, in both England and France, much of the overcrowding occurs in the older and more decrepit part of the prison stock. But issues of regimes, staffing and resources have latterly been hived off into a separate debate. In England, the initial point was one of the cost of the system, driven by the levels of overtime worked by prison officers and local restrictive agreements on staffing levels. When the Prison Department 'bought out' the overtime, and implemented new staffing arrangements and working patterns under the Fresh Start programme, it was claimed to be an opportunity to make significant improvements in regimes. None the less, in general, improvements did not materialize (McDermott and King 1989). In the Netherlands, budgetary constraints starting in the mid-1980s led to a change in regimes, with prisoners being expected to work half rather than full days. The move towards less work was not in itself considered a bad thing, since the intention was anyway to move towards more social and educational activities. However, the extent to which such activities have been implemented has been affected by the budget cuts.

The Reach of the Law

All four countries have at least one primary piece of prison legislation, even if it has been extensively modified by subsequent laws. In England and Wales there is the Prisons Act 1952; in France, the Code de Procédure Pénale (CPP); in Germany, the Strafvollzugsgesetz (StVollzG) of 1976, implemented in 1977 (extended to the old East Germany after reunification); and the primary Dutch legislation on prisons is the Prisons Act 1951, modified most recently in 1976.

Yet many matters pertinent to prisons and prisoners are given force and shape through other legislation. In England, the Health and Safety at Work Act 1977, and the Criminal Justice Acts of 1967, 1982, and 1991—to name but four—affect prisons in various degrees. The first cannot be enforced in prisons, though Richardson (1985) notes that the Home Office expects its provisions to be observed; the second is the legislative foundation of the parole system; the third, *inter alia*, made radical changes, to the forms of youth custody; and the last included legislative provision for private prisons. In France, legislation on public health, for example, is applicable to prisons, and the minister of health has rights to inspect establishments in this connection. The StVollzG is the major legal source dealing with adult prisons in Germany, though it is only one of half a dozen laws that between them deal with the major aspects of custody for most groups of inmates. Other legislation deals with parole, young offenders and so on.

What the laws actually deal with is also variable. The English Prisons Act is the least detailed. In essence, it gives control of the prisons to the Home Secretary and empowers him to open and close prisons and to make rules for their governance. The regulation of prisons is thus largely left to delegated legislation—the Prison Rules (and similar rules for Detention Centres and Youth Custody Centres)—and administrative documents known as Standing Orders and Circular Instructions. The French and German legal systems differ from the English model more than they differ from each other. They are, compared to English law, more likely to enunciate rights, including inmate rights; although the extent to which rights can be effectively claimed is a topic to which I shall return in Chapter 8.[9]

The third section—'Décrets'—of the French Code de Procédure Pénale (CPP) contains detailed regulations for prisons. 'Décrets' are issued by the government, and although they are a kind of subsidiary law they are not delegated legislation in the English sense; they have the full status of law (it should be borne in mind that the law-making procedures in France differ from those in England). The CPP itself thus contains the kinds of details which in England are left to the Prison Rules or to administrative directions. The German StVollzG is broadly similar in scope to the French CPP, though the specific provisions differ. It does, though, adopt a form of language not seen in the French CPP. While the CPP often states

[9] Comparisons of Anglo-Saxon and continental law have occupied many legal textbooks. But in brief, the English, common-law, system is based less on substantive legal rules and more on procedures, and directed towards the resolution of disputes rather than the establishment of rights. There has been, historically, no clear distinction between private and public or administrative law. Continental (or Napoleonic or Romano-Germanic) laws are codified and substantive in orientation, and provide models for social relationships couched in terms of rights, admitting of a distinction between laws regulating conduct between citizens, and laws regulating the relationships between citizens and the state (David and Brierley 1985).

that 'the Governor shall be responsible for . . .', thus providing inmate rights as the correlative of administrative duties, the StVollzG reverses this format. It frequently indicates that 'the prisoner has the right to . . .' or 'is required to . . .' or 'shall have . . .'. In general, and despite the lack of formal statement in the StVollzG itself, it seems to be accepted that the primary responsibility for ensuring that rights are observed and duties performed falls to the governor of each establishment.

The Dutch legal system is sometimes described as falling between the common-law and continental models. It makes a number of substantive provisions including several inmate rights. But many areas are left to the Prison Rules and Prison Regulations. In general, the Dutch orientation towards the law seems to have been that it should remain flexible, and responsive to specific grievances which can be dealt with as they arise. Thus, while most rights given to inmates are qualified—prison governors may withdraw rights on specified grounds—the arrangements for challenging such withdrawals are relatively sophisticated.

Administrative Rules

Laws, even the comparatively detailed French and German laws, do not say what they mean in so many words. They are expanded upon and implemented through administrative rules which, in the case of prisons, provide a detailed framework regulating the lives of inmates, the work of staff, and procedures for dealing with both normal and abnormal events. But since the reach of the law is rather different across the four countries, it follows that the scope of these administrative rules, and their legal status, varies also.

The lack of detail in the English Prisons Act means that the Prison Rules assume a correspondingly greater importance from a managerial point of view. Even so, they offer a poor guide to life in prison, since they are not detailed and leave some important issues unaddressed. Other administrative regulations are, in consequence, of great importance. The two main sets of documents involved are Standing Orders and Circular Instructions (usually abbreviated to SOs and CIs). SOs comprise, broadly speaking, statements of policy objectives, definitions of administrative structure and function, and specific statements of the duties and responsibilities of staff and prisoners (Zellick 1981). There is an ongoing commitment to publish SOs as and when they are revised, which means that not all are yet published.[10] CIs tend to deal with transitory issues, though some of these may be of great consequence (such as the subject of CI55/84, which set out a new management strategy for the service). CIs are not generally published, though some appeared in the appendices to the Prior Committee Report on prison discipline (Home Office 1985); others have come into public view via court cases; and a few are well-known to outside observers even though in theory they are still not public.[11] But as Morgan and Richardson (1987) point out, even SOs and CIs do not necessarily help us understand how prisons are run, since often they lack explicit criteria or detailed provisions which would structure the way in which staff exercise discretion. In some cases they simply repeat statements in the Prison Rules without offering any more detailed guidance. However, the key issue in England with regard to administrative rules is the attitude taken by the courts towards them. In brief, the courts have largely taken the view that the Prison Department can only exceptionally be held to account for breaches of the Prison Rules, while internal administrative orders can only be challenged indirectly, on the basis of judicial reviews of the legality of particular actions taken in respect of inmates.

In France, despite the more detailed nature of the primary law, the prison administration provides amplification, clarification, and interpretation of legal provisions as well as direct instructions to prison staff through 'Circulaires de l'Administration de la Justice'. These are issued by the Director of the Direction de l'Administration Pénitentiaire (DAP), sometimes in association with other ministers. Selected circulars are made available through the

---

[10] The process of publication began following a European Court case dealing with prisoner communications, which resulted in the redrafting and publication of the relevant Standing Order, SO5. However, those not published are lodged with both Houses of Parliament, and some have been cited in court cases, so that there is some degree of public knowledge about their contents.

[11] Circular instructions published as appendices to the Prior Report (Home Office 1985) included CI58/76, dealing with the restoration of lost remission; CI14/80, on the investigation of allegations against prison staff; and CI48/84, on prisoners' contact with legal advisers. One which became known through a court case was CI51/83, on female prisoners with babies, cited in *Hickling* (1985). Well-known to observers, though unpublished, was e.g. CI10/74, dealing with the transfer of refractory prisoners, since replaced by CI37/90 in Sept. 1990, when prison reorganization took place. CI37/90 differs in some important respects from its predecessor and is discussed later in relation to prison discipline.

Bulletin Officiel de la Ministre de la Justice, issued quarterly. Others are summarized briefly in the DAP's annual reports. They concern a wide range of issues; for example, the procedure to be followed after an incident, correspondence between prisoners and the President of the Republic or his wife (traditionally the French may address grievances to her), or directions for collecting statistics on the use of disciplinary measures. Although the procedures for challenging administrative decisions are somewhat complex, for reasons explained in Chapter 8, the fact remains that the comparatively detailed nature of the CPP means that many more specific provisions relating to inmates have a legal, rather than purely managerial and administrative, authority.

Although the German prison law applies to all Land prison systems, the administrative regulations increasingly differ from each other. Feest (1982) notes that when the StVollzG was implemented, the Länder also promulgated joint administrative regulations. But these gradually diverged in content as each Land published specific and local directions. The extent of published information is, however, remarkable. For example, Baden-Württemberg promulgates rules in an official bulletin, *Die Justiz*. Some concern matters of purely local interest, ranging from prison budgets to security procedures for keys. Some deal with changes of function or procedure in only a single establishment. But a few deal with issues touched on in the law, such as house rules for prisoners' recreation, inmate clothing, and instructions on the use of firearms against escaping prisoners.

The Dutch Prisons Act gives prisoners only a small number of specific rights. However, the subsidiary rules—the Prison Rules and Prison Regulations—frequently make reference to inmate rights, such as rights to information concerning their detention, to keep various personal items, to purchase items from the prison canteen, to exercise and to participate in sports, to smoke, to correspond with others, and to have visitors. Most notably, though, the Dutch orientation to inmate rights is that their existence and maintenance is provided for through a right to complain. Since no list of rights can cater for all circumstances in which they may be claimed, or in which qualifications or restrictions may apply, the right to complain is the primary right and others can be established through this mechanism.

All these regulations and instructions are produced at central

level and passed down to individual institutions. But prison governors in all countries have some powers to make rules applicable to
the local situation, which may vary from one establishment to
another. However, the role and importance of these rules cannot
be measured in quite the same way across the countries. Two
examples, England and France, suffice to demonstrate this.

In England, Governors' Orders provide the mechanism for local
regulations. For example, they may allow inmates of one institution, and deny those of another, permission to take radios into the
exercise yard. Even within a single establishment, some inmates
may be required to make their beds with military-style 'bed-boxes'
of neatly folded blankets, and those in another part of the prison
may be free from this rule. If a prisoner were to refuse to comply
with such requirements he may find himself charged with an
offence against 'good order and discipline'. Such rules are often
posted, typed or hand-written, on notice-boards or pertinent
places—the door to the television room will have a note stating
which channel will be switched on at what time, the gate to the
exercise yard a note that prisoners may not take off their shirts
while on exercise, and so on. The 'house rules' in France, however,
are often quite substantial documents, comprising several sections.
A typical set of rules would cite the provisions of the CPP on prisoner communications, discipline, and hygiene, and then give
detailed information on the prison timetable (0700 rise; 0800–1100
work, etc.), instructions on how to apply for visits and details of
visiting hours, rates of pay and rules concerning the canteen (some
repeated from the CPP, others strictly local), and matters such as
the number of showers allowed per week and the arrangements for
kit-change. These house rules are also more carefully vetted
than their English counterparts, since they are submitted for
approval to the Regional Director and the Juge de l'Application
des Peines.

The Main Actors

Much of the discussion in later chapters will revolve around bodies
and mechanisms which are either highly specialized sections of the
prison services, or not part of the prisons structures at all. It is
worth giving a brief overview of these bodies before getting too
involved in particular issues.

Inspectorates

Only in England is the Prison Inspectorate formally independent of the prison service. It was separated from the Prison Department in 1979 and, like the Department, reports to the Home Secretary. It inspects individual establishments, produces 'thematic reviews'— surveys of particular functions or problems common to a range of establishments—and has been called upon to investigate major escapes and riots not only in England but also in Northern Ireland. The Chief Inspector's annual reports, which are laid before Parliament, have often been critical of the prison system and in particular of the level of overcrowding in local prisons.

In France, two main inspection mechanisms operate. The Inspection des Services Pénitentiaires (ISP) is a department within the headquarters organization, while the Inspection Générale des Affaires Sociales (IGAS) has also conducted work on prisons. The ISP divides its activities into two main types of inspection: 'missions de contrôle générale', directed at routine working methods, and 'missions d'enquête', conducted after important incidents— escapes, collective disorder, suicides—or following apparent management failures or lapses. But it has also conducted reviews of areas previously identified as problematic across the prison system as a whole, such as open visits and segregation areas, the utilization of prison space, and hygiene in overcrowded establishments. And it has been involved in at least one joint inspection, with the Inspection Générale des Services Judiciaires, of the assistance provided to released prisoners. The other inspection body, IGAS, has a tighter brief which concerns the health of prisoners and the hygiene of prisons (CPP: Art. D372). It reports to the minister of health, but its work is steered through an inter-ministerial committee. It studies individual establishments, but also prepares reports on specific areas of interest, such as medical prescriptions and AIDS.

Although there is an inspectorate within the Dutch prison department, it comprises only two inspectors and some advisers. Like the English inspectorate and French ISP, it investigate escapes; unlike the English Inspectorate, though like the ISP, it has a management function, advising on matters such as the staffing levels and authorized posts; and in addition it investigates certain kinds of allegations by prisoners, in particular those concerning maltreatment by officers.

Inspection is most devolved in Germany, largely because of the plurality of systems. The federal ministry of justice maintains a small department which appears (outside observers say) to concentrate on financial affairs. The Länder have small inspection, research, or inspection/research facilities—in some cases, one or two people—and the results of their work are not published. However, some systems give responsibilities for inspection to senior departmental managers. In such cases, each official will have responsibility periodically to inspect a group of establishments, or a couple of them will have, along with other functions, a general responsibility to oversee the entire system.

'Lay' Oversight of Prisons

This exists in three of the four countries, the exception being France. In England each establishment has a Board of Visitors. The membership of each Board is drawn from the area surrounding a prison, and at least two members must be magistrates (in England, the vast majority of magistrates are lay persons and not professional, full-time functionaries). The Boards' duties are: to inspect the prison, usually by having one or two members visit the establishment every fortnight; to hear prisoners' complaints, which are then taken up with the governor; and to authorize the segregation of inmates under Rule 43 (a governor may only segregate an inmate for twenty-four hours).[12] Despite their freedom to oversee their establishments—they may see any person, place or prison record—they have often been criticized as ineffective (see e.g. Maguire and Vagg 1984).

In the Netherlands, a 'visiting committee' (Commissie van Toezicht, or CvT) exists for each prison, with an advisory role somewhat similar to the English Boards. Since 1977, each Commissie has had a 'grievance subcommittee' (Beklagcommissie) which hears prisoners' complaints and seeks, broadly speaking, to mediate between prisoner and governor. It has powers to change decisions made by governor, although its substitute decisions may be appealed by inmate or governor to a higher-level committee, the Centrale Raad van Advies (Central Advisory Board). The Central

---

[12] The boards' powers to conduct disciplinary hearings against prisoners, so controversial in the 1970s and 1980s, were terminated on 1 Apr. 1992. A number of other changes concerning their role in handling complaints were also in train at the time of writing.

Advisory Board is also an independent and lay body, with a number of responsibilities in addition to the hearing of such appeals. It advises the Prison Department on policy proposals and new circulars (though many circulars, announcing, for example, changes in rates of pay for prisoners, will be dealt with in a very summary fashion). And it must also be consulted when new members are appointed to Commissies van Toezicht and when an establishment changes its function or a new establishment is planned.

Strictly speaking, the German scheme for local, lay participation in prison affairs functions less as a mechanism of oversight than as a welfare-oriented body for inmates. The local boards, or Beiräte, are voluntary bodies attached to each penal institution. Their responsibility is to the organization of the prison and the care of prisoners. They may receive requests, suggestions and complaints, and see prisoners in their cells without supervision. They may obtain information relating to cells, prison activities, vocational training, catering, medical supervision, and the treatment of prisoners generally, and examine the arrangements made in these respects. The booklet *Strafvollzug in Baden-Württemberg* (1983), produced by the Land Ministry of Justice, simply notes that members of Beiräte are often in contact with social and business organizations, and may help prisoners through these connections. It also suggests that they may act as 'middlemen' between prison officers and prisoners, and that their 'outsider' status means that their suggestions can be helpful to the prison authorities. A similar publication, *Strafvollzug in Nordrhein-Westfalen* (1984) merely records that members are appointed for a five year term, and that members of the Beiräte attend a yearly conference with the head of the Land prison authority. The rather sympathetic study by Shäfer (1987) points out that such bodies may be ineffective because they have relatively high workloads.

Routine Judicial Oversight

The French system of external oversight relies not on 'lay' members, but on local government and the judiciary. Each 'Tribunal de Grande Instance' appoints one or more Juges de l'Application des Peines (usually abbreviated to 'JAP') for each prison in its area. The JAP's role is, in summary, twofold; advice to the governor on the treatment of prisoners and on general issues within the prison, and decisions affecting the prisoner's release.[13] Article 722 of the CPP states that the JAP shall determine the 'principal forms of penal treatment' for each prisoner, including decisions as to leaves, semi-liberty, remission, parole, suspension and part-suspension of sentence, and special treatment. He is to advise on the transfer of prisoners (with an exception for 'urgent cases'), and more generally on 'house rules' (Art. D255); and he must be advised of any use of segregation (Art. D170), any disciplinary hearing (Art. D249), and any incident. Prisoners may lay before the JAP any 'useful information' concerning segregation or punishment, though this appears to be intended less as a complaints mechanism than as an an opportunity for an inmate to make a representation in relation to decisions about his 'mode of treatment'—that is, remission, home leave and so on. However, it is a balance to the provision of Art. D 250/1, which allows the JAP to adjust the treatment (including remission) following a disciplinary hearing.

The role of the JAP has been and still is controversial. Historically, it has undergone a number of changes. Many of the JAP's decisions must be made in consultation with a Commission de l'Application des Peines ('CAP'), comprising the JAP, the prison governor, one or more assistant governors, the chief officer, and members of the education and welfare staff. The balance of power between the CAP and the JAP tipped from the latter to the former in the 1970s and back again in the 1980s, with considerable debate about the extent to which the judiciary should have, and did or did not have, effective control over decisions that affected the internal order of the prison.[14]

Ombudsmen

The concept of an ombudsman—an official who deals with complaints about governmental maladministration—has grown in popularity over the last twenty or so years. However, with a few

---

[13] Article 727 of the CPP also establishes a Commission de Surveillance, comprising the JAP and other judicial and local-government officials who have a duty to visit the prisons in their area, to hear complaints, and to meet annually. This would be, in any organizational chart of prison administration, an important body; but in practice it exercises a kind of arm's-length supervision while some of its members are clearly much less involved with prison affairs than others. It tends not to be of much practical importance in the oversight of prisons.

[14] Changes recorded in a series of anonymous articles in *Justice* (1981).

notable exceptions, ombudsmen have not become important actors in the sphere of prisons.

Germany has no ombudsman. In the other three countries, ombudsmen deal with only a handful of prison cases, and though there have been calls to set up special prisons ombudsmen (see e.g. Hall-Williams 1984) it is only in England, and following the Woolf Report (Home Office 1991), that an ombudsman-like official with particular responsibility for prisons has been appointed.[15] The Dutch ombudsman deals with only two or three prison cases each year, and is limited in that he may not review general government policy, or matters where statutory remedies are available. His reports have been hard-hitting but have rarely resulted in reforms. The English Parliamentary Commissioner for Administration (or PCA) also handles only a small number of inmate cases—133 in total between 1967 and June 1980 (Birkinshaw 1985a). The constraints on his role have produced some controversy. He may not question the merits of policies or criticize 'bad rules', and his recommendations are not binding on the department concerned. Thus, in the case of *Kechtl* (1970), the PCA could not directly criticize a Home Office rule that inmates alleging injury arising from medical maltreatment should make a prima-facie case of negligence before being allowed to contact lawyers. He suggested instead that the Home Office review the rule. The review concluded, against widespread criticism, that the rule was satisfactory (Birkinshaw 1985a; 1985b). The French ombudsman has a wider remit, and may directly criticize 'bad rules'. But while he deals with a larger number of cases than the PCA (Neville Brown and Garner 1983) his reports indicate that virtually none of his cases involve prisoners.

Parliamentary Scrutiny

Some mechanisms whereby prison affairs are scrutinized are common across all four countries. Departmental budgets ultimately depend upon parliamentary approval, and specific line entries may be called into question. But other and more specialized arrangements exist in each of the countries for parliamentary oversight of prisons. In England, parliamentary scrutiny is primarily via two select committees of members of parliament, the Public Accounts

---

[15] This post, the Independent Complaints Adjudicator for Prisons, is considered further in Ch. 8.

Committee and the Home Affairs Committee, both of which have remits wider than prisons but have produced reports on various aspects of imprisonment. In addition, the Prison Department lays before Parliament an annual report which may prompt debate. In the Netherlands, the most important feature of parliamentary scrutiny is that the second chamber of parliament has a 'commission for requests' which will deal with inmate complaints. In Germany, members of legislative bodies may make a request (Antrag) for information from a government body. These may be extremely detailed. In Baden-Württemberg, an Antrag in 1986 asked for information on probation, privileges, budgetary measures, overtime, staffing, prisoners' work, the legal position of the earliest date of release from custody, and some dozen other points. The document even included instructions as to how the statistics should be tabulated. The reply ran to about 100 pages and formed the basis for a number of supplementary questions.

Legal Scrutiny

This may apply both at the level of law- and policy-making, and at the level of individual complaints. In France, the Conseil Constitutionnel can rule on the constitutional position in relation to proposed or actual changes in law.[16] The Conseil d'Etat exercises both an administrative function of a priori advice on bills ('projets de loi') and judicial review of administrative decisions, though the mechanics of the latter are somewhat complex. In essence, challenges to the legality of individual decisions go either to the regional Tribunal Administratif or to the Conseil d'État, depending upon the nature of the issue. The latter also acts as an appeal mechanism from the Tribunaux, but if the appeal is successful the case will simply be referred back to the competent authority for a fresh adjudication. One feature of the Conseil d'État is that while it holds both administrative (legal advisory) and judicial powers, it is staffed by civil servants rather than judges.

[16] Neville Brown and Garner (1983: 9-10) describe the Conseil Constitutionnel's functions as adjudicating on the validity of elections and referenda; expressing an opinion on 'organic' laws approved by Parliament; deciding on the constitutionality of ordinary laws, if challenged, prior to their promulgation or ratification; and ensuring that parliament and government keep within their respective domains with regard to legislative activities. A law may be referred to the Conseil for scrutiny by any of four specific state officials, or by any group of 60 deputies or senators.

In Germany, the main avenue for legal scrutiny is via special criminal courts, the Strafvollstreckungskammern. They have a remit to make decisions in the first instance over various matters regarded in England as administrative—parole being the best example. They also hear cases brought by prisoners, with the proviso that any case brought must allege a breach of the StVollzG. Prison service administrative directions are not considered part of the law, so that it is possible for the court to uphold complaints based on a point of law, even though the administrative rules were followed. In such cases the rules must be changed or re-interpreted to conform with the legal decision. Beyond this level, the Oberlandsgericht (Land supreme court) hears cases based on constitutional provisions, and in practice decisions made in one Land are regarded as valid for others, so that (notwithstanding my earlier comments on administrative rules) all the Länder are kept in step. There is also a federal court which will hear appeals from the Länder.

In England, the main mechanism for challenging administrative decisions is by way of certiorari to the Divisional Court, though its decisions may be appealed to the Court of Appeal and thence to the House of Lords. Other grounds (e.g. mandamus) exist, but are less used. Cases such as alleged negligence would be brought in the civil courts.

In the Netherlands, Ybema and Wessel (1978) note that 'since a general system of redress of administrative wrongs as existing in France is lacking, it may easily occur that there is no redress available at all, appeal lies neither to a special administrative tribunal nor to organs being part of the administration or to the ordinary courts'. This explains the importance of the Centrale Raad van Advies (CRvA) as a special tribunal. However, they note that the ordinary courts have gradually extended their powers and will in some circumstances act as an appeal mechanism from administrative decisions, while also hearing actions for damages as the result of administrative acts. Cases will be heard not only when decisions have been made, but also where administrative action is imminent, in order to restrain the administration from acting.

Concluding Comments

This chapter has been directed primarily at the organizational and legal aspects of imprisonment. At one level, the frameworks it

describes are key features of the four prison systems, and an understanding of them is essential to informed discussion of the prison
systems themselves. There are major differences across the four
countries in, for example, the opportunities given to inmates to call
for legal review of administrative decisions that affect them, and
this must be borne in mind when discussing matters such as inmate
rights, which depend crucially upon the ability of prisoners to
bring matters before a legal authority. But at another level, these
broad frameworks are a poor guide to the ways in which prison
institutions actually operate. I have not yet discussed in any detail
the extent to which the ability to bring an administrative decision
before a court is actually likely to result in a right being recognized, still less complied with in practice. And despite offering
broad descriptions of the prison systems, I have not yet discussed
issues that would be of immediate importance to inmates—or, for
that matter, prison staff—such as the degree of disciplinary control
that can be exercised over prisoners. Thus the following chapter
deals with the ways in which the four prison systems have changed
over the last twenty or so years, and the subsequent suite of three
chapters discusses the day-to-day life of prisons, as seen through
the eyes of inmates, staff, and prison managers.

3

Accountability and Change in Prisons: a Selective History

PEOPLE have different reasons for wanting to make prison systems more accountable. They may intend to make the system more responsive to inmate needs—a broadly humanitarian version of accountability. They may plan more restrictive regimes, for reasons that often have to do with wider penal policies or managerial aims. Or they may simply seek greater efficiency and effectiveness, treating accountability primarily as an issue of value for (taxpayers') money. This is not to say that all attempts at prison reform are attempts to make prisons more accountable, in whatever sense that word might be employed. It is to say that the idea of accountability is often relied upon, either implicitly or explicitly, to justify change.

Change can, of course, come about in several ways. Major reforms have taken place in England and France roughly every ten years, often taking several years to implement fully. In England, one can cite the 1979 May Report on prison staffing, and the 1987 introduction of the Fresh Start staffing package; in France, the 1974 and 1985 reforms of prison regimes. In between these events, of course, other significant developments have also taken place—for example, the 1982 changes to the English youth custody arrangements, and the 1983 adjustments to parole; and the tightening of sentences for serious offences in France in 1978, 1981 (mostly repealed in 1983), and 1986. Dutch prisons have seen a number of organizational changes—the 1977 reform in grievance procedures, and, throughout the 1980s, the introduction of standardized management structures, the move towards open management styles, and budget cuts arising out of the McKinsey management consultancy study in the mid-1980s. Only in Germany—allowing that unification resulted in a complete overhaul of the old East German prison system since 1989—has there been a more or less static situation in relation to prison regimes since the introduction of the 1977 Strafvollzugsgesetz.[1]

The changes cited above are clearly of different types, encompassing prison regimes, the staffing and administration of prisons, and factors such as early release and sentencing policies which influence the size and nature of the prison population. But such intended changes must also be seen in the context of unintended developments which have also had major impacts on the prison systems. Three of the four prison systems I shall discuss experienced severe pressure on prison places, and the fourth, Germany, having accomplished reductions in its penal population in the 1980s, began moving towards capacity in the early 1990s. England and France both had increasingly large levels of overcrowding throughout most of the 1980s, and regimes deteriorated as a result. They have both also had lengthy periods of industrial action by staff, and a series of riots. The role of designed changes—whether in terms of the liberalization or tightening of prison regimes, or attempts to squeeze greater efficiency out of resources—must accordingly be judged as only one part of recent prison history. I shall return to this theme at the end of this chapter.

The following four sections illustrate different kinds of linkages between accountability and change. The first deals with the effectiveness (or otherwise) of independent accountability mechanisms and procedures to deliver messages that change is necessary. Since the use of independent bodies has been most pronounced in England, this section is largely oriented towards English developments. The second takes up the theme of accountability as a concept used strategically in order to legitimate changes in prison

---

[1] Remembering, however, the changes in the use of imprisonment, especially for juveniles, discussed in the last chapter. One other point is worthy of mention. After the reunification of Germany in 1990, the laws of the Federal Republic were applied to the old E. Germany. In prison affairs as elsewhere, this created some problems. Many of the old E. German prisons were simply not physically capable of housing the kinds of regimes envisaged under the StVollzG. Most inmates were accommodated in large cells or dormitories, and it was common for prisoners to work outside the establishment (see Weis 1991). But at the same time, Weis notes that adult prisoners who fulfilled the work norms received 18% of the standard non-prison wage for their job. The implementation of the StVollzG meant that prison wages dropped to a figure roughly equivalent to 5% of the average adult wage (cf. Weis 1991 with Dünkel and Rössner 1991).

systems. It argues that, on occasion, ideas of increased accountability have been used in much the same way that a good deal of prison research is treated—to provide justifications for policy changes already decided upon for political reasons. The third addresses the extent to which mechanisms intended to make prison systems accountable have themselves been in need of reform. And the fourth considers the impact of political concerns about crime and justice generally on the prison situation.

England and the 'Independent Messenger'

The concept of independence is relatively straightforward, though the practical arrangements for independent bodies have proven complex. The essential argument is that any large bureaucratic organization, despite the presence of official aims and rules for processing cases, also has unofficial aims and rules, and perhaps also contains significant groups of persons whose interests diverge from those of the organization at large. Thus Weber's (1970) model of a rational-bureaucratic organization has to be modified in practice by acknowledging that individual workers use short cuts to achieve acceptable performance levels; that the organization as a whole has an inbuilt tendency towards its own survival, if not expansion; and that groups of workers (or organizational departments) may seek to influence other parts of the organization in ways that reflect their own official or unofficial goals.[2] This has a number of implications for management which will be discussed in later chapters. But so far as 'clients' of the organization are concerned—in this case, inmates—it means that the way they are treated in practice may diverge significantly from official rules concerning their treatment. In particular, complaints about their treatment may not be properly entertained, while processes of inspection can, if done in-house, be 'captured' by interest groupings within the organization, or act, in effect, as a means of furthering the interests of the organization as a whole.

One solution is to divorce the grievance process, or the inspection team, from the organization. To insist that all inspection work, or all grievances, be handled by an outside agency would of

---

[2] The issue of unofficial goals both by worker groups and organizational subunits has a large literature. See e.g. the edited collections by Pugh (1984) and Salaman and Thompson (1973).

course be impossible. Inspection is not only a matter of public accountability, it is also a part of routine management; and requiring all complaints to be processed by an independent agency would be to set up a separate but parallel agency. However, certain kinds of inspections, and serious complaints (or appeals against decisions in lower-level complaints), can be processed by an independent body. In principle, such bodies, being removed from the organizational environment and unofficial work norms, should be able to produce unbiased decisions. In addition, they should have fewer problems in defending their legitimacy.

In the context of prisons, an 'independent' agency might mean, for certain purposes, a court, as in the case of the German special prison courts, set up in 1976. But in England, the dominant strategy, for cultural reasons which extend beyond the scope of this book, has been to set up specialized bodies reporting either to the Home Secretary or to Parliament.[3] Currently, the two independent bodies most concerned with prisons are HM Chief Inspector of Prisons, and boards of visitors.[4] But the core problem of an independent agency is this: even if it can identify key problems in service delivery, it is often not sufficiently close to the centre

---

[3] Normanton (1966) proposes that the forerunner of modern accountability can be seen in the creation of special courts, reporting to the monarch (whose function was to provide a 'judicial audit' of tax revenues and expenditure), with powers to call government officials to appear before them. By the 20th cent. this kind of function had been devolved to officials such as the Comptroller and Auditor General, with a broader remit to investigate efficiency and effectiveness as well as the 'regularity' of expenditure. Rhodes (1982) offers a view based on the development of inspectorates, and suggests that a key distinction should be made between 'enforcement' and 'efficiency' inspectorates. The former usually have a range of statutory powers and may be entitled to conduct prosecutions; they would normally report to Parliament. The latter are often 'in-house' inspectorates within government departments, reporting to the relevant Secretary of State. In addition, of course, the UK makes extensive use of willing volunteers to sit on a variety of committees, of which the most important in prison terms is probably the Parole Board.

[4] However, a third body is likely to come into being in 1992-3. Following one of the recommendations of the Woolf Report on the Strangeways riot (Home Office 1991) that an independent adjudicator would introduce an element of justice being seen to be done to complaints, plans were formulated to create the office of 'Independent Complaints Adjudicator for Prisons'. The consultation document issued in Mar. 1992 proposed that the adjudicator's brief would be twofold; first, 'in relation to the grievance procedure, where his role would be to recommend, advise and conciliate at the final stage of the procedure', and second, in relation to disciplinary hearings, where it would be to act as a 'final tribunal of appeal'. However, a number of areas, including parole and the clinical judgement of doctors, would be excluded from his consideration. See HM Prison Service (1992).

of administrative power to be able to argue effectively for change.

English Boards of Visitors are sometimes described as the 'eyes and ears of the Home Secretary', and this is quite literally correct. They have no executive powers, but pass on their concerns to the appropriate part of the administration. The administration, however, has no clear and direct duty to act in response. Research on Boards of Visitors (primarily Maguire and Vagg 1984)[5] has suggested that the visible effects of their concern at local level are minimal—often because extensive consideration within the administration is not translated into concrete action. Small wonder that some Board members describe talking to senior officials as less useful than addressing fluffy white clouds (cited in Maguire and Vagg 1984: 120). And confronted with poor physical conditions in overcrowded prisons, often petty rules, and regimes obliterated by constraints on staffing, it is hardly surprising that Boards of Visitors often feel that sorting out an individual inmate's complaint that he was not allowed to change his library books is only marginally more useful than rearranging the deckchairs on the Titanic. The plain fact is that the Board of Visitors system is primarily intended to pick up routine problems and individual complaints in individual prisons. Where the problems are recurrent, major, and largely to do with the level and structure of resources, individual Boards can and do report them regularly to the administration. Faced with many such reports each year, the administration has, frankly, little option but to respond as soothingly as it is able. The Board of Visitors structure is built on the presupposition that the overall structure is working well. Where it is not, the boards are out of their depth.

The English prison inspectorate, being centralized and able to review the situation system-wide, is not in the same situation. However, as Morgan (1985) noted, it was dogged in its early years by delays in the publication of its reports, largely occasioned by the decision to include the Home Secretary's response in the published version and the period of time taken for that response to be written. Moreover, many of the responses were pious hopes, stating that improvements would be made if or when financial resources were forthcoming (see Chapter 9). The fact of the matter

---

[5] For earlier criticisms of the boards, see e.g. Martin (1975).

remains, however, that the inspector cannot do more than draw attention to shortcomings; he has no sanctions at his disposal, and no executive powers. It is the role of the Home Secretary, in practice on the advice of the Prison Department, to formulate appropriate courses of action. These may, of course, be pious hopes, or decisions that nothing can be done. The value of the inspector's reports, at this point, simply lie in their being sticks with which members of parliament can try to beat the Home Secretary.

The single major exception to this has been the impact of the Chief Inspector on the development of integral sanitation in cells. Judge Stephen Tumim was appointed Chief Inspector in 1987. From the beginning, he announced his intention to press for integral sanitation; following a review of the current situation and alternative sanitation systems, he made recommendations which envisaged the ending of 'slopping out' within about seven years (HM Chief Inspector of Prisons 1989). As the report was released, the Home Secretary announced a somewhat less ambitious programme designed to reduce the need for slopping out from 50 to 13 per cent of prison places by the end of the century.

The reasons for this general inability to initiate changes are unsurprising. First, the practical extent of independence is qualified by the nature of appointments, powers and responsibilities. Appointments are made, not by the prison administration, but by the Home Secretary (in the case of boards of visitors) or the Crown (in the case of the Chief Inspector). While this clearly leaves both Boards of Visitors and the inspectorate with considerable room for manœuvre, it does not mean that they are free of the constitutional and procedural restraints imposed on them by the appointing bodies. The Boards of Visitors are appointed on behalf of the Home Secretary and report to him. There is some degree of ambiguity in the situation, in that the Boards are not expressly prohibited from reporting, for example, to the press; and, indeed, some produce and publish periodic reports and give press interviews. But if it began to appear as though they regarded themselves as primarily reporting to the press and the public, the Home Secretary would undoubtedly see this as problematic. And while the Chief Inspector is required to present an annual report to Parliament, it would clearly be seen as a serious breach of convention if the minister responsible for the service being reported on did not have first sight of the document. The English doctrine of accountability also links together powers and responsibilities. Both Boards of Visitors and the inspectorate have broad powers of oversight in terms of their ability to conduct unannounced inspections, and have access to all parts of the prison, all staff, all inmates, and all paperwork (though the latter usually does make prior arrangements for inspections, given the size of the team and the length of the inspection—see Chapter 9).[6] But they have no executive power, the argument being that such authority also implies a line-management responsibility and would bring them directly into the management structure.

Second, the field of action of complaints and inspections arrangements is often limited to individual grievances or single prisons, even if the issues behind them are general properties of the prison system itself. The oversight function is often also restricted to reporting problems to senior management (or ministers), with no brief to take further action. Meanwhile, many policy changes are initiated 'top down' for political or managerial reasons, having little to do with the specifics of the prison situation. The end result is that reforms have frequently come about as a result of pressures put on administrators by the larger political environment, rather than by mechanisms intended specifically to identify problems in the prisons they are ultimately responsible for.

While the English approach conforms to the maxim of 'no authority without responsibility', the end result is that the authority of the independent bodies, derived from the nature of their appointment, is simultaneously undermined by their lack of direct powers to set right even those matters that could immediately and simply be remedied. But this does not have to be the case, as the discussion of the Dutch complaints system (below, and in Chapter 8) and the American use of prison standards and 'accreditation' (Chapter 9) indicate.

Accountability as a Resource in Reform: Accountable Regimes and Fresh Start

In this section I want to look more generally at how the rhetoric of accountability is employed, not least by senior officials and ministers, to justify changes in the prison system.

[6] There are some limits to this, however; inmate medical records, defined by the administration as medical rather than prison records, are thus not open to inspection.

One example, albeit at a fairly low level, is the history of 'accountable regimes' in England in the 1980s.[7] The May Committee, formed to inquire into the problems of the prison service following the August 1979 riot at Wormwood Scrubs prison in London, made a large number of recommendations concerning general features of prison management and staffing arrangements (Home Office 1979). One of its major concerns was the quality of prison regimes, and the administrative response was to develop proposals based on clear objectives and priorities, with formal mechanisms of accountability for the delivery of regimes. The intention was to provide management with a sense of purpose, staff with a sense of commitment, and inmates with richer regimes. In addition, the new proposals were to ensure that resources were being properly utilized. The mechanisms through which these would be done consisted of a modified form of management by objectives, with 'contracts' being agreed between regional directors and prison governors as to the objectives that would be reached within an agreed level of resourcing; and regime monitoring arrangements to determine whether objectives had in fact been reached. The whole package was labelled 'accountable regimes'.

The sense of accountability implied in accountable regimes was clearly that of value for money, in the sense of strong management being committed by contract to clear, agreed objectives to be reached through the efficient and effective use of agreed resource levels. The assumption seemed to be that this would weed out inefficiencies, which were largely discussed in internal papers in terms of regime elements which made competing rather than complementary demands for inmate time. And it seemed further to be implied that working in a well-led, purposeful, efficient, effective organization would improve staff morale.

Both Chaplin (1985) and Ager (1986) describe the mixed results of the pilot programmes in Shepton Mallet and Featherstone pris-

---

[7] 'Accountable regimes' was a project begun in 1981 within headquarters, and was intended to 'identify the separate activities which together formed the regime of the prison, then to describe the resources normally required, the minimum viable resources, the priorities, effects of shortfalls and constraints' (Ager 1986). This information could then be used to look at the relationships between resources and performance, and to suggest changes in organizational structures. A pilot project was carried out in selected prisons from 1982 onwards. Much of the discussion here also makes use of an unpublished paper by Brian Chaplin, one of the Prison Department officials originally involved in implementing the initiative.

ons. As Chaplin concludes, Featherstone was a large new prison which already had a number of special management features and sufficient resources, in principle, to provide a full programme of inmate activities. The scheme was on the whole well-received, and some of the inefficiencies associated with activities competing for inmate time were resolved. Shepton Mallet, a much smaller and older prison, was less well-resourced, and had a less strongly formalized management structure. The pilot scheme did result in more co-ordinated management, but the effects were more muted at the day-to-day level. The accountable regimes scheme was subsequently promulgated nationwide, though in somewhat different ways in different regions.

The idea of accountable regimes was, however, dramatically altered by the impact of Fresh Start, discussed below. Its forms—'contracts' and 'regime monitoring'—were turned to newer uses. But prior to Fresh Start, the main conclusions that could be drawn were these. First, accountability was primarily modelled on the concept of value for money, and principally applied to inmate regimes. But it became apparent in the comparison of Shepton Mallet and Featherstone that the best value for money could be obtained where an optimum level of resources were provided. In an environment where resources are insufficient, the concept of value for money is no longer an adequate yardstick for evaluation. Operating the regime becomes a matter of protecting core functions, with the structural inefficiencies that this can create.[8] Second, it became clear in the later stages of the 'accountable regimes' scheme that the contract was in reality a one-way document, with governors being accountable to regional (and, later, area) directors for performance without any corresponding accountability of area directors for the provision of resources. It became a way of driving

---

[8] e.g.s might be: the use of classrooms as dormitories in overcrowded prisons, thus squeezing out opportunities for education; the closure of workshops in local prisons in order to provide staff for court escorts; and the use of staff on 'detached duty', who are paid at a higher than normal rate, to work in establishments they are not normally assigned to in order to cover staff shortfalls: all situations to be found in some English prisons in the 1980s. Fresh Start was intended, among other things, to reduce the need for detached duty. However, Morgan (1983) challenged the general assumption of poor resourcing, pointing out *inter alia* that in the five years to 1982, more prison staff had been added to the staff complement than prisoners to the prison population, and that overtime working levels had also increased. Clearly some of this increase was used up by relatively inefficient working practices, particularly in relation to staffing escorts to court.

governors without making firm commitments on support. In practice, governors and regional directors would agree both objectives and resourcing, but for various reasons the regional directors often found themselves unable to deliver the agreed resource input. This was subsequently recognized in the Woolf Report (1991: para. 12.97), where it was proposed that such contracts should place clear obligations on both parties.

The Fresh Start model, which overtook the idea of accountable regimes, originated in 1985, with the then Home Secretary announcing at the annual conference of the Prison Officers Association that he proposed to study new schemes of staff pay and attendance. PA Management Consultants were used to identify possible new complementing and shift arrangements. The Home Secretary's speech referred to 'new systems and procedures' and 'pay arrangements which will continue to provide a fair level of take-home pay while allowing your members and management to escape from the shackles of dependence on overtime'. One of the key problems of the service had been its dependence on staff working long hours of overtime, with the consequence that staff had themselves become financially dependent on the overtime payments. The Home Secretary's words signalled an intention to buy out management dependence on overtime.

A second idea also lay behind this move. The prison service had long argued that locally-negotiated restrictive agreements as to the deployment of staff had resulted in substantial inflexibility, inefficiency, and unnecessarily increased staff costs, including overtime payments. PA Management concluded that task lists, that is, the lists of functions which staff are expected to cover at different points in the day, 'lack sensitivity to the inherent variability in prison activities, are influenced by shift system limitations, and in practice are contracted or expanded to match staff's overtime availability at the expense of service and regime quality' (Home Office/PA Management 1986: 5). The conclusion was that manpower utilization could be improved by 15–20 per cent, and that this time could be used either to enhance regimes, or to reduce overtime and manning levels (and thus costs).

In short, Fresh Start was framed in terms of management accountability for the efficient use of staff, and ultimately the rhetoric of the value for money obtained by the taxpayer. The principal components of the Fresh Start programme were increased basic pay, a unified staff structure, a group working system for staff, and progressively reduced overtime with 'time off in lieu' rather than overtime payments. Staff were co-opted into Fresh Start, partly (and obviously) because it offered them increased pay for fewer hours. But at the same time, as the Woolf Report makes clear, they were also persuaded that the end result would be to enhance regimes, their own motivation, and job satisfaction. They were also given to understand that the man-hours lost by eliminating overtime would be partly offset by the recruitment of new staff.

A rolling programme of 'Fresh Starting' prisons commenced in 1987. As it progressed, it became increasingly clear that whatever had been agreed between staff and governors on the one hand, and the central administration on the other, bore little resemblance to the ultimate aims of the latter and its agreements with the Treasury. Relatedly, the initial concept of efficient staffing leading to enhanced regimes, and 'value for money' was progressively displaced by a more single-minded insistence on economies. Much of this came to figure in the Woolf Report's diagnosis of the ills of the prison system following the Strangeways riot in 1990 (Home Office 1991).

The improved utilization of staff that Fresh Start was to bring about was, at least initially, discussed publicly as though it would be translated into improved prison regimes. Documents of that time optimistically discussed the improvements in regimes that would result. The Prison Department then proposed to phase in a 15 per cent reduction in staff hours over three years, though with the longer-term aim of recruiting new staff to fill half the man-hours lost. In other words, and despite claims about the enhancement of prison regimes, it opted to take the benefits in terms of immediate cash savings, with a broad commitment to improve staffing at a later stage. It subsequently became clear that this did not mean new staff would be deployed to existing prisons; rather, they would be used flexibly, but mostly to enable new prisons coming on-stream to be staffed. Once reductions in man-hours were achieved at existing establishments, other things being equal, they would be permanent. Meanwhile, the initial reduction in staff hours was made across the board, without any attempt to evaluate whether individual establishments were or were not inefficient. And in separate negotiations with the Treasury, the prison administration agreed to much greater cuts than those discussed in notices to staff, amounting to 24 per cent of staff hours achieved over a five year period. The resource implications were apparently not fully understood even by those training senior staff and governors for the implementation of Fresh Start, and the full extent of reductions in staff availability that the package demanded only became fully evident by early 1990. By that point there was widespread cynicism among staff about departmental aims.[9]

Fresh Start was not, in any obvious or immediate sense, 'about' accountability. In its initial phase, it was about buying management flexibility. It was intended to promote efficient management by reducing dependence on overtime and removing allegedly restrictive practices. The benefits were to be enhanced regimes for inmates, and a leaner, fitter, more motivated staff. Having said this, accountability was clearly a background issue in so far as its goals could be depicted in terms of value for money for the taxpayer. However, by the time Fresh Start was actually implemented, the benefits were conceived of in terms of staff flexibility and the 'management's right to manage', coupled with cash savings. While this might still be represented as value for money, it is clearly a rather different and short-sighted version of that concept. The implicit background concept of accountability had moved from a concern with overall, collective benefits to a simple issue of cutting costs and, in a sense, changing the balance of control to favour central management more strongly.[10]

Reforming Accountability

Systems of accountability, like the prison systems they oversee, are sometimes revised or reorganized. This can happen for a range of

---

[9] For a fuller description of the development and implementation of Fresh Start, see the Woolf Report, ch. 13, on which the comments here are based.

[10] Jones (1993) discusses some of the background to this. Her argument is that while, in the 1980s, government doctrine was built around the concepts of economy, effectiveness, efficiency, and value for money, in practice the only goal actively pursued was that of economy. Kirkpatrick (1986) has also criticized the 'mechanistic and ultimately cynical approach' of the Treasury, while Metcalfe and Richards (1987) identified a more subtle problem. They argued that the Financial Management Initiative was a victim of its own success. It identified areas in which cuts could be made in public expenditure, but appropriate management styles failed to evolve and a longer-term strategy based on concepts such as the value of particular 'outputs' was not forthcoming. In short, while civil servants may have better control over costs, they cannot make assessments of value for money.

reasons, and one should not be too cynical about them. The 1976/7 revision of the German prison law, the 1977 introduction of a new complaints system in the Netherlands, and the 1992 removal of disciplinary powers from English Boards of Visitors, all testify to the ways in which mechanisms of accountability, when they become problematic in their own right, can be reformed. At the same time, all three examples clearly show how proposals for change—whether intended to address prison problems or problems with the structure of accountability—become the subject of wide, and sometimes political and budgetary, debate.

The 1976 German prison law was a long time in the making. As Feest (1982: 10) notes:

Efforts to regulate German corrections in a comprehensive statute have been through a gestation period of more than a century . . . Several bills were drafted, but not enacted. In 1961, all German Länder adopted a joint set of regulations on the prison service (DVollzO), which did, however, not have the formal character of a legal statute . . . To redress these short-comings, as well as to 'reform' prisons, a national prison commission was set up in 1967. It submitted its recommendations in the form of a draft bill in 1971. This preliminary work was given additional impetus in 1972 by the Federal Constitutional Court's ultimatum to the federal legislature to produce a prison law in the near future. But it took another four years of deliberations and tug-of-war between Bundesrat and Bundestag until the Prison Act (Strafvollzugsgesetz) was finally enacted in 1976 and became binding in 1977.

The kinds of issues that the new law was forced to address are explained by Huber (1978), using examples from the extensive prison litigation of the 1960s and 1970s, including, for example, cases relating to the use of solitary confinement, overcrowding, and the censorship of correspondence with lawyers. In an effort to remove the need for such litigation, the new law included provisions for special courts to hear inmate grievances. The restrictions on grievances were that they must allege an infringement of the law, rather than of prison regulations; and, in some states, that the complaint must first have been ventilated through the prison administration, so that the issue before the court was one in which an official decision had already been made. Despite the provision of the European Convention on Human Rights that there should be no restrictions on access to courts, this arguably made sense in the context of the German inheritance of a largely Prussian legal tradition. The expectation was that they should review the legality of decisions already considered definitive within the administration, rather than acting in the place of the administration in a matter only halfway through the administrative process. On the positive side, since the court operated largely through the medium of written documents, and on an inquisitorial pattern, inmates did not require the services of a lawyer to initiate or argue cases.

Yet the 1977 grievance procedures contained a number of loopholes, chief among them being that no machinery existed for enforcing a court decision if the administration did not act in accordance with it. In the main, prison administrations did comply with legal decisions. After all, most prison administrators and governors were and are themselves lawyers, not disposed to treat the courts lightly. But the possibility always remained that if the issue were sufficiently important for the prison service, court decisions could be ignored.

In the Netherlands, the inmate-grievance system was overhauled in 1977. The need for reform lay in the provisions of the 1953 Prison Act, which created the possibility for inmates to complain to a visiting committee but provided only that the committee could advise the governor on the matter complained of. This arrangement, Nijboer and Ploeg (1985) note, led to a situation in which governors were at liberty to ignore the advice of the committees, while prisoners lacked confidence in the complaints procedure. The problem was identified by a state committee, sitting from 1964 to 1967 and looking at the legal status of prisoners. It took, however, ten years for relevant legislation to be brought forward. The 1977 reform required visiting committees (Commissies van Toezicht) of the prisons to set up grievance subcommittees. These could hear complaints from inmates, provided only that they related to specific areas set out in the law; and they could substitute their own decisions for those of the governors. This meant that, in effect, the committees were given executive powers to intervene in prison management. But the question of executive intervention was fudged by the convenient fiction that such decisions were 'substituted for' those of the governor and did not overrule them. The new system attracted roughly four times as many complaints as the old, and thus quickly developed a body of 'jurisprudence'. One has to use this word carefully, since decisions of the grievance committees, and the body established to allow inmates and governors to appeal their decisions, are not, strictly speaking, legal decisions. None the less, the grievance system—and indeed the prison system itself—operates in all important respects as though they are.

The system is not without its problems. Nijboer and Ploeg (1985) reported that a significant minority of inmates did not know that this system existed. More recently, Mante has argued that the system is much less formalized than it appears, with many committees having moved towards a mediation-based role. Others have pointed out that the system is still weak, because it is geared only to breaches of individual rights and not to the correction of systemic flaws in the prison administration (Kelk and de Jonge 1982). But the key to its success seems to be the power of the Beklag commissies to substitute their own decision for that of the governor, a power which, if not often used, at least makes their opinions weigh heavily with prison management. The end result, judged from the vantage-point of some fifteen years' experience, is that the right to complain has become the single most important right of Dutch prisoners, and the complaints system is, as Kelk (forthcoming) suggests, 'relatively excellent' even though practical results have often been small-scale.

In England, as noted above, arguments about the effectiveness of external oversight have frequently revolved around the issue of 'independence', and the independence, powers, and effectiveness of the Boards of Visitors has been at the heart of a major argument that has, after almost twenty years, only just been concluded. This argument was that the independence of the Boards of Visitors as inspectors and hearers of complaints was compromised by their power to conduct the more serious disciplinary hearings against inmates, and to give punishments. The specific allegation was that, in carrying out this disciplinary role, the Boards acted as a part of the administration, and were, in practice, prepared to give punishments based not only on the specifics of offence and offender, but also on the basis of the message which they sent to the prison as a whole. In other words, they were acting as though they were members of prison management. It therefore followed, it was said, that if they were not independent in relation to discipline, they were unlikely to be so in dealing with their other concerns, complaints and inspection.

The Boards had originally been given disciplinary powers because—ironically enough—their independence was supposed to guarantee the fairness of their disciplinary adjudications. But certainly by the 1980s, and probably much earlier, they were clearly perceived by inmates to have been co-opted into a managerial point of view. Many Board members in the early 1980s were reluctant to give up disciplinary hearings; they cited reasons including the respect accorded to them by staff because of their disciplinary powers, and the argument that the information they obtained through disciplinary hearings was useful also in the context of dealing with inspections and complaints. There was, to be sure, a fair degree of specious logic in the argument that Boards could not separate the two roles, just as the arguments deployed by Board members were rationalizations of a deeper attachment to what they saw as one of their 'power bases'.

The public debate on the issue arose primarily out of the disciplinary hearings following the riot at Hull prison in 1976, when it became known that disciplinary hearings following that incident had seriously lapsed from any reasonable standard of fair conduct (Taylor 1980). From that point on, the Boards and the Department, which was put in the position of having to defend the Boards against accusations of poor conduct, were put on the defensive. There were numerous court cases, and for several years they carried out what with hindsight we can describe as a spirited rearguard action. None the less, a large number of inmates were successful in attempts to have their disciplinary hearings quashed.[11] As a result, Board members underwent increasing amounts of training for adjudications. However, once the courts had accepted the principle that Board procedures were subject to judicial review, and more especially once the principle was established in *Tarrant* (1984) that inmates at such hearings could be legally represented, the game was increasingly seen as not worth the candle. Acting on a voluntary basis, Board members could none the less find themselves embroiled for lengthy periods of time in legal and procedural arguments; and the delays in hearing cases meant that whatever the situation has been in the past, disciplinary hearings by Boards had become a problem for management rather than an aid to it. There is some evidence to suggest that the governors of the large London prisons, at least, began to use their discretion to limit the numbers of cases appearing before the Boards after the *Tarrant* decision.

[11] For details up to 1984, see Fitzgerald (1985), Richardson (1985); for cases up to 1988, see Treverton-Jones (1989).

The Prior Committee was set up to consider the future of Board involvement in disciplinary matters, and its report (1985) recommended various improvements in procedure and limitations to Boards' powers, although the abandonment of Board involvement in discipline was ruled out on cost grounds. The resulting White Paper (Home Office 1986) diluted these proposals, envisaging only limited legally-qualified assistance to Boards, again on grounds of cost. But the next move followed the 1990 riots. The Woolf Report observed in 1991 that in all probability many Board members no longer wished to maintain the role, and gave credence to the argument that whether or not Boards were in fact independent, the fact that they were not seen as independent because of the disciplinary role was sufficient reason to abandon it. From 1 April 1992, the Boards' adjudicatory role ceased to exist. What impact this will have on inmate perceptions of the Boards, of course, is a question that only time will answer.

Politics, Accountability and Parole

Penal policy is, of course, ultimately in the hands of ministers and members of parliament, whose interests usually lie not with the detail of prison regimes but with broader questions of public security. At this level—in so far as the executive does what government wants—the issue of accountability is largely overshadowed by party politics. In the 1980s, the issue of parole demonstrated this quite clearly in both England and France.

In England, a 1982 proposal to reduce overcrowding by increasing the use of parole was announced by the then Home Secretary to the Conservative Party annual conference. It attracted such criticism that it was almost immediately withdrawn, and this played no small part in the subsequent replacement of William Whitelaw by Leon Brittan as Home Secretary. A careful rethink was undertaken, and in December 1983, Brittan implemented a new series of proposals designed to have much the same effect, but using a double-track approach. On the one hand, most prisoners serving short terms would be automatically eligible for parole after one third of the sentence. On the other, selected groups—persons who had murdered police or prison officers, murdered in the course of an armed robbery, or committed sadistic or sexual murders of children, could expect to serve at least twenty years before parole would be considered; persons sentenced to more than five years for violent offences would be unlikely to receive parole at all; and the procedures for determining parole were changed to provide a stronger input from the judiciary.[12] These provisions have been challenged in both the domestic courts and the European Court of Human Rights, thus far with rather limited success on narrow issues.[13]

The significance of this history lies in Stern's (1987: 162–3) observation that prior to 1983, and at least since the 1960s, penal policy had been decided on a broadly consensual basis, with cross-party support for new measures, an all-party committee proposing new initiatives, and few differences of principle between the two major parties as to the kinds of measures they had introduced (both, in other words, had introduced both restrictive and liberalizing measures). The 1982 Conservative Party conference and the 1983 policy changes on sentencing policy, however, clearly signalled that penal policy was now driven by party politics.

The French experience is even more striking. As Faugeron (1991) notes, penal policies bifurcated in France in the latter part of the 1970s. A 1975 reform of imprisonment was:

confronted with press campaigns, especially against the most visible outcome of the reform: the practice of granting home leave to prisoners. A law and order discourse gradually superseded the liberal discourse. Police criminal statistics, reformed in 1972, started being used as unquestionable evidence of the rise of delinquency and of the need for a law and order policy. The publication of these statistics was accompanied by the denunciation of the judges' leniency. From then on, these parallel arguments, about the rise of crime and the judiciary's leniency, constantly set the background for the political debate. (Faugeron 1991: 250)

Other factors which also put the government on the defensive included prison riots in 1974, a series of sensational crimes, an economic recession and increasing intolerance of immigrants. Between 1978 and 1981, the rightist Giscard d'Estaing government introduced a number of more stringent sentencing arrangements. A law of November 1978 instituted a 'security period'—that is, a period during which day release, home leave or parole would not be granted—which was to last for the first two-thirds of a sentence of

---

[12] These measures were set out in Home Office (1984), and commented on in Maguire, Pinter, and Collis (1984).
[13] This litigation is more fully discussed in Ch. 8.

three years or more, and half of any sentence of ten years or more. A more controversial measure, however, was the 'security and liberty' law of February 1981.[14] Those convicted of a fairly wide range of violent offences or some serious types of property offence could now be given longer sentences, which would be served under less permissive conditions as regards day release, home leave, and parole.

The 1981 law was partly revoked by the newly-elected left-wing government in 1983; yet the 1978 law, which was arguably more repressive, remained in place. On the one hand, penal policy—especially with respect to parole and temporary release—was a matter of cross-party dispute. On the other, political perceptions about the 'penal climate' and public attitudes inhibited the removal of the more stringent measures, even though the lesser (but more controversial) ones could be disposed of.

Discussion

One view of the procedures for accountability which I have discussed—inspectorates, inmate grievance procedures—is that they should oversee the routine or ordinary aspects of prison life. Much of what they do relates to the correction of wrongs which, while perhaps important to individual inmates, are relatively minor from the perspective of the system as a whole. On this view, such mechanisms are neither intended nor likely to press for wholesale reform of the prison system. They operate within a domain assumption that the general scheme of things is acceptable. Yet in the English prison system, and at times in both Germany and the Netherlands, this domain assumption has been quite clearly false. And under such circumstances, in England, while both Boards of Visitors and the inspectorate have extensively documented the major flaws in the system, their messages—for they are in effect only messengers—have only very infrequently been heeded.

In practice, it seems, pressures for change in the prison system only rarely originate 'bottom up'. Most are designed and imposed 'top down', based on the concerns of elected politicians, ministers,

---

[14] Faugeron (1991: 251) notes that despite the repressive policies pursued in respect of imprisonment, the 1981 law also sought to reduce reliance on pre-trial detention, and make a number of more or less liberalizing changes in criminal procedure; it thus represented a kind of bifurcation in criminal justice policies.

and senior administrators. Fresh Start, in England, is an example of such a reform brought forward primarily on the grounds of the cost of the prison system and, in particular, of the cost of staffing institutions. When such top down reforms are implemented, it is not unusual to see 'increased accountability' being floated as one of the arguments to legitimate change. However, overarching this, one must also add that the penal policy context became increasingly conservative throughout the 1980s. While such policies have had to take account of very real problems, such as prison overcrowding, they have also been used to signal tough action against serious offenders—examples being the changes in parole in both England and France. Penal policy changes instigated at the party political level have not been concerned with accountability even at the level of rhetoric or symbol. The only sense in which one can speak of accountability in this context is that of the accountability of ministers to the dictates of their parties.

Finally, it seems that the leading edge of change instituted bottom up has been through a mechanism not explicitly discussed here, namely the courts. The majority of changes in prison conditions, in three of the four countries (France is the exception) have arisen out of court cases brought by inmates. In England these have been somewhat restricted, since the courts have, by and large, only seen fit to intervene in relation to procedural matters, in order to enforce the application of natural justice in areas such as disciplinary hearings. In Germany and the Netherlands, however, the scope of court-ordered reform has been wider. In the one exception, France, the dynamics of reform seem to have more to do with political wills and inmate protests, matters taken up further in Chapter 11. So far as inmate litigation is concerned, however, a fuller discussion is reserved for Chapter 8.

4

Inmate Experiences of Prison

THE previous chapter discussed accountability mechanisms and their significance for larger-scale prison issues. This chapter begins afresh with an examination of the everyday life of prisons that mechanisms of accountability are, in part at least, designed to oversee. It provides some sense of the day-to-day prison life that senior prison officials are ultimately accountable for, and the kinds of problems that might be expected in the practice of accountability. I shall concentrate here on the perspective of the inmates, and the two following chapters expand on this by looking at prison staff and governors.

Some Observations and Caveats

Sykes's (1958) study of an American maximum-security prison suggested that there are five 'pains of imprisonment': loss of liberty, loss of social acceptance, loss of personal autonomy, loss of security, loss of possessions, goods and services, and loss of (hetero)sexual relations. He also noted, and later researchers have borne it well in mind, that prisoners outnumber prison staff and that the latter depend on the co-operation of the former to keep the prison functioning in an orderly fashion. Clemmer (1940; 1958) and Irwin (1970; 1985) both deal with, though take different sides on, the view of inmate subculture as either an adaptation to the perceived constraints placed on prisoners by institutional life, or an adaptation to the more general conditions of their lives which they bring with them into the prison. Subsequent studies suggest that, insofar as the inmate subculture can be considered an 'adaptation' in the first place, both are probably right, though to varying degrees in different instances.

Cressey's (1961) edited collection takes the concept of inmate subculture further, by likening it not only to informal social control mechanisms, but also the problems of institutional change. Inmate subcultures in general, and the inmate power structure groupings in prisons, are seen as intimately linked with contraband goods and services, informal communication systems, and social-control mechanisms. On the one hand, the degree of unpredictability in prison life can be used as a means for staff to control inmates. On the other, institutional change which threatens the position of entrenched inmate interests can result in disturbances.[1]

Jacobs (1977) also discusses the linkage between inmate groups and institutional change in Stateville, Illinois, a prison which developed over a period of some fifty years from being virtually the personal fiefdom of the governor to being run as a rational, bureaucratic, professional administration. One key point here, which is significantly different from the arguments made by McCleery (in the Cressey collection) and by Clemmer, was that relationships in prison were radically changed not so much by the imposition of bureaucratic administrative structures and the employment of treatment staff—though these caused friction in different periods—but by the linkages with other levels of government and the increasing legal challenges to the system that formed the wider context for the changes within the prison walls. A second point, drawing on both the importation and adaptation models of inmate subcultures, was that while the major inmate groupings largely revolved around Chicago gangs and ethnic power blocs, the gangs on the whole acted more in consort against the administration than in opposition to each other. This may, however, have had more to do with the particular circumstances of Stateville than the general context of prisons.

Taken as a whole, the implications of these points can be summarized as follows. First, imprisonment is largely, for inmates, an experience of loss of individual autonomy; access to virtually all goods, services, and activities is highly controlled. It is hardly sur-

---

[1] Problems of institutional change can occur because the balance of power between staff and different inmates can be altered by organizational changes, as where treatment-oriented professionals are introduced into the structure. Informal social control mechanisms include e.g. the ability of staff to delay or deny access to routine goods and services as an unofficial sanction, or providing or allowing unofficial 'perks'. The ability of prisoners to control other inmates' behaviour through manipulation of access to contraband, and ultimately through threats and violence, has also been documented. See esp. the chapters by McCleery and Galtung in Cressey (1961); other studies on these topics include Garofalo and Clark (1985), Kalinich and Stojkovic (1985), and Los and Anderson (1976).

prising that the interstices of the official regime constitute a significant area of inmate social life; this is where they can regain a limited degree of power over their own lives. Nor is it surprising that staff, who in practice are in a minority and who have to get things done on a day-to-day basis, seek to manipulate the informal as well as formal elements of prison regimes. And finally, both staff and inmates, and the official and informal aspects of prison regimes, are linked together in a situation which is structured by broader organizational and institutional factors. The problem, then, is how all these areas interact in practice, and, ultimately, how the formal processes of accountability bear down on, and are managed at, a face-to-face, day-to-day level.

A great deal of other research, both qualitative and quantitative, has gone into the description of day-to-day prison life and prison conditions in England.[2] The work of King and Elliott (1977), King and McDermott (1989; 1990), and McDermott and King (1989) provide what is probably the most consistently fine-grained account of life across a range of establishments. The King and McDermott papers, for example, outline in some detail the ways in which staff and inmates can manipulate each other (refusing to open doors to allow inmates access to toilets, 'shit parcels' thrown out of the windows, parties of inmates detailed to collect them), and, more quantitatively, point out that despite the massive increases in prison resources in England since the early 1970s, regimes have if anything deteriorated, with less time out of cells, fewer activities, and so forth.

Work on practical aspects of prison regimes in France, Germany, and the Netherlands exists in varying quantities, though it tends on the whole to be positivistic or journalistic rather than qualitative in its approach. Formal research in the Netherlands includes studies of inmate subcultures and adaptation, evaluations of special regimes, and the relationship between management models and regime 'service delivery'.[3] A number of French studies have looked

[2] Much of the Anglo-American literature on prisons, with particular emphasis on aspects of control, is reviewed in Ditchfield (1990). While my intention is merely to highlight a number of key themes, Ditchfield's work is an exceptionally full account of our stock of knowledge in these areas.

[3] Empirical studies published in English include Grapendaal (1990) and van der Linden (1987) on inmate adaptations to regimes and subcultures, Berghuis's (1987) evaluation of a special regime, and the links between management models and regimes (Vinson et al. 1985; 1987).

at different aspects of regimes, including work, isolation, and establishments' house rules.[4] Other works, including Delteil (1990), Favard (1987), and Pauchet (1982), discuss prison conditions, though from a relatively unstructured, journalistic viewpoint. Taken as a whole, however, their comments are heavily critical of the overcrowding, physical conditions, and lack of regime activities. Pauchet's thesis, in particular, is essentially that prisons are insecure in two senses: first, the number and type of escapes, and second, the nature of the regimes, with the uneven application of control by staff and the level of inmate-on-inmate violence. In Germany, perhaps the most comprehensive study of prisons is that by Dünkel and Rosner (1982), though it relies heavily on statistical information such as disciplinary and other measures, self-mutilation and suicide rates, and information on personnel such as the numbers of treatment staff in different institutions. Beyond this, most of the empirical studies of regimes appear to have been conducted in social-therapeutic units or prisons, and are oriented towards questions of treatment.[5]

One point that emerges clearly out of this broad range of material is that, whatever comments might be made on a particular prison system, the diversity of regimes and conditions across prison systems needs to be stressed. In discussing, for example, dependence on staff, inmates in English prisons often make reference to the consequence of lack of sanitation in cells. This is that staff may ignore inmate requests to use a toilet, thus forcing them to urinate and defecate in a bucket in the cell. This is almost unheard-of in the other countries, where in-cell sanitation is the norm.[6] In relation to Sykes's point that inmates are deprived of possessions, being allowed only those items which the institution allows them to keep, there is a world of difference between England and the Netherlands, the two systems which are furthest apart in this respect. The difference is one of the history of the administration's attitude towards inmates, and we can see the historical accretion of

---

[4] e.g. Bibal *et al.* (1983) on prison labour, and Fize (1984) on isolation. A study of 'house rules' was completed in 1985 by the Ministry of Justice (*Rapport du Groupe de Travail sur les Règlements Intérieurs des Établissements Pénitentiaires*), but remains unpublished.

[5] But see e.g. Schick (1986), which is an account of imprisonment in Hamburg by a serving inmate, and presents a number of examples (possibly unrepresentative) of poor decision-making, restrictive rules, and lack of regime planning.

[6] Though it is lacking in some of the older Dutch prisons, including Den Haag.

such attitudes reflected in matters such as prison design. For example, prison cells in the Netherlands almost all have electric sockets supplying standard mains current. English cells, with limited exceptions, do not. For this reason alone, it would not be possible, even if it were culturally acceptable, for inmates in most English prisons to have electric typewriters, coffee filter machines, stereo systems and so forth in their cells (this is not to say that all Dutch inmates have them, but certainly most have at least one consumer item which requires a normal electrical-mains supply). And on another front, there is a world of difference between, for example, the rather tough disciplinary regime in Fresnes, a large French maison d'arrêt near Paris, and the way in which Dutch prison staff talk and drink coffee with inmates—and on one occasion when I was present, sit down together to repair the inmate's walkman. This is not to say that the Dutch attitude towards inmates and regimes is liberal, loose, or undisciplined. Inmates talk about the 'repressive tolerance' of the institution, that is, its paradoxical ability to absorb and smother inmate disaffection and deviance by tolerating it. And they point to occasions on which the staff consciously and, in a way, ceremonially demonstrate that they are 'taking control'.[7] But quite clearly, administrative and staff attitudes towards inmates, staff-inmate interactional styles, and the ways in which control is exercised, can and do vary widely between different prison systems.

Differences exist between systems, but also within them. To the extent that establishments are designedly differentiated by function, contain discrete and specialized units, and take different categories of inmates, this might well be expected. To the extent that institutions build up their own traditions and solutions to practical organizational problems over time, it might also be expected. Yet some

---

[7] Two examples of Dutch practices in this regard can be offered. One is that if it is thought that inmates about to be taken to the segregation unit or to be transferred will offer resistance, a large number of officers—up to 20—will assemble to carry out the move. As one inmate noted, the large number of staff on hand is intended to demonstrate from the outset that resistance is futile, thus minimizing the risk of violence. The second is that systematic cell searches are carried out, not by prison staff, but by separate tactical squads who carry out the searches in riot gear. Staff were happy with this arrangement since much of the friction associated with cell searches was diverted from them. These squads are also sometimes paraded at prisons for reasons which inmates were typically unaware of, though they assumed the intention was to remind them that an iron fist does exist, despite the more common wearing of velvet gloves.

of the differences within prison systems are as great, or greater, than the overall differences between them. In France, for example, there is a world of difference between the strict regime of Fresnes and the much more relaxed atmosphere of Melun, an institution for long-termers (inmates and staff attributed this largely to the different staff unions that predominate in each establishment).[8] In the Netherlands, which has a generally modern prison estate, one can still find institutions such as Doetinchem, originally a sanatorium and now a semi-open prison, in which inmates sleep in poorly appointed 'chambrettes', small bedroom units separated from each other with wooden partitions. And in England, inmates are well aware of differences in the privileges available at different establishments of the same kind. Such differences tend to be obscured by talking of prison systems as a whole and making comparisons between them.

Having offered an extensive list of caveats, I shall now turn to 'my inmates', interviewed between 1986 and 1988, their views, and some specific issues which seem to me to have some bearing on the idea of accountability.[9]

Dependence and Infantilization

Inmates everywhere are dependent upon prison staff for the organization of virtually every aspect of their lives, from meals to visits, and from obtaining library books to being protected from attacks by other prisoners.[10] French inmates had a very descriptive word for this aspect of their dependence on staff. As one observed,

'Infantilization' is important. If I want to send something to my family I need authorization. A lot of little things like this. A double visit, I have to ask. If we ask correctly it's given, but we have to ask.

The concept of infantilization points first to the need to ask permission for virtually everything they wished to do on their own

[8] One social worker at Melun, for example, used the word 'traumatized' to describe the state of mind of inmates arriving from Fresnes; prisoners I interviewed who had been transferred from Fresnes said that after that experience, being in Melun was 'like heaven'.
[9] Further information on the numbers of interviewees, the interview schedules, etc. is provided in App. 1 on data collection.
[10] The extent of this dependence and the linkage between dependence and control is widely recognized and has, indeed, fuelled much of the academic discussion on prisoners' rights which is further discussed in Ch. 6.

initiative. Second, it refers to the manner of asking: one has to ask 'correctly'. Third, it was sometimes used to point to the rigidity of staff control over inmates, with the need, like young schoolboys, to wait, to queue, or to be supervised or checked at so many points in the day, as one activity gave way to another—for example, as work finished and the exercise period began. But equally significantly, it encapsulates the notion that the creation of dependence is also the creation of a means of control. If inmates depend upon staff for visits, the threat of refusal to allow visits is also a means of ensuring compliance.

This experience was by no means confined to French prisoners. It was reported in all the prison systems, but with different colourings. Inmates in Dutch jails typically felt that although in some respects it was easy to make requests, there was no form in which a request could be couched that would guarantee its being treated seriously, i.e., considered and answered. On the other hand, there were a number of significant ways in which Dutch inmates could make—indeed, were forced to make—their own decisions, for example in deciding how to exercise their limited entitlement to make phone calls.[11] There was, however, comparatively little formal 'queuing and checking' of the kind common elsewhere. In Germany, inmates mentioned delay in the consideration of any request as a significant factor, a kind of ceremonial underlining of the point that the inmate is dependent on the administration for an answer, and thus dependent on the staff's sense of time and bureaucratic procedure. In England, infantilization came about through bureaucratic delays in handling requests, and the dependence on staff for access to toilets. However, in England, infantilization seemed to take second place to another problem, that of the perceived arbitrariness of decision-making.

Arbitrariness, Predictability, Rationality

In principle at least, one feature that marks prison regimes is their predictability. Prison life is marked by timetables, by the

[11] House rules concerning the frequency and length of phone calls varied; but the essential point was that in many institutions, inmates had to choose whom to phone, and when; and had to book the call at the office. This procedure was often regarded as unnecessarily inflexible, and likely to result in a significant proportion of unanswered calls; and it was felt by inmates to be one of several ways in which they were placed under pressure.

predictable ways in which problems of various kinds are responded to, by established rules and norms, and by an official hierarchical and bureaucratic structure which defines and allocates responsibilities, rights, duties, and capacities. Inmates are, in theory, treated in a systematic and even-handed way, according to established criteria—though those criteria often, of course, include assessments of the inmate's behaviour and personality.

Prison regimes do indeed have some of these characteristics, although prisoners themselves were often unclear quite how assessments of behaviour and personality were made.[12] One example from the English training prison was that once the the SOs (senior officers[13]) could decide whether or not to allow prisoners to make telephone calls, inmates said they found no discernable pattern as to who was allowed to make calls, when, to whom, or for what reasons. This may be a situation in which the rationality of the decision was not apparent to the inmates, who could not therefore predict the outcome of any given request to make a call. But the end result was that inmates assumed that 'if the face fits, they let you make the call'. On other occasions, however, it appeared to inmates as though there was a rationality to the way they were treated by staff, although a rationality influenced by wider social factors in the prison, such as the degree of surveillance of officers by senior staff:

[12] As a result, there is a good deal of cynicism about decision-making, and a number of apocryphal stories esp. about psychological testing are persistent. I have heard the following story retold by inmates in different parts of the English prison system, with names and places altered to suit the convenience of the storyteller. At X institution for young offenders a psychologist would routinely conduct a batch of tests on newly received inmates, in which one question was: 'what would you do if you saw a battleship coming down the road [usually named as the main road of the local town] towards you?' The new arrivals, having been briefed by already-tested inmates, would routinely answer: 'I'd sink it with a torpedo from my submarine'. The psychologist's response would be: 'where do you get the submarine from?' to which the standard answer would be 'The same place you got your battleship from'. Although this is often used as an example of psychologists' craziness, it seems significant that no inmate I have heard the story from appeared interested in what such a question would be intended to show.

[13] Since the research was done, the staff ranking structure has been changed as a result of 'Fresh Start'. In this discussion I have retained the staff designations as they were used at the time of my interviews, and as they are still sometimes used by inmates. The old designations 'Principal Officer' and 'Senior Officer' are now replaced by Unified Grades 6 and 7 respectively. Assistant Governors grade 1 and Chief Officers grade 1 both became Unified Grade 4, and Assistant Governors 2 and Chief Officers 2 became Unified Grade 5.

You get different things on different landings in the same wing. You can get away with more on the 2s [the first floor] than the 3s [the second floor] because there's more activity, people walking round etc. But the 4s are furthest away from the office, so it's often the easiest—officers will let things go. (inmate, English local prison)

The kind of problem that many inmates pointed to, however, was the unpredictability and apparent arbitrariness of staff behaviour in enforcing or not enforcing rules. This was mentioned in a number of connections, which can be illustrated in the words of some (English) inmates:

On exercise, they stop you taking your radios with you, but it depends on the staff. Some will stop you and others won't. It's the same for wearing slippers to work. Some guards are unpredictable, some are nice but have off days, some are helpful but strict, and so on. (inmate, English training prison)

At times the staff will pull you up for silly things. For example, cassette tapes. You might lend them out, and sometimes the officers will take the tape to another prisoner for you, and sometimes if you ask them to do it they'll nick you. It is a rule, don't lend out tapes, but they ignore it cos it's not important to them. (inmate, English training prison)

It's OK if they'll nick you [report a disciplinary offence] all the time, you know where you stand. The ones that let things go now but not tomorrow are the problem. (inmate, English local prison)

Moreover, inmates were aware that staff did not always agree among themselves about actions taken in relation to the inmates:

The staff bicker amongst themselves. For example, last week one sacked a cleaner but the SO [senior officer] said he shouldn't have been sacked, so now he's back on. (inmate, English training prison)

These kinds of issues are not, however, unique to England. Observations by two Dutch inmates indicate a similar perception of arbitrariness:

The behaviour of the staff is very important. They are too arbitrary, so you have to know what's possible and what isn't. (inmate, Dutch semi-open prison)

If you want an extra phone call it just depends on which guard is on duty if you'll get it. (inmate, Dutch HvB)

In Germany, a different problem seemed to emerge. The inmate consensus seemed to be that the rules were not applied arbitrarily, but in a highly formal and bureaucratic manner. This created a somewhat different situation. Even though many 'ordinary' things were easy to get, they none the less had formally to be applied for, and deadlines for applications scrupulously observed—as one inmate put it, 'the time of the filing of a claim is important'—while decisions that could not be made on the wing did not come back quickly:

Relaxation measures are dealt with in a more slow-moving manner. There is no problem with house internal things, but dealing with applications is still too slow-moving. (inmate, German young offender establishment)

The distinction made here is between matters regulated within the accommodation unit and matters regulated centrally and according to legal provisions, known as relaxation measures (Lockerungs-massnahmen). However, the latter category was held to be overly restrictive of everyday affairs, such as access to sports facilities, and to be so slow that short-term inmates could be released before decisions on requests were reached.

A general sense of overly strict internal divisions within establishments, and their 'unintended consequences', also seemed to apply in all prison systems, and across a wide range of areas which had to do with the regime. One example, discussed below, is that cell changes may have consequences for work allocation and vice versa. But other examples can also be adduced:

The ateliers [workshops] have washing machines: new inmates have to wash their clothes in the showers until they're allocated to an atelier. (inmate, French maison d'arrêt)

In a Dutch half-open prison, in relation to home visits:

At the moment the situation is like this, that you'll get your travelling-money back if you can show the train or bus tickets. A few months ago the institution's van brought us to the station. They don't do that any-more. This implies that you have to wait for almost an hour for the bus. So what most of the boys do is that they take a taxi to the station, which costs 10 Guilders. If you don't get any money from outside this means that you've only got 15 Guilders left—your earnings minus costs for recreation etc., minus money for the taxi. These 15 Guilders are not enough to buy a ticket to Amsterdam for instance. That means that the prisoner won't pay for his train, gets caught, has to pay but can't and gets a warrant [i.e. a demand for payment within a specified time]. He can't pay it so starts stealing or gets extra days for it.

In both English and German prisons, any mismatch between individual status and the general conditions of the prison could become a cause for complaint. For example,

I, as someone who is serving a long sentence, cannot take part in jogging, swimming etc, because I am not yet allowed to have relaxation. (inmate, German young offender establishment)

I'm after home leave. I'm entitled to it, or would be if I was in a training prison. I'm told I can't have it because I'm in a local—but I'm only in a local because of medical disability. (inmate, English local prison)

The English inmate's complaint is that although his sentence length would in principle mean that he should be allocated to a long-term prison, his medical state means that he is kept in a local prison (which has extensive medical facilities) and therefore he is subject to the same regulations that apply to short-term inmates—thus denying him the home leave he would be able to apply for (by calling it an entitlement he is in fact overstating his case) were he in a training establishment. In the German case, a parallel situation exists; the inmate was a long-termer held in a largely short-term unit. He therefore did not get access to facilities available to his co-inmates.

Relations with Staff

One key factor in the question of how far inmates feel they need to interact with staff is that of how much decision-making power staff have. In general, basic-grade staff usually claim that they have little or no decision-making power, though this is to gloss over what counts as a decision in the prison context. As mentioned earlier in this chapter, the question of whether or not to open a cell door to allow an inmate to the toilet can be a significant 'decision' in many English contexts. None the less, prison officers do typically have few formal powers which they may exercise directly. In the Netherlands,[14]

[14] This quote, while illustrative of the general position, cannot be taken as indicative of the exact situation in all Dutch prisons. In e.g. one unit of the Over-Amstel prison in Amsterdam, an inmate pointed out, 'Staff can't decide on anything. For example, they need the permission for extra phone-calls', while in an

The guards can only advise on things. They can't decide on anything themselves; the brigadier decides. The governor can decide on anything. He has the absolute power. Only on things like an intermission of your sentence, the department decides but the governor still advises. The social worker can only give advice. He can't decide on anything. (inmate, Dutch semi-open prison)

In Germany, the situation is much the same:

The officer in charge of floor is involved in every decision. However, this authority to decide is formal and can be undone any time. At last, the governor decides. Everything concerns, is related to, the security of the institution and is therefore under the control of the governor. (inmate, German young offender establishment)

In France, because of the strict controls on discipline staff by their superiors, inmates characterized most officers as turnkeys, who had virtually no discretion in their work and who were afraid of making any decisions on their own initiative. As one pointed out, 'The guards are afraid of the system more than us, they can do nothing'. This is perhaps not a wholly accurate statement; the French prison service is highly unionized and staff unions have exerted considerable influence on penal policy in general. But it may still be substantially true at the level of staff–inmate interaction.

Getting Things Changed

Inmates are not, despite the points made above, passively dependent on staff. Nor do they launch their requests into a void; they usually try to find out who has the authority to make a decision and what kinds of criteria will be applied. The problem with the kind of arbitrariness outlined above is that inmates find their powers of prediction (of what will be allowed if requested, what will be *de facto* allowed if possessed, or which members of staff are likely to be more or less amenable) greatly decreased. But since inmates usually seek to reduce the 'unknowns' in making requests, and to maximize their chances of getting what they want, they are usually well aware of what is easy to obtain and what is not, of what kinds of consequences follow from particular requests being granted, and of how to set about getting the things they want.

adjacent unit, another inmate indicated, 'Staff can only decide on very little things like having an extra telephone call'.

These factors varied across the different systems and sometimes within them.

In France, for example, arranging to change one's cell was difficult in the centre de détention (long-term prison) studied, but easier in the maison centrale. The former operated a progression system, in which inmates were first allocated to double cells on the ground floor and literally 'moved up' the prison, with the 'best' prisoners on the top floor (all the cells from the first floor up were single cells). Moving from one floor to another was therefore a mark of a change in status within the establishment. To change from one atelier (workshop) to another was, however, more difficult in both cases:

To change cells is quite difficult. You go to the surveillant-chef, who looks at your dossier and finds someone to change you with. Within an étage [floor] it's easy, otherwise not. (inmate, French centre de détention)

To change atelier is difficult. I have to wait three to six months for other work if I stop in this atelier, and would have to change cell too. (inmate, French centre de détention)

To change cells isn't difficult. You write a note to the chief of security, or ask to see him. You ultimately do need to write a note. I've changed my cell three times. If you're not difficult you get treated well. To change workshops is more difficult, people there [i.e. the new workshop] need to agree, to want the person. (inmate, French maison centrale)

To change atelier is more complicated [than a cell change]. You have to ask more strongly; ask the chef de détention or direction. You must present a reason. (inmate, French maison centrale)

The key point, in France, was that one had to write to the relevant officer or governor in the establishment, and to justify the request. But some things were more easily accepted than others, partly depending on the nature or the request and partly on the nature of the decision-maker:

It's easy to get an extra telephone call. More difficult to get an extra visit. Very difficult: home leave [permission de sortir], because the decision is for the JAP and the governors. (inmate, French maison centrale)

To get anything, personal goods, you need permission. Study books are difficult enough. You need to write to the educators. It's not easy, it may take one or two weeks. They don't like stuff being sent in, don't like parcels. And with for example a home leave, there's always a cell-search afterwards. (inmate, French maison centrale)

In Germany, inmates generally saw little difficulty in arranging to change cells within an accommodation unit, though responses to requests were often said to be arbitrary. Requests to move to a different accommodation unit, however, although easily made, were frequently denied.

In the Netherlands, much depended on the nature of the prison. The situation in the closed prisons was one in which 'extras', or matters such as cell changes, were not difficult to obtain (notwithstanding the comments on arbitrariness above), though they were often the product of negotiation and the striking of 'deals':

For a special visit you apply. You do have to explain, and you can do deals, for example if you apply for six hours, you can have five hours over three days.

None the less certain decisions, such as those of the prison doctor, could not be effectively challenged. And matters such as allowing several inmates to associate together in a cell were strictly forbidden, because—inmates presumed—the staff were worried that homosexual activity would take place. And not everything was easy or simple: as one inmate noted, 'In almost any case you have to fill in a note—the guards don't dare to take the responsibility'.

This issue of guards not wishing to take responsibility is a rather complex one, because although Dutch prison staff on the whole are expected to have a lot of contact with inmates, they also seem to feel that they are not always able to exert their authority as a consequence of the rather non-confrontational day-to-day style of prison management. Thus some inmates may get what they want because they are prepared to imply (though not explicitly make) threats to staff. One inmate had the following story:

The fact of being liberal makes the staff lazy—for example not opening doors, tolerating pot. You have to threaten them to get things. Once I wanted to go to the toilet and called the guard. I went in the pot before the guard arrived and then went to empty the pot in the recess. The guard stopped me and said it wasn't permitted to empty the pot in the day because of the smell on the wing. I said the pot smelled worse in the cell and opened the lid to make it look as if I would empty it over him. He backed down. (inmate, Dutch maximum security prison)

The comments above relate to inmate perceptions in closed prisons. In the semi-open prison the situation was seen rather differently, with more strict rules on the one hand, but less systematic enforcement on the other:

Changing your cell is quite difficult; at least if you ask it. If you just do it they won't say anything of it . . . Changing of possessions is also officially not allowed, but everyone does it.

For home leave, you could get the extra leave but in that case your next regular leave lapses. That means almost four months no leave. It would be better if we had a weekend leave every 14 days.

In England, perhaps the best summaries of the situation were the following, offered by two inmates at the local prison:

There are few things you have no control over. You can ask for most things. The main things you can't control are allocation, pay, unexpected things, e.g. if you miss a film and want to go another time. (inmate, English local prison)

You can get a welfare visit on compassionate grounds. They're very few and far between. We get two visits per month. You can have a VO [Visiting Order] early, but not an extra one; you can get longer visits, but not more. You show your case to the welfare and the AG [Assistant Governor] okays it; welfare aren't so powerful now because everything is more liberal. (inmate, English local prison)

Other areas of rigidity were in relation to bathing and work changes. For most inmates, one shower a week was the norm and extra showers were virtually impossible unless the inmate had some special grounds for them, such as working in the kitchen. The situation in the workshops was partly conditioned by the degree of training necessary for any particular job. As one inmate in the training prison reported,

Change of work: depends where you're working. Once you've been trained up for the tailors or plastics shops they like to keep you there. Cell changes, I've never known them say no.

Two general axioms that emerged from inmate interviews were that the more offices are involved, the more difficult it is to secure agreement to something; and the further up the hierarchy one has to go to have a request considered, the more likely it is that the request will be dealt with in terms of formal criteria, and thus 'the less you can get away with'. Certainly there were situations in which inmates were able to achieve substantial improvements in their conditions largely because their requests were handled by wing staff. In one case, an inmate was able to remain out of his cell, rather than locked up, all day because the wing Principal Officer agreed it on his own responsibility, without consulting his superiors. The inmate was medically unfit for work, and a strict application of the rules would have meant his being locked in the cell at worktimes—which, combined with other periods when all inmates are in their cells, would have meant being locked up for close to twenty-three hours a day.

Regime Problems

Although inmates in all four countries were critical of the physical conditions in the establishments, they often saw little point in complaining about them. For example, in virtually every prison I visited, inmates commented adversely on the quality of hygiene. In England, one comment was, 'The hygiene and toilets are bad, and the new kit can be almost as bad as what's handed in'. In the Netherlands, it was argued that any serious study of institutional hygiene and cleanliness 'would expose bureaucratic incompetence and disregard for the law'. Similarly, the quality of institutional food was criticized everywhere, with inmates (except in England) claiming that they had to supplement their diet by buying fruit and fresh vegetables.[15] Such criticisms also encompassed the operation of the prison canteen. In Germany, a common complaint was that the prison canteens sold goods at higher prices than ordinary shops, and which had exceeded their sell-by date. In England, one major complaint among the inmates I interviewed concerned the need to buy toiletries from earnings rather than personal cash. Earnings ran (at the time of my study) from about £1.40 per week for a prisoner without employment to about £4 per week for the higher earners. At the lower end of the pay scale, inmates are forced to choose whether they will buy tobacco or toiletries (soap, toothpaste, shampoo etc.). Many inmates smoke, but also view such items as 'essential to keep clean' or as one of the ways of asserting individuality. Interestingly, this was not a system-wide rule, and it was this that fuelled the complaints: inmates named

[15] There was, however, a great degree of variation in what inmates bought and in the availability of facilities for self-catering. In both France and the Netherlands, many (though not all) inmates had access to small cookers—often, in dormitory accommodation in France, in the dormitory itself. In both countries it was common to see cardboard boxes of vegetables in inmates' cells and dormitories.

other establishments in which they claimed this rule did not exist, and claimed that their own establishment was 'out of step'.

However, the major complaint everywhere concerned visits, and in particular the lack of opportunities for heterosexual contacts. Visits were regarded as a key aspect of the regime, but also as lacking in privacy even where sufficient visiting time was allowed:

Visiting rights—that would be the number 1 situation to cause friction, it is very important. (inmate, English local prison)

Visits are very important; once a week, it's enough but there isn't enough privacy. (inmate, Dutch semi-open prison)

In France, this criticism was often bound up with the question of sexual freedom and the inconsistency of tolerating homosexual behaviour in prison while denying heterosexual behaviour:

There are four important things; sexual liberty, visits, privacy, home leave. The visits here are open, everybody can see. You should have more privacy and more time. (inmate, French maison centrale)

For us, what could we have? Only sexual liberty. They know there is homosexuality, they see it in the spyholes, but ignore it. But if for example my wife visits and sits on my knee, it's forbidden. Most of the inmates here are married or have girlfriends. (inmate, French centre de détention)

In sum, inmate views on regimes were not that different between the countries, despite the different levels of actual provision. None the less, there did seem to be differences in the perceptions of order in the establishments. French inmates stressed their 'infantilization', and noted that the high level of supervision of staff by their superiors led to a similarly high level of control on them by staff. Yet they also perceived the structure of most French prisons as ill-suited to the prevention of the kinds of minor everyday frictions associated with moving the inmate population through activities such as mealtimes and exercise, and as not offering a great degree of protection from other inmates:

Most important for the inmates is that 50 per cent are considered foreigners even though actually they're French Arabs. All go their own way, there's no discipline in the prison. The one who yells the loudest gets his way. It would be better if we weren't mixed together, it causes tension. The staff can't handle trouble, their duty is to get out and call the gendarmerie if the inmates get out of hand. (inmate, French maison centrale)

A slightly different point was made in Germany. Inmates had a general sense that many provisions and services were tightly controlled—although, as with France, the law was still balanced by areas in which administrative discretion needed to be applied. Much of the application of discretion was characterized, not as arbitrary, but as too highly rational; one's legal/administrative status determined the facilities available in a very tight way which did not always accord with common sense. One further consequence of the degree of organization was that any request which went even slightly outside the established routines and procedures—that is, which required an original response, an addition to the 'stock of interpretations of the regulations'—had to be fought for. And despite the appearance of order in German prisons, inmates observed that there was a high (though not necessarily serious) level of inmate-on-inmate aggression, often triggered off by trivial things. In the Netherlands, as one inmate put it, 'It's not the States over here. Only the guards will find it interesting if you're a "heavy" criminal. There's no right of the strongest.' But despite generally relaxed regimes and the relative openness of staff to requests, it was said to be necessary to make sure that requests were dealt with and not sidetracked into endless consultations with other staff that form part of the managerial culture.[16] And in England, inmates were more preoccupied with the apparent arbitrariness of staff in applying rules than with what the rules were.

Finally for this section, the situation with regard to contraband was complex, and I can only make a few general comments. In most prisons, apart from unauthorized trading of personal possessions, the primary forms of contraband were cash and cannabis, both of which were effectively inmate currencies.[17] Much of what inmates said conformed with the received views within sociology about prison contraband. On the one hand, staff were perceived as tolerating a limited degree of soft-drug use, often on the grounds that inmates were more relaxed and easier to deal with. What counted as a 'limited degree', of course, varied between open and closed institutions, and between countries. In France, for example,

---

[16] One inmate offered an alternative formulation: 'They [staff] mustn't promise things they cannot fulfil and afterwards put the blame on the department.'

[17] This is not to say that other forms of contraband did not exist. They did. But apart from the Netherlands, where the extensive availability of heroin in prisons is well-known, hard drugs tended to be a minority interest. Although earlier literature often refers to tobacco as an inmate currency, and although it is still sometimes used for this purpose, it seems to have been supplanted by cannabis in many institutions.

the 'official' position is that possession of drugs constitutes a criminal offence for which a court case would be brought. Yet despite inmate assertions in both prisons where I conducted interviews that drugs were common, the staff I interviewed could not recollect any such prosecutions ever having been brought. But having said this, staff and inmates everywhere were concerned about the informal power that dealing in drugs brought, and about the tensions arising when supplies ran short.

It seems that inmates, faced with a system in which they are more or less powerless, none the less can and do seek individual privileges and small improvements of various kinds in their day-to-day life. This they do by trying to understand the rules of the system and, equally importantly, the characters of those who run it; who has authority to make decisions, who makes decisions in principle and who makes them arbitrarily, what can be got easily and what cannot, and thus what is thus worth asking for and what is not. But this relates in the main to privileges which individuals can obtain for themselves. And to say that inmates can improve some aspects of their prison life easily is far from saying that they are satisfied with prison conditions in general.

Rights and Videos

One Dutch inmate pointed out that, in general, inmates are more interested in having access to good videos than they are in having stronger rights in relation to their treatment in prison. There is undoubtedly much truth in this; and while it may be surprising, in view of the stress placed on rights in recent academic literature on prisons, there is some rationality to it. Rights are essentially enforceable claims that the inmate has on the prison authorities in relation to the way he or she is treated. Academics may regard them as providing ground rules for inmate–prison relations. Yet they tend to be, for most inmates, and most of the time, background considerations. They only come to be of direct relevance where inmates wish to use them in order to require the administration to follow or abandon a particular course of action.

Moreover, having recourse to rights is often problematic. First, in most jurisdictions, not only do rights tend to be procedural rather than substantial (inmates can require that decisions are based on specified criteria, or in conformity with some specified procedure, rather than that a particular outcome is reached), but the procedures for enforcing rights tend to be cumbersome and lengthy, relying ultimately on the courts. Second, the point at which one reaches for rights is often the point at which there is a high level of conflict between the inmate and the administration, and calling on rights can give an inmate the reputation for being troublesome, with possible negative consequences for his or her ability subsequently to manipulate situations by informal means. Third, many of the things inmates seek cannot easily be obtained by rights. In the case of visits, for example, although cases exist in which inmates are denied visits, they are few and far between; in most cases the issue is one of obtaining an extra, or longer, visit, thereby rising above the provision specified in the regulations or laws. Clearly there are cases in which rights are at issue; but for most inmates, most of the time, what is being sought is an extra provision, having the nature of a privilege, and determined by institutional staff or governors. For all these reasons, on the whole, inmates probably prefer to seek some informally negotiated outcome, or at least have the decision made by officials they have direct access to, rather than 'demand their rights'.[18]

The attitudes of the English inmates are not untypical of all the inmates I interviewed. A question about whether inmates should have special rights pertaining to their treatment in prison met with agreement, though there was no unanimity about which rights these should be. The specific suggestions made were conditioned by their recent experiences in prison rather than by any reflection on the nature of imprisonment. It was also pointed out that many of the matters about which arguments arose were, at root, either very petty, or linked to conduct and attitudes, for which no clear rights or standards could be set or effectively enforced. Inmates tended to take the view that if facilities were *de facto* available (visits, letters, gym, films, etc.) that was sufficient—though there was always room for improvement. The only right for which serious arguments were raised (and this probably reflects the timing of the interviews, with a UK general election in the offing) was the right to vote. A similar question on whether prisons should be required to meet minimum standards for regimes similarly met with scepti-

---

[18] The question of inmate rights is discussed further in Ch. 8, which also describes a number of legal cases and analyses the extent to which a rights-based approach to questions of prison regimes and conditions is sustainable.

cism. It was pointed out that while current conditions left something to be desired (food, cleanliness, and supply of hot water were mentioned), at a realistic level, minimum standards were unlikely to achieve anything. In the case of hygiene, for example, it was said that the principal factor affecting hygiene was the attitudes of other prisoners to keeping themselves and their cells clean, and that this could only be improved if prisoners took more responsibility in such areas.

Conclusions and Implications for Accountability

Most of the ground covered in this chapter has had only a tenuous connection with the concept of accountability. But it does indicate the kinds of situations in prisons to which abstract ideas of accountability have to be applied.

Take first the notion of 'infantilization'. It is often observed that inmates are dependent on staff for many things which in normal society one can do for oneself. It is also observed that inmates must often comply with staff definitions of good behaviour in order to obtain some of these things as 'privileges'—in short, that they must behave themselves and must 'ask nicely'. And equally, it has been observed elsewhere that the fact that such things are defined as privileges rather than qualified entitlements (that is, that inmates meeting specific and stated criteria are guaranteed them) means that staff, or at least senior staff, retain substantial discretion which grievance procedures do little to address. The very fact that privileges are discretionary means that a complainant must demonstrate that the exercise of judgement was unreasonable—an argument which can at best be tenuous, since staff can always retreat behind a shield of 'security considerations'.

Prisoners are not completely at the mercy of the system. They can and do take steps to increase the probability of a decision being made in their favour. But the kinds of decisions we are talking about here are typically only at the level of getting a cell change or an extra telephone call. Other, more major, problems such as the poor fabric of the buildings, overcrowding, and so forth can hardly be dealt with by inmates through the kinds of practices outlined in this chapter. Moreover, even in the small matters, inmates are limited in the extent to which they can maximize their chances because—certainly as they see it, and probably to some extent in reality—the decision-making procedure has an admixture of arbitrariness. The result is that in an inmate's day-to-day life, whatever the formal programme that the institution is committed to delivering, there is no clear link between behaviour and outcomes. It is not that inmates experience insecurity because prisons cannot protect inmates from each other—for example, from inter-inmate assault (though this is, on occasion, true). It is not that inmates experience psychological insecurity because there are no links between behaviour and outcomes. In most prisons the likely outcome of attacking an officer, attempting to escape, and so forth are fairly clear. But at the level of routine questions such as whether one can change library books, get an extra visit, take one's radio into the exercise yard or swap cassette tapes with other inmates—which, to most inmates, most of the time, are the important issues—there is a lurking unpredictability.

Procedures for ensuring that the legal requirements are met and that a stated regime is operating can detect such problems through inspections—for example through talking to inmates—in exactly the same way that researchers can discover them. In England, the bread and butter of complaints to Boards of Visitors consists of exactly these types of complaints. But the problem of how one sets about remedying the host of minor grievances, the small resentments, the 'interactional order' of the prison, is permanent and intractable.

One argument that has been brought forward on occasion is the 'empowerment' of inmates through the provision of clear and precise inmate rights to particular services and facilities. This, it has been argued, will ensure that an inmate will know whether or not he or she is entitled to something, and a grievance procedure cast in terms of rights ought to be able to establish the facts of the matter and reach a decision in a straightforward manner. But, as I have tried to indicate, two problems exist. First, much of the dispute is not about 'basic' provisions which would be covered by rights, but about 'extras' or concessions such as cell or work changes. And second, the framework of rights is itself a rather cumbersome process which can generate bad feeling quite as effectively as it can right wrongs. Relationships are important in prisons, especially given the simple fact of infantilization that I started out with. The key to securing good relations is not the application of sophisticated legalisms, but a sensible approach to negotiation and mediation within a framework acceptable to both parties.

Finally, it is important to make a cautionary observation. Many of the quotations and comments in this chapter have supported three general conclusions about prison systems. First, they are internally diverse. Second, many of the fundamental inmate experiences—of dependence on staff and so forth—are similar across systems. And third, inmate views about their situations are, surprisingly, rather unrelated to the actual conditions they experience. It would, however, be a mistake to assume that because all systems share these characteristics, they are all the same. They are not. Two points in particular should be singled out. One is that, in comparing establishments across systems, one can begin to see clear variations between the systems. Perhaps the simplest example in this regard is that the English failure to provide electrical sockets in cells (with only a very few exceptions) in and of itself leads to a massive differential between conditions in England as compared with the other countries. The other is that the English system appeared to produce much greater levels of unpredictability in staff behaviour than the others, and, for inmates, a higher level of perceived arbitrary decision-making. It is probably this, rather than any particularly poor conditions, that is the hallmark of the 'English prison experience' for inmates.

5

Staff Views

THE front-line, basic-grade discipline staff in a prison are the low-est-ranking members of an official, bureaucratic hierarchy. Their job is straightforward in principle, though more complex in practice; the extent to which they are accountable for their work is similarly both simple and complex. The job is primarily to carry out the instructions of those above them, and they are accountable to their superiors for what they do (or don't do) and, more crucially from their perspective, for any mistakes they make. Most of their work concerns day-to-day routines and security, for which the most relevant written regulations are the prison rules and the orders and instructions that modify them. So far as they are concerned, no-one is ever disciplined for breaching the European Prison Rules. And if an inmate makes any request which is outside the routine provisions of the regime, the safest course is often to refer it to their immediate superior.

This fundamental simplicity does, of course, translate into a considerably more complex day-to-day reality. This can be described briefly along three dimensions: terms of service and the role of unions, relationships with inmates, and changing management demands and expectations.

In Germany and the Netherlands, prison staff are appointed, like other civil servants, on terms of service which amount to a virtual guarantee of a job for life, barring gross negligence or major criminality. The strength of their contractual position is reflected in the frustration sometimes felt by prison governors, who repeatedly told me how difficult it was to impose any but the most minor disciplinary warnings on staff members (this is discussed in more detail in Chapter 6). Prison staff unions, however, are relatively weak in these countries when compared with those in France and England. In these latter two countries, although staff have secure contracts—in the sense that they cannot have their employment terminated arbitrarily, formal procedures cover disciplinary punishments, and so forth—there have been bitter and protracted disputes. However, these have taken on a rather different character in France than they have in England, and this is discussed in more detail towards the end of the chapter.

So far as the character of prison staff work is concerned, probably the best known work is that by Lombardo (1981). Though describing an American prison in the 1970s, in which matters such as staff training were a good deal less sophisticated than in most European prison systems, the general descriptions of work in the prison provide a valuable orientation. Lombardo categorizes prison officers' work as involving human services, order maintenance, security, supervision, and rule enforcement. 'Human services' encompasses a range of issues, from counselling inmates on problems arising both inside and outside the prison, to setting up interviews with social workers and other staff to deal with such issues, ensuring that inmates get the food, clothing, and so on that they need, and identifying and dealing with self-destructive behaviour by prisoners. Order maintenance refers to a policing function in which staff are expected to prevent fights, gambling, homosexuality, and a variety of other activities; in general it can be described as the task of keeping the inmates safe from manipulation, intimidation, and assault by others. Security is often said to be the single biggest task of an officer, and basically refers to the prevention of escapes. This is both a passive task—watching and counting inmates—and an active and sometimes contentious one, inasmuch as it involves cell searches and the like. Lombardo describes supervision as a matter of ensuring that inmates are ready to do whatever they are supposed to do, or be wherever they are supposed to be, at the appointed time. In addition, he includes tasks such as ensuring that cleaning and food distribution take place efficiently and effectively. Lastly, rule enforcement is described as encompassing both citing inmates for rule violations, and detecting illicit activities.

Lombardo's study, and indeed most research on prison regimes, concludes that the practical performance of these functions is rather more complex than any job description might suggest. It has often been observed that since prisoners have no inherent motivation to co-operate with staff, incentives for them to do so must be structured into the prison regime. The practicalities of being a prison officer thus have much to do with maintaining some kind of day-to-day relationship with inmates, in the context of which the latter—or at least most of them, most of the time—are prepared to do what the staff want. In terms of line accountability, then, staff are often placed in positions where they must act in ways that, formally speaking, could be defined as infractions of staff discipline.

In addition, changes in management philosophies in recent years have begun to alter the demands of the job. In England, Fresh Start has seen, alongside statements about the 'management's right to manage', greater devolvement of some functions to lower-level staff and the creation of teams working in different areas of the prison. It has made major changes in some areas (for example, staff shifts and overtime working). It has probably also produced some rather subtle shifts in staff perceptions of what it is their senior officers expect them to be paying attention to. In the Netherlands, the restructuring of management over the last ten years has placed a similar emphasis on devolvement of responsibility and the use of small teams, which again has produced some changes in the ways that staff relate to each other and to inmates.

The position of chief officers in France, Germany and the Netherlands is interesting (the position was done away with in England as a part of Fresh Start). In these three countries, there remains a division between governor-grade staff and officers which roughly equates to the 'commissioned/non-commissioned officer' divide in the military. The chief officer is the head of the 'non-commissioned' side, the uniformed officers. Although it is possible for chief officers to become governors, it is comparatively rare, because of the structuring of entry to governor grades. Often, therefore, one sees chief officers, having reached their position of seniority after many years' service, working under comparatively young but more highly-educated governors. But in practice, their influence in the prison is very extensive. While governors may pursue a career path that will move them from one establishment to another every two or three years, chief officers tend to be in post at a single establishment for much longer periods. They can afford to take the long view; the incumbent governor does not have the same 'deep knowledge' of the ways of the particular prison, nor is he (or she) likely to remain for long.

A number of other staff, usually specialists in medical or human sciences, also work regularly in prisons. Their numbers vary considerably from one prison to another, while the specific structural arrangements of their employment differ between prison systems. But one would normally expect to find medical staff, and sometimes nurses, social workers, or probation officers, education staff, psychologists, and representatives or ministers of religious bodies. All these staff have one thing in common; while they work in a prison, it is simply their place of work. They are usually employed by non-prison agencies and report to superiors in those agencies. Yet they must also find some accommodation with the lines of responsibility and the day-to-day routines operating in the prison.

The following discussion deals first with basic-grade discipline staff, and then turns to senior officers and professional staff. Industrial relations, in particular in France and England, are then briefly considered.[1]

The Basic-Grade Staff

So far as the lower-ranking disciplinary staff are concerned, an initial cautionary note is in order. The experience of day-to-day work in prisons seems to vary rather substantially across the four countries: and even within a single prison system, there will be large variations. In all countries, prisons have fixed posts, in which staff will be placed for a period of months; these typically include the front gate of the prison, some administrative functions, and the prison canteen. Staff in these posts either have relatively little direct contact with inmates, or work with a small group of inmates over a relatively lengthy period. Officers whose primary responsibility is in the prison hospital or some other specialized location will have different experiences to those who are primarily 'discipline officers'. And in France and England, some relatively junior staff may find themselves fulfilling quite important roles within the administration, for example as the governor's adjutant, or organizing social-cultural activities for inmates. In such posts, they may routinely be dealing directly with persons several grades above them in the hierarchy; have much more contact with the governor than the average prison officer (who may never have any occasion to speak with him or her); and perhaps also have relationships with officials outside the prison, for example in the regional

[1] Quotes from staff, and much other information in this ch., come from interviews with prison staff in the 4 countries. App. 1, on data collection, provides further details as to the numbers of staff interviewed and the topics covered.

administration. The following notes are not, therefore, intended to represent the full range of prison staff experience, but to underline some large-scale features of each of the systems.

1. There were differences between countries in whether staff were allocated to tasks on an individual or group basis, and in the management philosophies which gave rise to such differences. Both management philosophy and the patterns of staff allocation that followed from them had an effect on the way staff saw their job. France and the Netherlands were the two extreme positions.

In France, the surveillants are typically allocated tasks on a daily basis. Although they would be allocated to work with a small group of colleagues under the direction of a Premier Surveillant, they do not work regularly with any individual Premier Surveillant, or with a specific group of colleagues. The French staffing system, unlike the Dutch (and, since Fresh Start, the English), does not rely on building up small teams which work together under a single senior officer; while the rudimentary committee structure structure meant that basic grade officers were rarely involved in any meetings other than those in which the governor would address the whole staff.

In the Netherlands, by contrast, the 1985 reorganization of the prison service introduced several fundamental changes in staff work. The basic unit within a prison is the team of eight to twelve guards, commanded by a brigadier.[2] Each team is typically in charge of an accommodation unit (wing, pavilion, etc.), but also works with 'its' prisoners throughout most daily activities, on or off the wing. Each team has regular meetings to discuss its approach to its inmates, and these are discussed in more detail below. One guard summarized the changes that had taken place in this way:

In the past, there was a special staff for each function but now each task is combined with social things, for example making phone calls; prisoners no longer have to see the social worker for this—the guards deal with it. On visiting days [for the inmates of his wing] one of our team's guards goes to the visits with a guard from the other wing [that has visits] . . . We also go with prisoners to work. A lot of guards pre-1985 were just to

[2] Officially these staff are now described as team leaders; but, as in England with Fresh Start, the older designations remain in common use (e.g. as I never heard a team leader described as anything other than a 'brigadier' I have retained this term).

lock prisoners in—now they have to work with the prisoners. (closed establishment)

The reorganization also attempted to 'change the ground rules for the interaction of staff and prisoners' (Vinson et al. 1985), broadly speaking by encouraging greater inmate-officer interaction in order to break down inmate subcultures. At the bottom line, this simply means that officers drink coffee and play table tennis with inmates. As one brigadier said, 'The officers should be with the inmates, not by themselves. I don't care if it's chess, billiards or talk—it makes a nice atmosphere.'

Beyond this, it seems that staff in the Netherlands now spend much of their time talking to prisoners rather than simply guarding them. It is common to see inmates encouraged to talk about their problems to staff. Not that all inmates welcomed this; some felt that their only 'problem' was being in prison. Yet it would be common in most prisons to see staff and inmates, if not sharing personal troubles, at least talking and drinking coffee together.

2. Despite these differences, the work of the basic grade prison officer is always and everywhere fundamentally concerned with two issues: the institutional timetable, and security.

Prisons run on fixed schedules for meals, work, exercise periods, visits, and so on. Obviously in some institutions there may be more flexibility about some aspects of life, such as times for reporting sick, the possibilities for taking a shower, and so on. And obviously, from the staff point of view, team working provides a slightly greater degree of flexibility. But in large institutions where small delays in one part of the daily schedule could have a knock-on effect for the running of the whole institution, staff prided themselves on their efficiency at maintaining the timetable. Equally, if not more importantly, security was a major consideration and great value was attached to the observation of inmates. In France, officers were expected to record any significant details of inmate behaviour in their 'cahier d'observation'; each basic-grade officer was issued with a notebook and expected to make daily entries, which were scrutinized daily by senior staff:

There are 'observation books'. You note down that such-and-such an inmate did such-and-such a thing. You give the book to the 'primo' [premier surveillant] who signs it, then it goes to the surveillant chef, then sous-directeur, then directeur. If an inmate is in a group, aggressive, etc. you note it down. (surveillant)

Each officer has an observation book. Any remarks go in it. It is obligatory—Code de Procédure Pénale, section D276. You hand it in at the end of the shift. It's seen by the 'gradés', surveillants chef, the direction. Certain information is photocopied, placed in the inmate's dossier. (premier surveillant)

In Germany, as one officer pointed out, 'first comes security—the number of people present on floors'. But beyond that, staff were expected to make comments and notes, not in a notebook issued to them, but in the 'hand file' (Handakte) kept on each prisoner. In England, such information was recorded in memoranda written by staff and collated by the prison's security officer. King and McDermott (1990) note that there are occasional glitches in this process, and that staff have to be reminded to keep up a certain level of production of such reports. In the Netherlands, much of this material was entered in the major working document at wing level, the wing log book. If an inmate was, for example, getting up in the mornings, associating with others outside his usual circle, or appeared depressed or aggressive, it would be written down. The wing log books would be seen by the brigadier and governor, usually daily, and would be used as the basis for the discussions in the wing meetings (discussed below).

One feature of accountability worth mentioning at this stage, though I shall enlarge on it later, is accountability for security in relation to such observations. The noting of observations in the cahier d'observation, the inmate files, the wing log book, and so forth was taken as some indication of the level of staff vigilance. As one French governor pointed out, if a notebook contained no observations after a shift he would check to see what duty the guard had been doing; and if in his opinion the officer was in a post where there would have been things to note, he would pass back (in the cahier) a note asking for an explanation of the lack of observations. Yet it is also the case that security reports provide a means for staff to render suspicious virtually any behaviour: one might imagine a note stating 'inmate appeared to be behaving normally today' as giving rise to the question of why he appeared to be behaving normally—what was he up to? In such ways security notes can be turned into vehicles for staff to affect, perhaps in small but gradually cumulative ways, the kind of treatment an inmate receives.[3]

3. Discipline staff typically saw themselves as having no authority to make 'decisions'. At the same time they had and used various means to exercise control over their daily work and to reach some accommodation with inmates.

Basic-grade staff in all countries were quite clear that their work did not entail making decisions, in any significant sense of that term. Decisions were the province of senior staff and prison governors. The lack of formal decision-making power by lower-ranking staff is hardly surprising. Yet it was tempered by different considerations in each of the countries. In France it came about because of the strict hierarchy. In the Netherlands, it was in part a function of the team meetings, the increasingly frequent regime changes, and the need to consult colleagues: as one officer in a closed prison put it, 'Most of the things you do after consulting your colleagues. The only things you can decide on by yourself are extra air or extra sport for inmates'. Another explained:

There's almost nothing you can do on your own. If you do you get irritation from the other team members. We consult; sometimes we say 'do it like we do now but bring it up at the Thursday meeting'. It used to be that if someone changed something it was changed for five years; now we change some things every two weeks if they're not working out. (women's establishment)

Yet despite the hierarchy and the regulations, or at least between the lines of the regulations, staff did have some kinds of control over their own work; answering cell buzzers quickly or slowly, addressing inmates formally or informally, abiding by or ignoring petty regulations, or including or omitting small pieces of information from verbal and written reports to superiors.[4] A number of enabling and constraining factors structured this low-level

[3] In principle, it is possible to argue that the specific form of security record-keeping may affect the operation of security; e.g. the French system, based on notebooks issued to individual staff members, places an onus on all staff to make daily security reports, while the English system, based on memos, does not place so much pressure on staff to make regular reports and may thus allow greater discretion as to what is reported. However, this is a complex area and proper investigation would require much more detailed information on the nature of security reports.

[4] See also the examples quoted in Ch. 4: in England, staff refusing to unlock inmates wishing to use the toilet, or agreeing or refusing to allow inmates to swap cassette tapes.

discretion. At one level, establishments appeared to have unwritten rules of various kinds, which relaxed some aspects of regimes. In one French prison the longer-term inmates, grouped on the upper landings of the wings, were tacitly—though not officially—allowed greater freedoms to spend time in each others' cells. Staff at all levels effectively turned a blind eye to the regulations, though undoubtedly if any trouble emerged they would have been held accountable for it. And despite this apparent formal inability to make individual decisions, and (in conversations with staff) their frequent citation of regulations as unambiguous and comprehensive reference points, staff were, as one French officer put it, not cloistered behind the regulations. They had to work with the inmates on a day-to-day basis. The fact that staff are acting in accordance with regulations does nothing to lessen bad feeling when inmates perceive the regulations as repressive. As one French officer said, 'In cell searches, there is no conflict, but there is a tension and a feeling against staff.' In England, too,

The days are gone when you could say sod the inmates. You've got to cater for them and they're more challenging than they used to be. There are comebacks off prisoners and other officers. You have to think of the outcome.

The idea of accountability in day-to-day prison life thus had much to do with the recognition that what officers do is to a large extent transparent, in that they cannot control what becomes part of the stock of common knowledge about the way they work. Information about their actions can, and does, pass between inmates, between inmates and staff, and among staff. Thus some degree of freedom but some degree of caution go hand in hand: as one officer put it,[5]

I run the landing as I see fit. Any problems I refer to the PO [Principal Officer]. Otherwise prisoners would see the PO or someone would tell him. (England)

In Germany, the tensions between different aspects of the staff role, the formal regulations, the lack of formal decision-making

---

[5] The pre-Fresh Start designations have been retained here, since my interviews with staff were done shortly before Fresh Start was implemented (and staff often continue to use the old designations, since it is somehow dehumanizing and unwieldy to refer to someone as e.g. a 'Grade 7').

power but also the possibility of influence were summed up by one guard in the following apparently contradictory series of thoughts:

The scope for action is predetermined to the greatest possible extent . . . In dealing with the prisoners we are free to organize things the way we want them to a large extent. We are supposed to give our time and attention to the prisoners, build up a relationship of personal trust but at the same time be responsible for security. One must not, oneself, have any illusions about that . . . Through the ward conference [Stationskonferenz], however, we have means to achieve something for the prisoners. If I, for instance, want to push through something for one of the inmates that I believe is necessary, then my superior says no, and this only because he is superior to me. Decisions on relaxation, it always depends on how you put forward your arguments.

The Dutch wing meetings deserve some further brief comment, because they constituted a major means of decision-making about the treatment of individual prisoners in which staff participated. The tenor of the meetings depended to a large extent of the personalities of senior staff; but the following selection of quotations illustrates the range of matters discussed.

Most important is the meeting every week. We talk about people's behaviour, guards working—are they at the right place at the right time, the quality of their work. And the social worker asks for observations on prisoners—we have a book. The meeting includes the director (head of wing), brigadier and staff. We can ask the brigadier and director to leave if we want to, and we can criticize them. (closed establishment)

There is a team meeting every week; we report on half the inmates every week in a paper, and discuss how to handle them. For example we may decide we have to persuade a man to take a shower; it's very behavioural. (closed establishment)

We talk about the situation on the pavilion, inmates' behaviour, who takes a special interest in which prisoners, who is new. It is to keep each other on the same lines. Three quarters of the input comes from us, one quarter from the brigadier and the director. (closed establishment)

In short, much of the management of the prison at brigadier and governor level was in fact the management of individual problem prisoners, though it was effected through discussions with teams.[6]

[6] Despite the democratic image of wing meetings presented here, research by Kommer on prison staff suggests that only some staff attend the meetings, and that prison governors frequently take a more directive role than that suggested here (Kommer: pers. comm.).

This approach was clearly helped by the fact that staff worked regularly with a small group of inmates. It is by no means clear that in the context of, say, the French staffing arrangements and the size of wings, similarly fine-grained information could be obtained on inmates who did not present control problems.

One final issue in connection with 'non-decisions' is the way in which staff addressed inmates. The issue in England largely revolves around the mode of address to black inmates, and the use of common, though derogatory, terms in conversation (for an extended discussion, see Genders and Player 1989). In France, a slightly different concern emerged. The French, Dutch and German languages all have informal and formal ways of addressing others (tu/vous, je/Uw, du/Sie). In the Netherlands it was not an issue; in Germany it was rarely remarked upon; but in France there were strong feelings about the use of 'tu' and 'vous'. One surveillant chef in France noted a problem with:

. . . the way of talking to people; the staff call inmates 'voyou' ['thief'] etc. They 'tutoyer' inmates—I can understand that if an inmate is here for 8–10 years he becomes 'tu', but it shouldn't happen. It's OK for prisoners to 'tutoyer' staff but not the other way round. With the maghrebiens [Algerians, Moroccans, Tunisians], it can be a problem because of the staff. Staff tend to use 'tu' though they shouldn't, because they know the prisoner well because he's long-term. But the maghrebiens, especially, resent being called 'tu'. Equally, staff shouldn't accept cigarettes from prisoners though it's OK to give them.

The basic issue was that the Arabic inmates perceived the use of 'tu' by staff as analogous to a teacher talking to a pupil, or a parent to a child; it was one more brick in the edifice of infantilization, discussed in the previous chapter.

4. Staff are often required to manage tensions whose origins lie beyond their capacity to control.

The various features of the staff role mentioned above—security, rehabilitation, lack of power, but the need to retain some influence and control in minor issues as a way of securing a relationship with inmates—also need to be seen in the light of the fact that prisons are relatively complex institutions where decisions made, or problems arising, at one point in the organization can have a knock-on effect elsewhere. A French officer observed that the operation of the social-educational service in his establishment created frustrations—because of long delays before inmate requests were dealt with—which he was expected to smooth over, although he had no means of solving the root problem. In Germany, guards mentioned decisions on relaxation, and the mindless and tedious nature of some prison work, as matters in this connection.

5. Despite the comments above, to say that staff have few formal powers may be to overstate the case.

Staff have, after all, some powers which directly affect inmates, including that of making a formal disciplinary report. Yet where such powers were used it was by no means clear that staff would know the outcome of their actions. Although in England, the officer making a disciplinary report would routinely be present at the subsequent adjudication, this was not normal in any of the other countries. In France, for example, it was said in relation to discipline that:

If you receive an insult, you report it to the chef. Depending on the gravity, there is an avertissement [warning] or a prétoire [disciplinary hearing]. There are no systematic sanctions [i.e. no 'tariff' of punishments]. I only know the result of a prétoire because the inmate isn't in the cell block and there is a mark on the register. I assume he's in the punishment cells. (surveillant)

In other kinds of situations, where influence rather than the exercise of power was the issue, there were again differences in the extent to which staff knew what weight their observations or proposals has carried. Dutch prison staff were able to get a large degree of feedback (if only in the form of instructions) via wing meetings, and probably also felt that their ideas were more positively valued. One example mentioned by staff in the Hague was proposed by staff, and adopted, partly because it was a good idea in itself, but also because it provided a privilege for prisoners over which staff had control:

Guards and prisoners have more responsibility [since 1985]. We have been responsible for some things, e.g. setting up the garden was the idea of the guards and it was set up after a meeting with the brigadier. But the decision to close it would also be us and the brigadier. It does work on our wing, prisoners respond and don't abuse anything because now they can lose the vegetable garden, the chickens. (closed establishment)

6. Finally, there remained several areas in which staff were more or less explicitly exonerated from responsibility for inmates' reactions to their treatment.

Staff duties were fundamentally concerned with security, and this was the major area in which staff were held responsible for lapses; the concept of security is wide-ranging, touching on almost all aspects of prison life. Although some other issues of interaction with inmates emerged (such as the tu/vous distinction in French), inmates are, after all, seen by prison managers as people who may react unpredictably, who may have a stake in disrupting the regime, and who are essentially responsible for their own actions. Thus, though all prison systems have some rhetorical commitment to rehabilitation, and staff in the Netherlands in particular were now expected to interact much more with inmates than was the case pre-1985, in no system were staff regarded as responsible for the reactions of inmates to such social work. A Dutch officer pointed out:

We haven't experienced being made responsible for social work. For example, if we talk to the prisoners and they smash up the cell, it's not our responsibility. It's a new area, problems have not yet been discussed. We're only responsible for security. For example, if we're in the office and there's a fight it's our responsibility. (closed establishment)

It seemed to be the case in all the prison systems that even if staff were provoking an inmate by talking to him about personal or sensitive matters, and the inmate reacted by assaulting staff or smashing up his cell, the responsibility for that reaction would be put squarely on the inmate. Even though his senior officer may well give him a severe talking-to for his irresponsibility, it is most unlikely that any formal action would be taken against an officer in such circumstances.

Senior Officers and their Relationships with Other Staff and Governors

In the Netherlands, the roles, functions, and responsibilities of brigadiers seemed to have changed markedly under the reorganization. This can best be illustrated by two quotations, the first from a brigadier, the second from a guard:

All the little things come to me—extra telephone calls, extra visits etc., complaints about 'I've ordered something from the canteen and it's not come through'. I have control over inmates—it's mainly behavioural, for example I can appeal to their common sense. If they break the rules I report it, they see the director. There are cases where we have to put people in the punishment cell first and then see the director. Also if there's an injury I would send them to the hospital first and then notify the director. I can only advise, since the responsibility is the director's. (brigadier, closed establishment)

The management will only ask the brigadiers about something if it goes wrong. The brigadiers don't have much control—they know less about their floor than the team. Where it goes wrong is that he has to take responsibility but he can't remember the particular thing happening . . . We also have responsibility for what happens. No brigadier wants to carry your mistakes on his back, he will say 'why didn't you ask me?'. But with the reorganization, the brigadier will take the blame for the team. (guard, women's establishment)

In France, the two ranks above that of 'surveillant' were premier surveillant (or 'primo') and surveillant chef. The following quotes illustrate the range of their work:

There are five surveillants chef, that's normal for the size of establishment. My special responsibility is security. Others are responsible for the central living units; the workshops; finance; and one is both adjutant to the deputy governor and 'chef de détention'. The job is to take decisions and implement them. To follow through. You're responsible for one section, so you have to co-ordinate with the others, so no wires are crossed. But you're responsible to the chef de détention. (surveillant chef, maison centrale)

There are two or three 'primos' each shift. Broadly speaking their jobs are co-ordination, control, supervision . . . For 'primos', there is a responsibility for order and for discipline. To apply the general lines of operation. To ensure good execution. The 'chef de détention' co-ordinates—is responsible for problems of functioning—so one always has to co-ordinate with him and the governors. (premier surveillant, maison centrale)

I have to be there and the staff have to explain to me. I control relations between inmates, deal with problems, execute rules. I am basically responsible for inmates getting up, going to work, and following orders. I have no other powers—changes of cell goes to the chef de détention, changes of workshop to the governor. I don't have the right to change anything, but yes, it is possible to change workshop allocations, visits—and I can issue 'notes de service' [internal circulars] . . . I don't have the possibility to take initiatives. For example if someone asks for isolation I can't put them in, I refer it to the governor even if the request is directly to me. I can put someone in isolation overnight, pending the governor's decision, if it touches on security—this is a power given to the premier surveillant on night duty. . . . If prisoners complain about the food to me, I report it to the 'econome' who gets on to the governor and the 'chef de cuisine' and they discuss it. I don't discuss such things with the governor, always with the responsible service. (premier surveillant, maison centrale)

It is quite clear, on these accounts, that the 'chef de détention', though of the same rank as the other surveillants chef, is in fact a first among equals; though it is also common in prisons for the one with responsibility for security to have, de facto, a large say in the running of the establishment. And it is also fairly clear that despite the theoretically limited executive power of the premier surveillants, they are in fact in a powerful position, because the very fact that they decide to refer a request to the next higher level means that, from the perspective of the surveillant chef or governor, the request comes with an implied recommendation that it should be granted. At the same time, given the legal framework in which French prisons operate, many of the details dealt with by the premier surveillants and the surveillants chef are technically legal matters rather than internal issues covered by circular instructions or standing orders. Thus, on the one hand, 'One has to know the law when asked, because one is argued with. And one should not exceed the exact limit of the law' (premier surveillant, centre de détention); though at the same time, 'Because one ensures the law is applied, one can also change small things in the manner of application of the law' (surveillant chef, maison centrale).

A chief officer (Vollzugsdienstleiter—literally, a prison service leader) in Germany occupies a position of immense importance because of the way governors are appointed. Prison governors are legally trained and qualified, and for legal officials in the civil service the post of prison governor is only one step in a career which may include working in the procurator's office, conducting prosecutions, governing a prison, a period in the Land prison administration, and being a judge. In consequence, a governor may come into the prison service with relatively little prior knowledge of its day-to-day workings, and spend only two or three years there, then move into another legal post. Prison governors of long experience and a broad knowledge of prison can be found; mostly they have decided (or their superior has determined) that they are particularly suited to this particular niche in the civil service. But it is commonly the case that a relatively inexperienced governor will have to rely heavily on his or her chief officer; and, because most Land prison systems are small, and the possibility of working in a number of institutions is therefore circumscribed, he may have worked in the same establishment for many years. Realistically, most chief officers can (and some do) take the attitude that governors come and go, none of them know too much about the prison, and so the major influence on the running of the institution will be their own. In this kind of situation, chief officers were very much involved in institutional and regime planning issues such as planning for refurbishments and developing facilities for inmates.

My interviews in England spanned the beginning of Fresh Start, a process in which the fine distinction that officers were under the command of a chief officer and only indirectly controlled by governors was dispensed with. It had been the case for many years prior to Fresh Start that lines of responsibility were relatively complex, with distinctions between ranks and roles meaning that officers would sometimes report to one more senior officer in general, but to another officer of their own rank, or an officer senior to them, or indeed a governor grade, in respect of a specific job. Thus the general pattern of being responsible to 'next one up the line', or, indeed, the view of more senior officers that they were in charge of a group of staff was recognized everywhere as being of limited application. This became more pronounced after Fresh Start (because of functional group working practices) yet probably also more routine and easier to operate (because staff were regularly working with others in teams with designated responsibilities), despite the staffing problems caused by the changeover.

In some respects the work of the more senior staff was more varied; they were, it is true, often in charge of wings, but would also be detailed to run specific functional units—reception, the 'Rule 43 wing' (prisoners under segregation) and so on. In terms of their relationship with their staff, they said, much of their day-to-day work consisted of supporting basic-grade staff in various ways. One way was simply to ensure that information was channelled up or down to the right people about what prisoners were doing, or about developments or changes in routines or procedures. Another was making sure that basic prison routines for food, canteen, letters, and visits were operating (there was wide agreement about these as the 'basic' areas), because if these areas did not function smoothly, there would be inmate unrest and security would become problematic. A third was to use their rank to solve emergent problems in prisoner relations. A senior officer provided this example; one prisoner wanted to share a cell with a second (instead of his current cellmate), but the second did not want this arrangement. The senior officer told staff to get the inmate to make a formal request for a cell-change, and then refused it, accepting responsibility for the decision and stating when challenged that the grounds for the refusal were confidential. This solved the problem—the inmate could not change cell, while the heat was off the basic-grade staff. Finally, and unlike the French situation, while basic-grade staff and some senior staff were more guided by internal regulations than laws, a few did have to worry about the legal consequences of their decisions. This was so, for example, with the 'security officers' (a role, not a rank), who dealt with cell searches, collation of security information, and so forth. As one security officer pointed out, given the legal pressures on inmate disciplinary hearings, issues such as the proper presentation of evidence, arrangements for drug tests, and ensuring that orders to inmates had been lawfully given, had to be attended to carefully.

The 'Professionals'

In addition to the uniformed and managerial staff, prisons have within their walls the representatives of various non-prison service professions. Principally, these are doctors and nurses, social workers or probation officers, educators, and religious ministers. All of them in effect serve two masters. On the one hand, since the prison is their place of work, they must follow similar practices to uniformed staff in relation to security. They are in some senses responsible to the governor and must follow, if not orders in the formal sense, then at least his suggestions and proposals. And their relations with other officers can greatly influence the way in which they do their work, since they must make arrangements either to have the inmates brought to them or to be able to see inmates at their place of work or on the wing. On the other hand, groups such as social workers or probation officers are formally employed by non-prison-based agencies, their superiors are based outside the prison, and their professional responsibilities and doctrines may at times be difficult to reconcile with the practical requirements of prison life.

The German prisons studied each had a full-time psychologist on site. One problem they experienced was that there were no hard and fast descriptions of their functions. Neither, however, were routinely involved in any clinical psychological testing, which was done only in exceptional cases. One had become involved in regime development through his proposals for sentence plans, though prison staff found the idea attractive because they felt it would give them a greater decision-making role in the treatment of inmates. Most of his work consisted of advising management on relaxation measures—home leave and so forth—and, to some extent, on the treatment of violence and drug addiction.

The German social workers similarly found themselves in a position where, with no direct management authority, their role was one of advising on the treatment of individual inmates. As one observed, their influence was much greater on matters concerning individual inmates than it was on wider regime matters. The same social worker also commented that much of their work had to do with counteracting the negativism of staff views. Prison staff on the whole were interested in, reported to seniors, and noted on inmate files, problems with the control of inmates rather than the inmates' positive achievements. The social worker saw his work as trying to indicate positive developments in the inmates which would to some extent provide a more balanced view.

French social workers tended to occupy a double role, as one explained:

I am a go-between. I give advice on subjects—family, law, social and psychological problems. That is 80 per cent of my work. I also take care of the cinema, theatre, concerts—20 per cent of my work. I provide reports for the JAP [who is involved in decisions such as home leave]—they are taken notice of by the JAP and the Procureur here. One has a professional responsibility to society and to the inmates. But we don't make decisions ourselves. One isn't able to change anything without the authority of the director. But internal regulations can be changed. For example I can ask to have half an hour extra for a particular film, but the director decides.

In England, one of the most interesting developments both pre- and post-Fresh Start was the increasing decision-making power delegated to officers in charge of wings. The repercussion on probation practice (social work in prisons is provided by probation officers) was quite noticeable. Previously, for example, inmates who wanted to have a phone call made to their family would have to approach the probation officer. Probation staff thus had significant power as the 'gatekeepers' to facilities desired by inmates. But as wing staff were given increased authority to make such arrangements themselves, probation staff found that they were no longer such important players in the prison drama as far as the inmates were concerned. They held the key to fewer of the facilities that inmates wanted. Some welcomed this move, saying that such facilities should anyway have been available to inmates, and that this meant they could concentrate more effectively on inmates' social problems.

The Implications of Unionization in France and England

Questions of accountability are somewhat complicated by the presence of strong prison-staff trades unions in both France and England.[7] Trades unions are an important consideration in both countries, though for different reasons. The English Prison Officers' Association (POA) can be thought of as a more or less traditional union whose appeal lies in its ability to negotiate work-related and wage agreements for members. In this sense it enters into questions of accountability as an interest group whose co-operation with management must be elicited if prisons are to function adequately. And at the same time, administrators can depict it as an impediment to goals which have already been set, or argue that they will form a stumbling-block to policies under consideration. In France, although the unions have a similar function, the situation is more complicated. There are three major trades unions involved; there are differences between the unions as to their roles vis-à-vis the state; and the prisons 'syndicats' of these unions have anyway taken different stances from those of their parent unions on penal issues.

The English situation at the time of the 1986 riots was summed up by the then Chief Inspector of Prisons, Sir James Hennessy (HM Chief Inspector of Prisons 1987). Over the late 1970s and early 1980s prisons had become increasingly dependent on staff to

---

[7] Although there are a number of unions in both countries, covering e.g. trades and clerical staff as well as discipline officers, I have restricted this discussion to the major unions for discipline staff.

work overtime, while staff had become dependent on overtime payments. But staff had also become increasingly suspicious of what they saw as a management concentration on economy as a primary aim. By 1985 this had become focused into a dispute about manning levels. Despite a number of local disputes, Hennessy notes (para. 10.02):

at a local level, relationships between staff and management appear more often than not to have been maintained at an acceptable level, each side recognising the difficulties and working sensibly to achieve a better understanding; but at national level a gulf appears to have developed between . . . Prison Department Headquarters and the POA.

The POA took manning levels to be a matter of staff safety, and insisted not only that they had the right to negotiate safe levels, but that while negotiation was taking place, the Department's decision should not prevail. In addition, the POA did not see itself as bound by the outcome of arbitration procedures. The Prison Department was prepared to discuss manning levels, but regarded them as, in the last instance, non-negotiable. In April 1986 the POA national executive received a mandate from its members to call industrial action (of an unspecified kind) should the national executive think it necessary. And when a local dispute at Gloucester prison emerged in late April, the hard-line management strategy prompted a POA call for a national overtime ban. In addition, no paperwork would be completed, alerts (to call staff to the prison for security reasons) would not be recognized, communication between union branches and management would cease, and all previous agreements were considered void.

The outcome—a series of major riots—is considered further in Chapter 11. The subsequent development of Fresh Start, which began as a way of using staff more efficiently but also of buying out dependence on overtime, was mentioned in Chapter 3. But the key point is that whatever the rights and wrongs of each party's case, the dispute can be dramatized as a clash between two versions of accountability. The management view cohered around the slogan of the 'management's right to manage', that is, control. The POA view, although linked to traditional labour concerns about the situation of the workforce, encompassed the idea that accountability was not purely a matter of public policy and managerial control, but extended to limits on management control where it would put employees in unsafe situations. To exaggerate slightly, one might see it as a quasi-militaristic versus an industrial view of the prison staff position.[8] Undoubtedly a considerable amount of humbug was generated by both sides. But these contrasting positions on the legitimate scope of management in moving towards its objectives, capture the essential details.

French discipline staff belong not to a single union which dominates the prison field, but to smaller syndicats affiliated to one or other of the three major, multi-industry unions; the Confédération Générale du Travail (CGT), Force Ouvrière (FO), or the Confédération Française du Travail (CFDT).[9] FO, for example, has three syndicats affiliated to it, representing discipline staff, administrative staff and governors. Moreover, a situation existed in the early 1980s in which the key CGT and FO officials in relation to prisons were also senior members of the prison administration. Favard (1987) lists Hubert Bonaldi (FO) as ex-director of La Santé prison, and sometime head of a special internal commission within the administration; Aimé Pastre (CGT), though junior to Bonaldi, was also a senior prison governor.

Both CGT and FO prison syndicates, whatever the provenance of their leadership, have a 'law and order' position which, perhaps ironically, resulted in strong differences with the left-wing govern-

---

[8] And ultimately, it seems, the government objective was not to reach some common agreement with the unions but to undercut their power. This is at least part of the agenda of privatization, discussed in Ch. 13.

[9] Although not immediately relevant to this discussion, it is worth noting that these unions are divided on the question of the extent to which they should be 'political' organizations (that is, intervening in political policy questions) as opposed to 'class actors' (that is, representing the working class). French unions have for many years been criticized for having adopted a stance within the broad political consensus rather than being part of the 'workers' movement'. For a more detailed discussion of this, see Touraine et al. (1987). But beyond these ideological differences, it is also possible that unions make a practical difference to the ethos of a prison. The social worker quoted in ch. 4, who saw very clear differences among inmates interviewed on reception, depending on which maison d'arrêt they had previously been allocated to, attributed the differences primarily to the ways in which union syndicates influenced the prison ethos. Inmates who had been in FO establishments, he claimed, had typically experienced harsher discipline than those coming from CGT institutions. This is clearly not conclusive. It can be argued that such differences may come about because of traditions unconnected with unionization, or indeed that the choice of union may have been connected with such traditions. But the idea that an association exists between union membership and the nature of prison regimes is certainly worth further investigation.

ment which came to power in 1981.[10] These differences were basically over security issues, and resulted in 1982 in a strike called by all the syndicates, despite a 1958 law prohibiting strikes by prison staff; and industrial relations were to be disrupted, on and off, for some three years.

Although the demands of the strikers—a thirty-nine-hour week, additional holidays, and pay and conditions comparable with the police—were relatively straightforward labour issues, the background to the strike lies in the differences between the left-wing government and the left-wing unions on penal policies. Although the 'Security and Liberty' law was intended to increase sentence lengths for serious offenders, the special units to hold disruptive prisoners, instituted in the 1970s, had been abandoned. This had left prisoners, the unions alleged, in a powerful position and staff were being assaulted with impunity. The end result was not, then, that staff were suspicious of the administration's intentions towards them. It was ultimately a strike over conflicting views about how the prison service should be run, and the nature of the discipline that should be imposed on the prisoners.[11] Unlike the English situation, the unions were lobbying on the political issue of penal policy rather than the narrow one of prison management. They were staking a claim to a political role.[12]

[10] Favard (1987: 73) tells of some incoherence among the unions, when it was 'particularly difficult' to fully understand the 'positions taken by the CGT, of which the president, Aimé Pastre, called successively for votes for Valéry Giscard d'Estang in 1981, resigned from his union positions but then took them up again, and then supported the left in the legislative elections before finding himself, in February 1982, a member of the Council for the Future of France created by Giscard d'Estang, which did not prevent his from telling Robert Badinter [the left-wing Garde des Sceaux—minister in charge of prisons] that he had for a long time belonged "to the same community of ideas as him"'. At about the same time there were several splits within both the CFDT and the CGT, with the creation of competing syndicates; the CGT had a 'CGT Syndicate of Penitentiary Personnel' and a 'National Federation of Penitentiary and Justice Personnel'. It was also in 1982 that the FO prison syndicate split into 3, dealing with guards, administrative staff, and governors.

[11] Favard's (1987) account of this is quite short; but see also the article 'Les Lieux du Lobby', *Justice* No. 92, October 1982. These events are also discussed, albeit in the context of an argument about prisoners' rights, by Fagart (1982). One claim that has consistently emerged about the 1982 strike, however, is that prison staff more or less systematically attempted to provoke riots in order to strengthen their hand, while the nascent prisoner unions appealed for calm in the face of this provocation. This is best documented in *Justice*, 92.

[12] It would be unjust to leave this discussion without mentioning the role of COSYPE—Coordination Syndicale Pénale—an umbrella group for administrators

Conclusions

Despite differences in management style and philosophy, and despite official rhetorics of rehabilitation, basic-grade staff in essence are concerned with two things. One is the institutional timetable; a delay in the daily routines in one part of a prison can knock on to create serious problems elsewhere. The other is security, which is routinized either through observation books, a system of memos, or meetings. On the wings, however, staff have to manage inmates in the context of prison regimes which many see as having become more liberal in the long term, with consequent increases in the number of requests and complaints, and requiring a more sophisticated approach to inmate management. They have to manage tensions that have often been created by such long-term trends, and over which they have little influence. The consequence is that as and where possible—probably to a greater extent in the Netherlands and England, and least in France—staff carve out for themselves areas of 'non-decisions' in which they can influence inmates positively, with privileges, or negatively, with punishments. Although refusing to allow an inmate to go to the toilet may seem a juvenile punishment, it was at such levels that informal punishments for behaviour that did not warrant an official report could take place.

Although staff are, at least in a general sense, responsible for the inmates in their care, in some specific senses they are not. If an inmate insults or attacks someone, smashes his cell or creates some disturbance, the general principle seems to be that the inmate is responsible for his own behaviour. There would have to be gross

and professionals which was formed by 9 syndicates in Apr. 1981. These were: the national association of prison teachers, the CFDT syndicate of Ministry of Justice central administrative staff, the CFDT syndicate of Ministry of Justice medico-social and social-educative personnel, the FEN syndicate of penitentiary education and probation staff, the FEN syndicate of Education Surveillée staff, the multiprofessional group of Paris prisons, the Advocates' syndicate, the Magistrates' syndicate, and the syndicate of prison doctors. Several well-known penal researchers, including Michel Fize and Claude Faugeron, were members from the beginning. COSYPE denounced the major discipline staff syndicates for 'political adventurism', while offering a more humane and reductionist vision of prisons. However, and despite the seniority of many individual members of COSYPE, the group has acted primarily as a talking shop and umbrella for publications; its role as a force for change in prisons has thus been, understandably, rather limited. More detailed descriptions of COSYPE appear in *Actes* (*Cahiers d'action juridique*), 37 (June 1982), text A ('Dix variations sur le thème du changement: COSYPE').

and clear provocation or tactless handling of inmates before staff would be brought to account for the inmate's acts.

In Germany and France, since the law was the source of many specific statements about the treatment of inmates, prison staff were aware of the legal framework of their work. In the Netherlands and England, where much is left to internal service instructions, it was these rather than the law that constituted officers' basic working guides. However, in no country was the law (or the internal regulations) constantly in the forefront of officers' day-to-day practices. Accountability for actions primarily means, for basic-grade staff, carrying out the instructions of the relevant superior officers.

Finally, although professional staff work in the prison, they are not of it; and the demands on them are framed not only by the prison but by the agency that employs them and its professional standards and doctrines. They are, inevitably, caught between the two structures and seek to find ways of doing their professional work within the constraints imposed by the prison organization. This is not, it must be said, always difficult; but it has to be considered and dealt with, day in and day out.

6

Managerial Accountability

AN organizational chart of the interface between a prison and a prison administration looks more or less like an hourglass, with the governor occupying the central position. Below him or her, the institution fans out, with its functions, staff, budget, and operational issues. Above, the regional and headquarters departments and bureaux fan up, with departments structured to reflect their own, rather different, concerns. Yet although the governor is in overall charge of the institution and its various parts, he or she must satisfy a number of masters over a wide range of issues; and in specialized segments of the prison—for example probation welfare work, and medical matters—professional staff may work as closely with co-professionals in headquarters or other organizations as they do with the governor.

Cressey (1959; 1965) repeatedly stresses that prisons have multiple and sometimes contradictory goals. Management is often, therefore, less concerned with end products—for example, treatment or recidivism rates—and more with what we might broadly define as 'regime delivery', covering all aspects of inmate life and staff performance from food, visits, exercise, and the prison library to workshop safety, medical facilities, censorship of letters (where it is still practised), and security. But delivering the regime also means the negotiation of budgets and staffing levels within which the prison must then be operated; dealing with matters such as inmate disciplinary infractions and staff discipline; local trades union negotiations, and the implementation of national agreements. In all these matters, although there is these days substantial delegation and a large amount of committee-sitting, the governor is the person on whose shoulders responsibility will ultimately rest.

Literature on prison management generally identifies four management models (Barak-Glantz 1981, also discussed in Ditchfield

1990).[1] An 'authoritarian' model is one in which the governor has complete power and there are few restraints on its arbitrary use, over staff as well as inmates. This was, Barak-Glantz argues, replaced after the Second World War with a 'bureaucratic-lawful' model, in which general principles and rules are applied by the central administration to the governor, and by the governor to his or her staff. The result is that the governor is simply one link in a bureaucratic chain. The 'shared-powers' model refers to a situation in which rehabilitative goals lead to the democratization of the prison, with some power being given to inmates, who form identifiable pressure groups. The fourth model, of 'inmate-control', is one in which inmate groupings, perhaps based on subcultures outside the prison, become sufficiently strong that they effectively control the prison through sharing power and settling disputes among themselves, with the staff and governor having little ability to influence the situation.[2]

These typologies, generated initially to explain shifts in prison

[1] Jacobs's (1977) study of Stateville prison is not inconsistent with this 4-phase model, though he distinguishes only 3 periods, with some transitional states: the authoritarian regime under the personal dominance of Warden Joseph E. Ragan; the emergence of a rational-bureaucratic administration headed by educated and often liberal professionals; and a period of legalistic rule created out of inmate legal challenges to the system, coupled with the penetration of street gangs into an informal prison power structure. This expands Barak-Glantz's definition of 'inmate control' by linking its emergence explicitly to external legal pressures on the administration, and also suggests that if there was ever a 'shared powers' period at Stateville, it was a transitional situation quickly superseded by the growing influence of the gangs. McCleery's (1961) earlier study of maximum security prisons, though concentrating on the problems of a rational-bureaucratic, pro-treatment regime following on the death of an authoritarian governor, is also consistent with Barak-Glantz's observations.

[2] DiIulio (1987) casts his models slightly differently, and refers to 'control', 'responsibility', and 'consensual' models, (e.g. the Texas, Michigan, and California prison systems respectively). The control model emphasizes inmate obedience, work, and education, with a paramilitary style of organization. The responsibility model places less stress on paramilitary organization and more on classification and grievance procedures; the assumption is that inmates can, within limits defined by security needs and imposed largely through classification procedures, be given some responsibility for organizing their own daily lives; there is more inmate counselling and much staff decision-making takes place as close as possible to the inmates involved, with order being maintained, where possible, through consultation rather than orders. The consensual model is a combination of both control and responsibility styles, with paramilitary staff who none the less deal with inmates informally where possible, and with extensive grievance procedures. The control model might thus be seen as a bureaucratic and hierarchical system under the control of an authoritarian chief; the responsibility model as roughly equivalent to Barak-Glantz's 'shared powers' model; and the consensual model as somewhere between the two.

control problems in America between about 1945 and the 1970s, may not be entirely applicable in the 1990s, or for that matter on the other side of the Atlantic. The 'authoritarian' model may well have been descriptive of the large prewar American prison farms, and indeed Jacobs's Stateville until the early 1960s, under the wardenship of Joseph E. Ragan. But it is doubtful that it really survived in Europe after the early 1900s. Thomas (1972), for example, records that many significant powers over appointment and promotion were removed from governors in the latter part of the last century, and that once control over prisons was centralized, governors were much less, if at all, involved in local pork-barrel politics, and were not political appointees (he does, however, record that communication problems between institutions and headquarters were both persistent and major). The inmate control model, equally, is probably most applicable to large American prisons.[3] Most of the Western European experience is with bureaucratic-lawful structures, somewhat modified by the gradual infiltration of shared-powers ideas. But this simply begs the question of what, in a European context, such models really mean.

Perhaps the best way to describe the current European position is to suggest that there is an interregnum in penal philosophy. While the bureaucratic-lawful model began to stress inmate rehabilitation, and the shared-powers model did so in a more thoroughgoing fashion, the empirical results of allegedly rehabilitative practices have largely been disappointing. Criminal-justice policies have moved, meanwhile, partly as a result and partly because of increasing crime rates, to the right; in the course of the 1980s they increasingly came to stress retributive concerns. Yet there has also been increased attention since the 1970s to issues of inmate rights, and the development of a 'justice' approach, in which the aim is to provide, at a minimum, a specified regime to which the inmate is

[3] Though it seems to be the case e.g. in Poland that a particularly strong form of inmate control emerged in several prisons in 1990/1. After the fall of the communist regime, and the election of Solidarity, the prison service was faced with a number of problems, one which was that a large number of staff resigned or walked off the job. This, coupled with a more general uncertainty among staff about the limits of their powers and the implications of the rule of law, allowed groups of inmates in several establishments to seize effective control, to the extent that certain parts of these prisons became 'no-go' areas for staff, while in other areas, inmates decided which staff they would allow in. This situation appeared to be continuing as of early 1992 (presentation by Zbigniew Hołda and Andrzej Rzepliński to an international seminar on long-term imprisonment, Prague, Apr. 1992).

entitled. This may be slightly overstating the case, and I shall discuss this further in the next chapter. The overall effect, however, has been to create a situation in which the aim of many governors has been simply to keep the institution operating, without any officially sanctioned and workable long-term goals in mind.

At the same time, the long-term growth of government seems to have led to the sedimentation of regulations into a kind of florid bureaucracy which often appears excessively petty and internally inconsistent. At the same time, more sophisticated, or at least complicated, managerial tools have become necessary, often entailing the establishment of consultative or advisory committees at all levels. And more recent concerns about public expenditure have resulted in governors spending more and more of their time dealing with resource issues. The idea of the 'governing governor', who holds a tight rein on the prison, is constantly about in the establishment, and does his own troubleshooting, is now largely defunct—or at least, this position is relegated to deputy and assistant governors. The 'Number One Governor', these days, is more often to be found chairing institutional meetings, attending headquarters functions, composing reports, or devising budgets.

The remainder of this chapter is not intended to be a rigorous survey of the position in each of the four countries. What it does is to illustrate some aspects of prison governorship, which may apply empirically only to one or two prison systems, and yet have some wider ramifications.

Who Are the Governors and What Do They Govern?

Two points, although trivial at first sight, suggest that there are some significant differences between countries as to the role of a governor.[4]

The Dutch, English, and French systems all conceive of governors as career professionals who will spend most of their working lives in the prison system. In England, the apex of a governor's career, at least in principle, would be to hold the post of Director. The post of Director General—after 1993, redesignated Chief Executive Officer—has usually been filled by an appointment from

---

[4] Most of the information in this Ch. comes from interviews with governors and regional and headquarters officials in the 4 countries. Further details of the numbers of interviews, etc. are provided in the App. on Research Methods and Data.

elsewhere in the civil service.[5] But in the course of a career spent largely working in prisons, a governor may be attached to one of the area or headquarters posts, while a few are seconded for one or two years to the prison inspectorate, a post which in practice now seems to imply 'grooming' for a more senior position. Despite frequent complaints about the inability of headquarters staff to comprehend the reality of running a prison, there are governor-grade staff with previous institutional experience in senior managerial positions.

This is not the case in either France or the Netherlands. In the Netherlands, while there would not appear to be any principle preventing the appointment of prison governors to senior administrative posts, it is simply not part of the administrative culture to do so. In France almost all senior officials must be legally qualified, and so many governors are *de facto* disqualified from holding such posts. In recent years, and in a radical departure from previous practice, a governor was appointed to head one of the less sensitive headquarters departments, that dealing with building maintenance.

The system in Germany is rather more complex. As indicated in Chapter 5, the post of prison governor itself not only requires legal training, but is actually part of a legal career structure. Their relatively short prison experience means that governors in Germany must rely more heavily on their senior discipline staff than their counterparts in other countries, because the balance of power between governor and staff is unusual.

These different models of governors' careers clearly indicate different levels of integration between governors in the field and headquarters staff, and to some extent also different expectations about what role governors are supposed to fulfill. The English and Dutch approaches appear to suggest that governors are, at least in principle, integrated into the administration and part of the general pol-

[5] In 1992, there were 6 Directorships in the English prison service in addition to the Director-General: building and services, prison medical services, personnel and finance, inmate administration, custody, and inmate programmes. Governors 'in the field' might realistically aspire to the last 3 in this list, since the others require professional expertise in non-prison areas. The post of Deputy Director General, first established in 1980, was abolished in 1990. For the background to this reorganization, as well as further details, see the Woolf Report, paras. 12.8–12.25. The change in nomenclature from Director General to Chief Executive Officer took place in 1993, when the prison service became an 'executive agency'; and from this point on, it became possible to fill the post with an appointee from industry as an alternative to recruitment from the civil service.

icy-making process. They should have a good all-round competence in practical aspects of man-management both in relation to staff and inmates, while also having a broad sense of the policy questions involved in management. The French structure seems to regard policy-making as part of a legal and managerial expertise, reserved for headquarters. The French governors I talked to did not speak of 'managing' their prisons but 'administering' them, with policies imposed from above. This semantic distinction is significant, as I hope to illustrate below.[6] The German vision is if anything even stronger in this regard, and can be seen as the end point of a constitutional logic which gives the courts wide powers not only to declare new parliamentary laws inconsistent with the constitution, but also to intervene in the interpretation of law and thus in the making of policies in a way quite different to the general oversight of policy exercised by the courts in, for example, England. The positions of both governor and headquarters administrator in the German system thus become linked to the need for legal expertise.

A second issue is to do with who can be considered as filling a gubernatorial role. Most prisons have two or more governor-grade staff, one in overall charge and the others being deputy or assistant governors, whose roles would normally be to run specific units or functions within the establishment and to deputize for the 'Number 1 Governor' when he or she is away from the establishment. But the position of governor is a role that can occasionally be played by persons not appointed as governors. This is true in the sense that, in all four countries, senior officers were required to 'act up' when governor-grade staff were off duty or away from the establishment (though the governor would be contacted if any major developments occurred). But it was also true in small establishments, where the official in charge might be a senior member of the discipline staff, operating (in France, for example) under the

[6] Metcalfe and Richards (1987) claim that in the UK, 'civil servants are inclined to define management as an executive function, presupposing the clear definition of objectives, policies, and if possible, corresponding performance measures. They view management as purely internal to Departments, concerned with internal routines and procedures. They limit management to hierarchical relationships, neglecting others which may be more productive.' These tendencies do not appear in such a strong form in English prisons, though the vague nature of the prison task does tend to lead to the primary concern being the maintenance of routines; it could be applied much more aptly to French prisons.

periodic supervision of a governor from a nearby larger establishment. An analogous situation occurred in Germany, where 'satellite' institutions—usually open prisons—were administratively regarded as adjuncts of larger institutions and headed by the governor of the 'main' prison, but in practice operated much of the time with no governor grade staff on site.

Such arrangements, if complex in practice, are in principle quite straightforward. But the idea of 'governorship' became more complicated in the Netherlands with the introduction, just prior to my research, of the role of Hoofd begeleiding as an adjunct to the gubernatorial role.[7] Vinson *et al.* (1985) describe the post as one which brings together the authority for the design, resourcing, and management of programmes, and which is designed to increase the integration and co-ordination of staff functions. Appointees would usually be trained in the human sciences (sociology, psychology, social work, etc.), and governors were, at the time of the research, given wide discretion within these guidelines to make appointments to the new posts, the appointees then undergoing a centralized training programme. This approach was expected to make possible the recruitment of persons into senior positions who did not have a long experience of prison matters and who might therefore stand outside the viewpoint of the 'discipline staff subculture'. It represented, therefore, not only a way of freeing governors from much of the developmental work that went alongside the actual operation of the prison, but also provided an in-house source of social-scientific expertise.[8]

The role brought with it a number of complications. The intention was to appoint as managers, though not necessarily direct-line managers, persons who had no prior experience of prisons over those whose working life had been spent in jails. In some circumstances they would have a direct-line-management role, since in addition to development and co-ordination work, it was clear that the role also included responsibilities such as deputizing for the

[7] This term would translate literally into English as 'head colleague', but a less literal translation such as 'Director of Inmate Programmes' is probably a more accurate reflection of the nature of the post.

[8] However, in practice the posts were being filled rather slowly; in the 4 establishments I visited, only one Hoofd begeleiding was in post. She had begun to develop her role initially through embarking on projects concerning staff training and a new method by which guards could choose which teams they wanted to work on.

governor, for example, in dealing with inmate disciplinary matters. All in all, it was a brave step, and one running in the opposite direction to English practice, which now insists that its prison managers are recruited from among those who have long exposure to the staff subculture.[9] And it was counter to the English trend in another sense as well, since the Fresh Start reorganization has redesignated deputy governors as 'heads of custody', with responsibilities for the day-to-day management of the institutions, thus locating the main responsibilities for planning, development, and integration with the governors.

Fronting for the Prison

Governors spend a great deal of their time dealing with matters which can intimately affect their institutions but which originate with, or must be processed through, headquarters or regional bureaucracies. Such matters may be dry, technical, and often financial: budgets, expenditure, staffing levels, and so forth. Yet they are clearly of great importance, since the good functioning of the institution depends upon the governor demanding and getting adequate resources to operate the establishment, in competition with demands being made by other governors, and in the context of pressures imposed by government expenditure targets and the like. There is, in these matters, a very lively sense among governors generally that the relationship with headquarters is an unequal one in at least two ways.

First, governors and headquarters tend to operate with different value systems. The task of a governor relates to his or her prison; headquarters must deal with the generality of prisons, and with the tasks of allocating, for example, a fixed budget among prisons. It is hardly surprising that in most prison systems governors feel that headquarters staff not only do not understand them, but do not attend to the realities of dealing with inmates. As one Dutch governor pointed out, when the educational budgets for establishments were worked out, no allowance was made for the fact that his

[9] This observation should, however, be qualified by two observations. First, it clearly does not apply at the very top of the organization, since the CEO came from the private sector to head the prison system, while in years past the Directors–General were normally brought in from other areas of the civil service. Second, while it may well be true for the prisons that the Department continues to run, it is less clear how far it applies to the privatized establishments.

prison held a man who was regarded as a potential troublemaker, and who had been given access to a large number of educational courses in order to occupy his time and thus prevent the emergence of control problems. The management of a prison tends to include a vision of matters such as education and other regime activities as ways of keeping control or preventing trouble; headquarters sees it as a budget which can be allocated on a per capita basis or in fixed proportion to historical costs. One result of this discrepancy in views is that governors often feel they are left managing problems originating elsewhere in the system. As one Dutch governor said, 'Most of the time they only hamper you. They take wrong decisions, and they provide you with unnecessary work.' Many governors would share such a sentiment.

Second, it is an unequal situation in terms of information flow. Headquarters can demand information from establishments, and can demand that specific types of information are regularly reported or that occurrences of certain types of events are reported promptly. It is to some extent possible for governors to manipulate or, in some circumstances, choke off this flow of information, as will be illustrated below. But on the other hand, when headquarters offices pass out information it is by no means always as complete or accurate as it could be. The most spectacular recent example of this was described by the Woolf Report, discussing the implementation of Fresh Start. One governor, involved in training governors and senior staff for Fresh Start, was asked where the additional staff to operate the new attendance scheme would come from (Home Office 1991: para. 13.65):

Staff believed the assurances which he and others gave because they were trusted and because the information on which the answers were based came from the appropriate division of Headquarters. He himself had no appreciation of the true level of economies which were being demanded. He states that it was only later that he became aware, as the result of a change of duties, that it was intended that there was to be a 10 per cent reduction in the level of staff across the region he was then working in. As he put it, this had 'the effect of making a lie out of all the training that had been given to senior staff and group managers'.

The balance of power within the organization can mean that governors are given responsibilities but not adequate resources to fulfil them. Governors can and should be held accountable for the appropriate deployment of the resources they are granted. Yet if those resources are not sufficient for the tasks the governors are required to perform, responsibility can and should be placed on the headquarters organization. To be sure, in cases of major institutional failure, such as escapes, the political fallout will land initially on the shoulders of headquarters officials, whether or not their actions are ultimately seen to have contributed to the failure. None the less, in at least one recent incident, the 1991 escape of two IRA suspects from Brixton prison, claims of inadequate headquarters support for institutions appeared validated when a senior headquarters official decided to resign because of the incident.

Planning and Monitoring

Much of the business of running a prison is not so much concerned with day-to-day issues as with identifying and assessing or justifying historical performance and 'track records', and with making and pushing forward plans for the future. To some degree both these tasks are now routinized and bureaucratized.

In all four countries, weekly, monthly, quarterly, and annual reports on various topics were called for by headquarters offices. Outside of reports on inmate numbers, the extent to which these reports mattered—in the sense that they were scrutinized and discussed rather than just filed—clearly varied. Indeed, it appeared that even within a single country, such as England, prior to their reorganization into areas, different regional headquarters would call for different returns based on the perceived prison problems and the predilections of regional personnel. Moreover, since each group with the authority to call for returns wanted the information in the form best suited to its own interests, prisons could find themselves in the position of filing three or four returns on a single subject. In England, prisons would supply regular information on prison workshops in one form to the national headquarters branch responsible for workshops; in another form to regional headquarters; and include similar information in other documents they were also required to supply regularly.

Having made this caveat, some pen-pictures here might be helpful in identifying different national practices as to routine regime monitoring.

In England, a system of regime monitoring was instituted in 1987. Originally part of the short-lived 'accountable regimes' philosophy, the forms were intended to enable prison governors themselves and regional psychologists to keep track of what provision was actually being made for inmates on a regular basis.[10] The weekly monitoring form covered: the planned and actual number of inmate hours (number of inmates involved multiplied by number of hours in the activity) in each of the main workshop and recreational activities in the prison; the completion of a number of weekly routines, including canteen, exercise, association, bathing, kit change, inmate mail distribution, library, and visits; a range of other 'regime indicators', such as numbers of first reports sick, disciplinary hearings, and inmate petitions; a sample check of the number of hours inmates spent out of cell, obtained by asking some six or twelve inmates randomly and computing an average; and comments on the reasons for not completing planned activities and other major events in the prison's week.

One example of such a form from a training prison notes that in a specific week the prison workshops employed 104 men as opposed to their capacity of 120, and that the workshops were open for 25 hours out of a maximum of 27 hours 45 minutes. Thus 2,600 man-hours of work were completed out of a theoretical capacity of 3,330 man-hours. The form indicates that, on some days, one of the instructors was away at meetings; and a note at the bottom of the form indicates that one workshop was closed for half a day to search for scissors that had gone missing.[11]

The usefulness of this information was said by governors to be rather limited, because it merely summarized points that the governor already knew, and because the information was always a week in arrears. Within the institution there was little need to go to the file in order to check on recent trends in the prison. However, from the point of view of the regional psychological staff, the series of weekly returns was a useful research tool. It was possible, for example, to look at the extent to which planned activities were actually operated, and to compare this with the extent of reports sick, inmate complaints, disciplinary hearings, and so forth. Governors on the whole received this concept of 'scientific management' with some scepticism, since it could not tell them anything

[10] For details of the 'accountable regimes' arrangements, see Ch. 3.
[11] During my UK fieldwork I collected a large number of regime monitoring forms, principally from the local prison and the training prison in which I conducted staff and inmate interviews.

they did not already know about the atmosphere of the prison and the likelihood of control problems.[12] The psychologists, on the other hand, held that the figures collected could act as proxy measures for the prison atmosphere, and could be used to look at the medium- to long-term antecedents of incidents and other management problems. Both groups were right from their own perspectives—though at the time I finished my fieldwork with them, the psychologists had not been able to 'deliver', in the sense that their database was still too small to be of use in analysing the conditions prior to incidents.

In the Netherlands, the management structure required less in the way of routine returns, and that little tended to vary depending on the prison. For example, information on the numbers of inmates working and not working might be seen as more important in relation to open or semi-open establishments, where idle inmates were thought to pose more of a risk to the institutional order. At the closed prisons with longer-term inmates, levels of staff sickness were thought to be better indicators of possible tensions and troubles. But in general, the headquarters management style, much like that of the prisons, has a strong collegiate flavour. When a problem is recognized at headquarters level—as happened in the mid-1980s with the level of drug use in prisons—the response is to set up an *ad hoc* committee, often comprising members of several interested headquarters branches, to look into the situation and make recommendations. The first problem the committee faces is usually a lack of concrete information, and its response is to call for regular returns on the problem from all institutions. In the case of the *ad hoc* working group on drugs, for example, all establishments were asked to produce regular figures on drug discoveries and seizures from inmates.

The comments above relate to regular statistical returns. However, another form of document has assumed a much greater

---

[12] In one of the institutions I studied, an alternative form of monitoring had been designed by the senior staff in the early 1980s and remained in use alongside the new system because, as the governor said, it provided a means of continuous *financial* control over the running of the establishment, including components on staff wages, allowances, and overtime, which together accounted for over 80 per cent of his expenditure. At the time of my interview he was considering—with regret—suspending the system because he did not think he had sufficient resources to run it alongside the system he was required to operate.

importance, certainly in England and the Netherlands.[13] Prison governors in these countries are now required to produce a 'plan of action' for the following years. In the Netherlands this is known as the 'beleidsplan', a system introduced in 1987–8. In England, where the system was initiated in 1984, it is described as a 'contract'.

In the Netherlands, prior to 1987, plans for institutions were prepared centrally. During 1987, institutional governors were asked to provide draft plans covering a five-year period, with the general intention that such plans would be updated every two years. The plans were to deal with three areas: first, the status quo and problems associated with it; second, the specific aims to be achieved over a two-year period, and more general aims in relation to a five-year period; and third, the identification of organizational resources that would be needed and any implementation problems anticipated. Although there was some flexibility as to what areas should be included in the beleidsplan proposals, a checklist of areas which it should cover was produced centrally, calling for specific mention of issues including drug-use, visits, recreation, and education. The purpose of this was to avoid a situation in which governors could ignore regime areas they personally did not like dealing with, or found institutionally convenient to gloss over. At the time of my fieldwork only a few completed documents had been submitted to headquarters, and most governors I spoke to were in the process of completing their plans.

The requirement to produce a plan itself had a mixed effect on the governors. On the one hand it required additional work, and indeed in one prison an agreement had been reached (though not implemented) for an additional assistant governor to be appointed to help in its preparation. There was also a degree of scepticism about the utility of the plan, since governors were fully aware that headquarters would then tell them that they should not follow

---

[13] The situation in France at the time of my fieldwork was that several of the larger prisons, including Fleury-Mérogis and La Santé, dealt directly with the headquarters administration over their budgets while most other establishments dealt with the regional directorates. Although governors had substantial flexibility about how they spent their budgets, those I spoke to suggested that forward planning was centralized and hierarchical, according with the doctrine that the functions and responsibilities of establishments are laid down in the Code de Procédure Pénale, so that what is being resourced is a standard level of provision across all institutions of a similar type.

their own plans, but implement alternative developments instead (at this time the headquarters was trying to identify units which could be developed into 'drug-free' regimes). And as one governor said, 'We used to send statistics to Ministry, but there was no response so governors stopped doing it. The beleidsplan is to get back the information.' None the less, the discipline of having to think systematically about the nature of their prison, and to propose development plans which would be binding on future governors, was accepted, if sometimes grudgingly, as useful. As one assistant governor said.

For me it's a mistake that every time there comes a director in a prison he has his own ideas and guards should do what he is thinking. In my being here twelve years I've had four directors, four different regimes. So I was glad with the planning we have to make, not on a person but for two to five years.

In England, longer-term planning has remained the preserve of headquarters, which annually prepares a five-year plan starting in the current year. This plan has been characterized by Woolf as 'a document which establishes what Ministers expect of the Service and what they are prepared to pay for those expectations' (Home Office 1991, para. 12.87). Clearly the most detailed targets are those set for the current and following years, and within the context of this broad plan, governors are only required to think one year ahead. The essence of the contract emerged out of the move towards mission statements in 1984. Two documents were produced, which laid out between them the aims of the prison service as a whole and the functions of establishments.[14] More detailed arrangements for implementing these aims were set forth in Circular Instruction 55/1984, and they were designed, as the then Director General, Christopher Train, observed (1985: 179), as 'a piece of machinery which enables the Governor to be accountable for the operation of his own establishment'. The idea was for an agreement, covering some twenty-two specific items, to be reached between each governor and his regional director (area managers took over the regional directors' role in this process following the 1990 reorganization). The contracts were fairly detailed, specifying,

[14] The 'Prison Board Statement of the Task of the Prison Service' and the 'Prison Board Statement of the Functions of Prison Establishments'. Both are reproduced in Maguire *et al.* (1985); they are not very detailed.

for example, how long on average each inmate should spend out of his cell per day. The agreement would then form a benchmark against which institutional performance could be measured in the coming year, and this was in large part the role that the weekly 'regime monitoring' forms took over after the demise of the 'accountable regimes' doctrine.

The system had, however, two flaws. First, as one governor observed, the contracts for local prisons and remand centres specified resource allocations for court escorts, although court escort requirements were largely determined by the courts, and were outside the control of governors. Second, and more seriously, although the governors were required to set objectives and to identify the resources necessary to achieve them, there was no corresponding obligation for the department to provide the resources that the regional directors, and later area managers, agreed to make available. Governors, whatever their good intentions at the beginning of the year, were inevitably put in the position of needing a 'plan B' up their sleeve when promised resources failed to materialize.

Relations with Headquarters in a Time of Change

In 1986–7, the Dutch Ministry of Justice invited the international management consultants, McKinsey, to review departmental organization and functioning. The result was a recommendation that the budget could be cut by some 40 per cent without loss of efficiency. Some departments were to be reduced by only 10 per cent; others were identified for cuts of up to 80 per cent, or indeed for closure or merger with other departments. The result was a great deal of turbulence, since at the time of my fieldwork senior officials were simultaneously trying to run the existing system and to change many of the organizational structures, while remaining uncertain as to whether their own jobs would exist after the changes.

Governors in the field were less radically affected by this. The turbulence in headquarters made their jobs easier in some respects, since the degree to which they were being monitored had dropped. As indicated earlier, many had ceased to send in monthly statistical returns and no-one had pursued them about it. They were aware, because they were in contact with headquarters officials, that in some cases the purpose of the returns had ceased to exist, or that headquarters staff were too busy with the reorganization to pursue routine matters. But at the same time, trying to get decisions out of headquarters staff when they were needed had also become problematic. Some officials had effectively ceased to make decisions because they were no longer sure whether the reorganization had left them with the decision-making power, or because they anticipated that the power to make them would lie elsewhere within a matter of weeks. In this situation, many governors began to use alternative ways to get decisions made. The easiest route was to go, not to the person organizationally nominated as the relevant decision-maker, but to a friend at headquarters who could be asked to process the issue through whomever had to be consulted, as a personal favour.

One prison governor, during the course of my fieldwork, tried another way of changing departmental policy. He was concerned about the likelihood of HIV transmission in prison, and took the view that condoms should be issued to (male) inmates to prevent infection via homosexual activity.[15] He used an interview with a journalist to announce his intention to make such a scheme available within his prison. He stated that since homosexual activity quite clearly existed, inmates should be able to protect themselves against HIV infection in the same way that they would do outside the establishment. The pros and cons of such a scheme are not immediately at issue here, though the policy at that time was that prisons (including cells) are public places and that homosexual activity was prohibited; to allow the issue of condoms would thus require a change in the legal view of a cell, or would amount to formally condoning an activity which was officially prohibited.[16]

[15] It should be remembered that this incident occurred at the early stages of discussion about HIV infection, when the major concern was still about transmission via homosexual activity. Since that time, of course, the issue has shifted to infection via needle-sharing among drug addicts. In this connection it is worth bearing in mind that a very large minority of Dutch prisoners are heroin addicts at the time of reception (and a not inconsiderable number are alleged to begin using heroin in prison). It was estimated that 20–30% of the Dutch penal population were hard drug users in 1984 (Erkelens 1984, quoted in Meijboom 1987). A 1985 estimate was 35% (Kelk 1991).

[16] Ralph Vossen (who assisted me with the Dutch fieldwork) and I discussed this informally with several officers, prisoners, and headquarters staff. The positive aspect was thought to be that such a scheme would reduce HIV infection. The negative aspects, according to the prisoners, were that unless condoms were issued automatically to all inmates, individuals would have to identify themselves to staff to obtain them—and even if they were simply left in a box for inmates to help

Understandably, the content of the interview became an item on nationwide television by the same evening. The following morning, senior administrators—rightly thinking that the governor had 'gone public' in order to force their hands on policy—called him to a meeting from which he returned looking distinctly haggard. No public discussion of this issue seems to have occurred since, and no such policy has been adopted. But the key point seems to me to be that such a strategy for attempting to change policy becomes thinkable, and indeed stands some chance of working, in periods of rapid change or crisis. And as a sidelight to this episode, it is worth recording that when I asked a senior headquarters official what might happen to the governor, his answer was that he would probably now have to wait an extra year to get the posting to the governorship of another establishment he was known to be aiming for.[17]

Dimensions of (Lack of) Control

As many headquarters officials pointed out, governors have to be seen as professionals who should be given latitude to exercise their managerial skills. None the less, the latitude of a governor is limited by the internal rules, regulations, policies, and finances of the prison system, as indeed it is meant to be. In practice, as several governors said, the combined effect of these constraints was that their room for manœuvre in any given situation was often quite small. Moreover, at certain times, in certain prisons, or over certain issues, the grasp of the headquarters administration over the institutions would be tighter than in other situations. For example, headquarters had, by and large, a stronger interest in the Dutch high security institutions than in the other prisons, because of their concern about the political consequences of a 'heavy' criminal—whose offences may be widely known to the public—escaping.

themselves, staff would still be aware who was using them. Thus the scheme would be unworkable unless the prison administration was prepared to recognize cells as being the equivalent of private residences for certain purposes; and the end of this line could potentially be that staff would need search warrants in order to conduct cell searches. The prison administrators also made this latter point.

[17] This compares rather favourably with the fate of John McCarthy, the English governor of Wormwood Scrubs who wrote a letter to the press complaining about the state of his 'penal dustbin'. He was invited to resign shortly afterwards, and indeed may have written the letter with his resignation in mind.

Similarly, particular concern over specific issues may result in closer central scrutiny of all prisons with regard to those issues.[18]

What, then, did governors feel they had most control over? Asking such a question reveals the rather nebulous, but detailed, ways in which management of institutions is conducted. In the Netherlands, one assistant governor, in charge of one wing of a prison, said:

I have control over nothing and control over everything. Over nothing: every governor has control over his own section of the budget, but can't change [it]. Over everything: I have guards and team leaders. To my guards I say I want to listen to them and also allow them to make their own decisions on the wing. Decisions outside the wing are mine, therefore staff can't change, for example, the amount of exercise, but can decide on visits. It is my decision which guards do what, go where, on what wing. Team meetings can't take decisions but only advise me. If you give staff more decisions they get more pressure from prisoners. Having a distance means it's easier to see the whole picture and say yes or no.

Another Dutch institution provided, through its unusual management structure, a miniature illustration of the wider constraints on governors. Over-Amstel comprises six units, each with its own governor, and central services such as the kitchen and visits area under the control of a seventh. With no overall governor in charge of the whole site, decisions affecting more than one unit had to be made collectively by all the governors. Within any one 'tower' (i.e. individual prison), because of the high-rise prison design, inmates move about by elevator and some thought had to be given to the timing of the prison day in order to avoid large congregations of inmates needing to move about at any one time. But in addition,

[18] A Dutch example of this idea of 'administrative grasp' may be helpful. The remand establishments at the time of my study were expected to operate at 100% capacity, and one governor was under pressure to increase his population by utilizing a high-security unit at 100% capacity also. The governor argued against this, on grounds partly of principle and partly of pragmatism. The principled argument was that inmates should not be placed in high-security units on grounds of administrative convenience unless their behaviour warranted such action. Moreover, the high-security cells were actually split into three sub-units, each designed and operated for different types of high-security inmates. If unsuitable (e.g. normal) inmates were located in these units the regimes would break down. The pragmatic argument was that if all the cells were full there would not be sufficient regime activity to occupy all the inmates located in the unit. His arguments succeeded to the extent that the administration accepted that the high-security unit should normally be utilized at only 75% capacity.

A change in one activity will change others. We have to look at the complete activity programme and staffing. It's not impossible, but complicated. There's need for liaison between towers as well, over for example feeding times. Also with visits—not only liaison within a tower but with central services.

Within the French system, one governor of a maison d'arrêt who I interviewed saw his job as largely administrative rather than managerial. Although he has to make plans and budgets, the majority of his work in relation to inmates, he said, consisted of executing the decisions of others. This is so because many decisions about individuals, such as controls on correspondence, are largely judicial decisions. This view might be described as a 'traditional' one, since it accords with the doctrine that the Code de Procédure Pénale comprehensively states what should happen in prisons. Yet as we saw in Chapter 2, the CPP is silent on many important issues.

Another governor, in a centre de détention, did, however, confirm the other plank of the French system, its highly centralized control. He stated that little development planning took place at institutional level; developments were largely proposed by the ministry and implemented without local discussion. However, he did have a number of ways of changing the regime. He could, for example, schedule more exercise, change the variety of goods in the inmate canteen, invite more or fewer outside teams to play basket- or volleyball, and so on. Yet there remained aspects of the regime that he felt were difficult to tackle administratively. He pointed out that searches were conducted in a less 'hard' way when compared with the maisons d'arrêt, and that this was not totally controllable largely because staff needed to maintain a continuing workable relationship with the long-term inmate population.

One area in which all governors, in all countries, were agreed was that their powers in relation to staff discipline had become very weak because it had become centralized. Yet in a time of change this created problems. In the Netherlands, one regional director observed that some governors sought to bypass official channels for sanctioning staff. In principle, they were expected to write a letter to the central administration which would be copied to the officer concerned, and then to await a reply. However, since replies could take up to six months, these governors were now cautioning officers themselves and placing notes on the personnel files. The problem then arose with repeated disciplinary infractions, for which governors would use the official procedures; since no documentation on previous misconduct was filed with headquarters, the repeat infraction would have to be formally treated as a first offence.

Accountability and some Conclusions

The literature on prisons proposes a number of distinct management models—control, shared-powers, bureaucratic-lawful, and the like. These models are fairly useful in discussing trends in prison management over a period of, say, the last fifty years; but they are much less useful in distinguishing between the prison managements of English, French, German, and Dutch prisons in the 1980s and 1990s, for the simple reason that all four systems operated along relatively similar lines.

This is not to say, however, that differences did not exist. As I have indicated both in this chapter and the last, comparisons could be made between French maisons d'arrêt and centres de détention which indicated the greater or lesser presence of paramilitarism and emphases on control. Management structures in different countries gave different powers to discipline staff, with somewhat different results in terms of inmates' abilities to organize their own lives. But other factors, such as the background and qualifications of prison governors, and their relationships with headquarters, also made a considerable difference to the nature of prison management and accountability.

Prison systems nowadays are large bureaucratic organizations. The place of the governor in such organizations is both central and marginal; central because of his or her managerial and administrative role within the establishment, but marginal because in many respects he or she is little more than a conduit between the institution and the central administration. However, it is important not to obscure national differences as to the gubernatorial role. It is at its most administrative in France, where a great deal of power is centralized in the hands of legally-qualified administrators in headquarters, and the most significant decisions concerning individual inmates are made by the Juge de l'Application des Peines. The German system provides some 'equality of arms' between governors and headquarters staff, since all such officials are legally trained, but the consequence of the career structure is that prisons are in practice largely run by chief officers. The more 'managerialist' organizations—England and the Netherlands—give governors greater leeway, though much of this has been eroded in England by the fact that area managers and headquarters are not bound by the agreements over resources negotiated in planning cycles as part of a 'contract'. Ironically, in the Netherlands governors do wield substantial power, but this came about unofficially in the late 1980s as the central administration began to buckle under the weight of reorganization.

Aside from such remarks on individual countries, perhaps a general conclusion should be this. Governors are expected to manage their institutions within a set of constraints, of varying natures— legal, financial, organizational, and practical. Indeed, the divorce of the allocation of resources from their day-to-day management constitutes one of the major themes of accountability. Yet these constraints also divorce authority from responsibility, so that governors are often regarded as responsible for matters that they cannot wholly control.

7

Accountability: Some Sensitizing Concepts

PREVIOUS chapters have described various features of the prisons, prison administrations, and staff and inmate perceptions of the institutions in the four countries. Later chapters will discuss specific issues such as inmate discipline, complaints, and inspections. This chapter, which forms a bridge between the descriptive and discursive parts of this book, provides some theoretical tools which can be used to bring order to the wide range of materials discussed in the coming chapters. In essence, it argues that accountability can be seen as a series of games; and while some games have elements of confidence tricks about them, accountability as such is not a confidence trick. That is, accountability can be understood from both structural-functional and interactionist perspectives, but cannot be reduced to an argument about hegemony.

Hegemony and Accountability

Why not treat accountability as simply one instance of hegemony, which has proven to be one of the more durable concepts in left-wing political and social analysis? The answer is that the concept is reductionist.

Hegemony, a term first employed by Gramsci (1971) and developed by Hall *et al.* (1978) among others, refers to the consent of the ruled to the domination of the rulers. In essence the argument is that economic and political domination is best achieved, not by the use of force, but by co-opting the working class into accepting its 'dominated' position. This is done partly by giving real political and economic gains which, none the less, are not structured in ways that would undermine domination; and partly through the promulgation of cultural and ethical values which reinforce the need for consent. In consequence the working class accepts the domination of the ruling class because the latter is seen as the provider of material benefits, such as a rising standard of living and welfare and health benefits; because the working class believes that political structures ensure its representation in the society's power structure; and because those structures or representation and the mass media reinforce the notion that mass consent to existing power relations is both necessary and forthcoming.

On this argument, the need for prisons can be depicted as a need for a crime-control mechanism, the primary function of which is to protect the working class from victimization by its own deviant members. Yet at the same time, since persons sent to prison are predominantly members of the working class, there need to be mechanisms designed to reassure the working class that prison conditions are neither overly bad nor overly good (that is, neither inhumane nor inconsistent with less-eligibility arguments), and that inmates are fairly treated. In this sense, mechanisms of accountability would be designed both to ensure that prisons are operating in line with stated policy, and to reassure the public that this is the case.

There are two problems with this argument. One is that, saving the distinction of viewpoint between discussions of class and of the public, these functions are, of course, those that mechanisms of accountability are purportedly designed to fulfil; there is no 'hidden agenda' here which the concept of hegemony is able to bring to light. The key difference between hegemony and a liberal-democratic view lies simply with the motives attributed to the state. The second problem, which follows on from the first, is that using the concept implies seeing the state as both monolithic and deeply cynical in its motives. To speak of hegemony is to view the world as one in which there is, in the last instance, a subservience of the state not only to abstract 'interests of capital' but to a cabal of captains of industry whose broad demands on the state are complied with, even though in specific instances (planning permissions? financial regulation?) they may be opposed.[1]

---

[1] These comments are directed principally at the situation in liberal democratic states. In other states, and in other historical periods, the concepts of hegemony and the 'monolithic state' are likely to have greater purchase; for example, it is now widely accepted that in the eighteenth century, during the formation of the modern

Yet such a situation is probably best seen as one in which different power blocs, with differing resources and agendas, are competing for control over a state apparatus which itself has significant resources and its own agenda, and sees itself as a player in the power game, albeit that it is in principle subservient to elected representatives. On this view, the advances won by labour are real advances even if they are also strategies for co-option by capital, and there remains space for 'real' liberal reformers, even if they are members of the ruling classes, to be seen as acting from genuine humanitarian motives rather than cynicism or the misguided acceptance of ideologies directed at the working class. Such a situation is likely to be complex, and best likened to a series of 'games'; even if they are games in which the rules, the playing field, and the players can change over time. Accountability thus becomes a fragmented field, in which the word has multiple meanings and is deployed in a variety of ways for different purposes.

To pursue this 'game-playing' view of accountability, we must turn away from broader political characterizations of the state and look more closely at a range of sociological perspectives.

Sociological and Other Theories

Accountability is a diffuse concept, and one that is mobilized by different groups for very divergent purposes. In the discourses of prison reform it is often held out as a tool in the business of making prisons more just and more humanitarian. In the hands of governments it is used as the justification for increased cost-efficiency and control over staff. In all these uses, however, it is in practice closely associated with the concept of control, and usually denotes a concern with and scrutiny over the structure and exercise of controls.

There exists—with a few notable exceptions—a much greater literature on control and compliance within organizations than on

capitalist states in Europe, the creation of prisons and the forms of prison labour were greatly influenced by the desire of capitalist interests to shape a largely agrarian population into a more settled, urban, and skilled proletariat, and that some aspects of this—in particular the protection of local populations from vagrants and beggars by policies of incarceration—verged on the hegemonic (see e.g. Melossi and Pavarini 1981). In addition, recent work on contemporary Chinese prisons suggests that at the local level, at least, one is dealing with a monolithic state in which the close ties between the prison authorities and production companies making use of convict labour enable the latter to manipulate the former. On the Chinese prison situation generally, see Wu (1992).

accountability.[2] Yet since the two concepts are so closely associated we may usefully plunder the work on control and compliance in order to derive some perspectives on accountability. Three perspectives in particular may prove useful.

One approach can be derived from a broad consideration of functionalist sociology and studies of organizations. One of the effects, if not a defining characteristic, of modernity is the differentiation within, and complexity of, the social system. Governments have become large bureaucratic organizations which execute a variety of different tasks. A relatively simple way of looking at this situation is to consider government organizations, such as prisons, as sub-systems which have goals set by the polity, and which are achieved through processes of adaptation and integration. Organizations must acquire the human and other resources necessary for their work, arrange themselves in ways that enable such resources to be used, and overcome practical problems which can be seen as giving rise to sub-goals. For example, the custody of inmates creates a large number of sub-goals concerned with clothing, food, physical security, and so forth.[3] The primary, though not the only, way of carrying out such tasks is through the medium of a hierarchical structure in which decisions are made by relatively few people and carried out by the many, who must also provide information to the decision-makers. At the same time, the integrative problem of how to ensure compliance with organizational procedures and regulations depends upon the use of rewards—salaries or wages, for example, but also prestige—and sanctions. The problem for leaders is thus one of allocating rewards and sanctions in ways that ensure compliance. In prisons, where there is a captive 'administered population', the problem is also one of providing sufficient inducements and sufficiently severe punishments to ensure this population's compliance with the demands made on it.

The question of compliance is the central topic of Etzioni's *Complex Organizations* (1961). Its approach is typological. Power may be exercised in three main ways: as coercive force, through

---

[2] The exceptions include the now-extensive literature on accountability in relation to the police and the medical profession. On the former, see e.g. Lambert (1976), and Morgan (1987) and Baldwin (1987)—Baldwin's paper being a commentary on Morgan's argument. Unfortunately, these discussions are only of limited application in the particular area of prisons.

[3] See esp. Parsons (1952), and Parsons and Shils (1962).

remuneration, and through the establishment and maintenance of norms. 'Lower participants'—employees, inmates, and so on—may be involved in the organization in three corresponding ways: alienatively (that is, they are alienated from the goals and functions of the organization), calculatively (for anticipated reward), and morally (through acceptance of norms). Coercive force would therefore be a primary form of control over inmates, while remuneration and moral involvement would be the main reward for staff. There is a recognition that coercive force is not often directly employed, but used only where other means fail. There is acknowledgement that the forms of power are not completely distinct, as where a mental hospital relies primarily on normative controls over patients, using coercion only where these fail. And there is recognition that some specific techniques may have various symbolic values; medication can be perceived sometimes as treatment, sometimes as punishment, and sometimes as one by staff and another by inmates. But there is little discussion of, for example, the normative controls on staff designed to ensure that overt, physical, coercive force is not used without good reason, nor of the broader mechanisms that dictate what levels of coercion may be used.[4]

These issues have been addressed, though often tangentially, from within the field of management studies. In Pugh's (1984) edited collection, March and Simon draw attention to the development of increasing levels of rational-legal authority, and its consequences in terms of increased demands for control over work. They argue that frequent though unintended consequences are rigidity of behaviour, increased supervision which is felt as increased authoritarianism and punishment, and internal organizational conflict. Other points made elsewhere in the collection also indicate how complex the use of authority can be. For example, the rule-following nature of organizations often means that the formal issue of an order is often, in reality, no more than the authorization of a 'next step' in a series of actions which are equally well understood by order-givers and recipients, and where the

---

[4] One other problem in the public sector in particular is that policy-making is constrained by wider social, legal and political issues, so that policies which may be rational for an organization perhaps cannot be promulgated, or must be implemented step by step over a lengthy period of time, to keep in step with developments elsewhere. See esp. the chapter by Lindblom, in Etzioni (1969).

recipients may know that the authorization is normal and can only exceptionally be withheld. On the other hand, the efficient execution of an order may depend upon its being given in such a way that it does not seem like an exercise of authority, so that its recipients feel co-opted rather than coerced into compliance. While much of this may be characterized as pragmatic managemental knowledge, it has relevance for notions of accountability which rely on the idea of a 'chain of command'. It suggests that formal authority provided by the official hierarchy may be underplayed in practice, because too heavy a reliance on it may be resisted from the shop floor.

There is no particular reason to link functionalism explicitly to a view of accountability that is limited to formal structures and relationships. As later explorations have shown, a broadly functionalist perspective can happily incorporate questions about the dynamics of formal and informal relationships in organizations. Just as 'control' does not have to be a purely directive, authoritarian phenomenon, so 'accountability' might be conducted in a variety of styles. In Chapter 1, two other styles—the 'stewardship' and 'partnership' models of accountability—were also described. Yet such an approach remains rather abstract, assuming as it does that there are clear organizational goals, that common values and norms may be used to ensure the accountability of the organization, and that clear rules and supervisory procedures are capable of ensuring accountability within the organization. As we have seen in previous chapters, real life is more messy, not least because the very idea of accountability is a political tool that can be wielded for various ends.

In practical terms, we know that inmates and staff alike inhabit a world in which allegedly rational-bureaucratic procedures, while certainly bureaucratic, appear from the 'shop floor' to be rather irrational. Inmate dependence on staff is extensive, and staff often seek to maintain a margin between the goods and services they are obliged to provide to inmates and those they can provide selectively, on a discretionary basis. Institutional governors have much less latitude to govern than is often supposed, except in situations where the organizational structure is changing so quickly that formal procedures begin to break down. The processes of accountability themselves sometimes become part of the machinery for effecting change, rather than being simply arrangements for monitoring performance; and this is especially the case when the objectives of the service become politicized, in relation to concerns such as the reduction of union power, for example. In the face of such knowledge, a second approach, that provided by an interactionist perspective, may become more relevant.

An interactionist approach would, in general, be more concerned with the ways in which structures are constraining of action, or can be used as a resource in interaction; or are created out of routines in behaviour. It would be interested in the ways in which accounts of events are generated and evaluated. Above all, it would concentrate on the normal and everyday world of interaction, whether between inmates and staff, governors and headquarters officials, or senior civil servants and ministers. And it would ask how particular concepts—'the manager's right to manage', 'inmate rights', and so forth—are invested with symbolic significance and used by the various parties involved in the prison world. The idea of accountability would therefore be cognate with an ethnomethodological use of the terms 'account' or 'accounting': the production of recognizedly rational properties of actions (Garfinkel 1984). It would also, however, have to extend to the production of recognizedly rational *evaluations* of those actions.

Within this kind of perspective, one is more likely to be interested in presentational aspects of prison life, such as how inmate behaviour comes to be defined as in breach of prison discipline, and how concepts such as 'due process' and the 'rules of natural justice' are applied by the courts in determining whether that punishment was properly imposed. In addition, such studies also offer a way of showing how and why particular modes of accountability begin to falter. For example, they may indicate why it is that prison conditions which would be regarded as 'intolerable' from a humanitarian point of view cannot be translated into legal concepts capable of adjudication by a court.

We have, then, two broad but distinct perspectives within which to make sense of accountability. One might be termed 'macro-accountability'—that is, the forms and processes of accountability at the level of the state, law, constitution, and policy. It would be concerned with the forms and structures which are held to ensure that prisons are accountable for what they do. It would be primarily concerned with the nature of control exercised over the prison system by senior managers, the courts, parliament, and so forth. A second might best be described as a 'micro-accountability' perspective, which would be more concerned with the ways in which accountability is a product of social interaction. It would be concerned with the ways in which control is exercised within the prisons, though it would not exclude consideration of issues such as the quality of relationships and the manipulation of symbols at other points in the macro-processes.

While these two terms, macro- and micro-accountability, indicate different emphases of attention on the prison process, they are not wholly divorced from each other and neither do they mark out clear boundaries as to what kinds of acts can and cannot be considered within each perspective. Demonstrating that policy decisions were made on a sound basis, or with due regard for humanity, propriety, effectiveness, or value for money, is an enterprise utterly different from making prison staff accountable for their actions towards prisoners. Yet the two areas are linked, since information about general aspects of staff behaviour or the handling of specific incidents may prompt higher-level policy discussions or court cases, while the implementation of policy changes may require staff to behave differently.

There is, in consequence, a need for a third perspective which can link the macro and micro approaches. Several observations may be helpful in this connection. Giddens's (1986) theory of structuration offers one type of link. The theory asserts, *inter alia*, that human beings are knowledgeable social agents who can rationalize and offer explanations for their conduct. That conduct is, however, institutionalized, and institutional rules and structures provide actors with both constraints and opportunities in their action (for example, the constraint that rules must be followed, or the opportunity to decide whether or not to invoke a particular rule). Routine and recursive patterns of action ensure the maintenance—or, in some circumstances, change—of institutions over time. In addition, power is regarded as an elemental social fact, is differentially distributed within society, and different forms of power are given to different social agents; however, the exercise of power can often be resisted.

In this context, some of the concepts introduced in Chapter 1 come to be important. The distinction between different *modes* of accountability—stewardship, partnership, and direction—can be understood as different models of the distribution of power between agents. Different spheres of expertise—legal, political, managerial, professional—can also be considered as the 'power bases' of the various participants. Day and Klein's (1987) notion of 'languages of evaluation' becomes important, since it indicates that the various power bases can be used both to take control over issues, and to apply different sets of evaluative criteria to them. In consequence there may be disputes about the appropriate coin in which accountability can be paid. Finally, Foucault's running together of power/knowledge as a single concept becomes significant, since certain forms of power structure the knowledge that one has, while some kinds of knowledge also constitute a form of power. Foucault's own examples are those of the uses to which reports and files on individuals can be put, and the use of statistical returns in policy-making. Both remain as true today as they were in the early years of the total institutions Foucault describes. In addition, however, one might make a similar observation about the multiplicity of 'knowledges' embedded in the different power bases involved in prisons. The 'legal knowledge' and 'managerial knowledge' of prisons are likely to look very different, with entirely different frames of relevance.

Having described three broad approaches to questions of accountability, the following sections elaborate on the kinds of issues and propositions that might be made within each of them.

Macro-accountability

Any list of generalizations about accountability at the level of the state are likely to be banal. For example, we can fairly confidently assert that even though accountability and control are two sides of the same coin, policy planning for the prison system tends to stress control rather than accountability issues unless the latter have particular political significance, as with, for example, the question of privatization. It is probably also fair to claim that where questions of accountability do emerge at the political level, they have to do with *post hoc* justification and legitimation than with the creations of new directions in penal practice. And at a more 'operational' level, it seems likely that while certain classes of decisions—about inmate complaints, parole and remission, and so forth—are 'owned' by different bodies in different countries, the reasons for the differences lie more in historical and practical contingencies than in immutable principles.

Another banal generalization, but one we may be able to develop more fruitfully, is that accountability rests symbolically on two 'planks', control and independence.

'Control' in this context refers to a concern to ensure that the prison department is in command of its prisons, governors are in command of their staff, and staff have control over prisoners. Many of the internal mechanisms of accountability built into line management are primarily intended to assure senior officials and, ultimately, the system's political masters that, to put it bluntly, prisons are neither bear-pits, nor hotels from which inmates can check out when they feel like it. The infantilization felt by many inmates has much to do with the ways in which control is exercised through creating dependence on staff.

'Independence' alludes to a relatively complex set of issues that have grown up around the modern nation-state. Governments, we tend to believe, should not be allowed to do whatever they want. There should be checks and balances based on the concept of the rule of law; and there should be ways for the electorate to be assured that money is spent wisely, and that official practices and decisions are not made arbitrarily, capriciously, or in a discriminatory way, and do not result in inhuman or degrading conditions. There needs to be scrutiny of prisons by persons who are not involved in the chain of command, and who can make evaluations of prisons which are based on criteria derived from social discourses which are rooted in, for example, legal or humanitarian world-views.

The end result of this line of reasoning is that governments must on the one hand operate prisons, or see to it that they are operated, and on the other, authorize a body at one remove from the actual operation of prisons to oversee them and to comment freely on their performance. Such a division is made possible by the compartmentalization of government activities. A prisons department may run the prisons, while oversight either remains with central government or parliament (in the form of a select committee), or is entrusted to a separate governmental organization such as an inspectorate, or turned over to committees of lay persons such as Boards of Visitors. Not infrequently several or all of these mechanisms may operate. In addition, the concept of the 'rule of law' ensures that the operation of prisons is subject to legal scrutiny as to its lawfulness. In short, and recalling the categories of Morgan and Maggs, 'directive' line management has to be complemented by legal and political scrutiny of the 'stewardship' of prisons, and possibly a partnership at local level, in order to maintain the legitimacy and credibility of the prison system.

Direction, stewardship and partnership designate different forms of accountability relationships. Day and Klein (1987), in a discussion of several public bodies, suggest that these different forms are applicable to different kinds of situations. Thus the police are felt to need operational independence combined with policy and financial scrutiny, the end result of which is the use of publicly appointed police authorities exercising a stewardship role. In areas such as medicine, however, the expert and professional status of doctors means that accountability is exercised—in so far as it is exercised—through professional bodies, while broader public policy issues are assumed to be aimed at general welfare priorities best debated jointly with advisory boards in a partnership model. In the case of prisons, however, one is dealing with a national-level bureaucracy with some administrative and some policy functions; local institutions, where there is some need to monitor the quality of 'service delivery'; and a historical dimension which incorporates some vestiges of local control over establishments. All three kinds of accountability can be seen at different points in the systems, though their practical importance is not always strong.

The directive mode is a 'high-control' structure in which A is subordinate and obedient to B. In practical prison terms, this is in theory the relationship between front-line prison officers and their seniors. It is, in Germany, a relationship that can be imposed by headquarters on prison governors in relation to specific areas of management, and constitutes a kind of sanction (see Chapter 10). In principle it can operate at higher levels still, as where, for example, a minister instructs the director of the service as to policies, and provides a framework of rules to govern the exercise of discretion such that decisions are made in accordance with them.[5]

The stewardship concept suggests that the head of a service retains substantial discretion in policy as well as in day-to-day mat-

[5] The proviso 'in principle' is important here, because in practice ministers do not often issue such direct and detailed instructions—or at least, they only do so after consulting the officials to whom they give such instructions. One could, however, envisage situations which come close to such directive accountability; e.g., a minister requiring the head of a prison service to supply a plan to contract out prisons within a specified time-frame.

ters, but is periodically 'audited' by a competent and independent authority—whether an elected body or a professional organization appointed ultimately by that body—to ensure that what has been done complies with the prevailing standards of fiscal regularity, humanity, propriety, and so forth. In England at least, we might place the appearance of civil servants before select committees, and the submission of inspectorate reports to Parliament, under this rubric.[6]

The partnership model is one in which the head of service retains executive powers, but is expected to work closely with one or more representative bodies in identifying and addressing issues such as policy priorities, service delivery, and good practice. What interests might be represented on the partner bodies, and how individuals are selected for membership, can be fairly variable.[7] In so far as one can speak of a partnership model applied to prisons, it is probably best seen—though in an admittedly attenuated form—in England and the Netherlands, where the Boards of Visitors and Commissies van Toezicht hold discussions with governors about the running of their prisons, and in France, where the governor is obliged to consult the JAP on various aspects of prison regimes. This is not to say, however, that the formal relationship between the Boards and their governors, or the JAP and the governor, is in any formal sense a partnership; nor that the relationship is necessarily a strong one in which advice given would be treated as having any particular weight. Perhaps the best generalization one can make in advance of a more detailed discussion is that many members of such boards feel that they have some partnership relation to the governor, while the governor may be less eager to see it that way.[8]

[6] Much of the discussion of police accountability in England has revolved around the concept of stewardship. In this context, the key relationship is between the chief constable and the police authority to whom he provides (or in some cases has not provided) accounts of policy and other decisions. See e.g. Day and Klein (1987), Jefferson and Grimshaw (1984), Lambert (1986), and Reiner (1985). However, in England prisons are, unlike policing, a central state function. Since the local accountability that applies to the police is absent in the case of prisons, the accountability of the prison system is very largely a matter for parliament, select committees, the inspectorate, and the courts.

[7] Day and Klein (1987) suggest that the partnership model is a reasonably good reflection of the realities of accountability in areas such as health care, education, and social services in England.

[8] These relationships are all at the level of individual institutions. The extent of openness to partnership at headquarters level is a different matter. Here, the

Morgan and Maggs note that each form of accountability has inherent limitations. The directive mode concentrates power at the top of a hierarchy, whether in an individual official or a committee. It offers few safeguards to those not represented by, or who cannot influence, those at the top. In policing terms this may mean minority groups of all kinds, including ethnic minorities. In prison, it means the prisoners and lower-level staff. The stewardship mode is a *post hoc* exercise in which there is little room for intervention in the operational, day-to-day exercise of discretion and in which there are corresponding difficulties in remedying that which has already been done; its strength lies in the audit of track records. The partnership mode leaves effective control in the hands of a service which may thank the partners for their advice, but proceed to ignore it. Such observations enable us to look at the correspondence between different components of a service and the forms of accountability applied to them.

These systems do not, by any stretch of the imagination, constitute a 'net' of accountability covering every aspect of imprisonment. They might better be characterized as a small fleet of trawlers which may catch different but not by any means every kind of fish that might swim in those waters. To translate this analogy into practice, we might observe that legal, managerial, and political terminology surrounding prisons tends to differ sharply, and to focus on different aspects of any particular question. Take, for example, prison conditions that could be described by a lay person as 'unacceptable' or 'inhumane'. In the political sphere, the issue might well become one of 'less eligibility', that is, that improvement may be necessary but could not be considered while other groups such as the elderly, the ill, or schoolchildren have a greater moral claim to government resources. In legal terms, however, concepts such as negligence and 'cruel and unusual

strongest example might be the Dutch Centrale Raad van Advies, which deals with appeals from decisions of the Commissies in institutions, but also offers advice on national prisons policy. In England, the nearest to partnership one gets is probably not the National Association of Boards of Visitors, which is not necessarily seen as a source of advice on prisons policy, but the informal relations between the prison service and the Home Office on the one side, and the National Association for the Care and Rehabilitation of Offenders (NACRO) on the other. Much is discussed between these parties at informal meetings, dinners, etc.; but the relationship has largely been forged at an individual level, and has never been formalized.

punishment' would be employed to make a judgement on the issue.[9] Such differences in terminology are not always disadvantageous. They may serve to widen the scope of possible remedy. Complaints about prison conditions may be taken to different bodies which scrutinize them from different perspectives, while only one such challenge needs to be successful in order for some improvement to take place.

One way to characterize this state of affairs at the macro level is to suggest that the business of accountability, while clearly differently constructed in different countries, looks a little like the children's game of scissors cutting paper, stone blunting scissors, but paper wrapping stone. Each of the major players—parliament, ministers, departments, the judiciary—has formal power, or expertise and information, which the others do not. Each has a venue in which the proceedings are conducted according to its rules rather than those of others. And each has a sphere of interest which, despite some overlap with others, is also in part distinct. The questions that logically follow such a characterization are: what is the balance of power between the different groups? and who has the ability to review whose decisions or policies? It is difficult to produce a thumbnail sketch of the balance of power between the various players in each of the four countries, but the following remarks on England and Germany illustrate the kinds of differences that exist.

The judiciary are key participants in many processes of accountability in Germany. Courts may rule on the constitutionality of new laws, and are thus in a position to block unwelcome policy changes. Courts are built into the fabric of inmate complaints; and many, if not most administrators and prison governors are legally trained. The ethos of the rule of law is strong. One result is that

---

[9] See e.g. R. v. *Home Secretary ex parte Herbage* (No. 2), 1987, in which the applicant argued that the conditions of his incarceration were 'cruell and unusuall' (in the terminology of the 1688 Bill of Rights) because of his location in a prison hospital along with mentally ill inmates. This case is discussed at greater length in Treverton-Jones (1989: 26–7). In another case, R. v. *Metropolitan Police Commissioner ex parte Nahar* (1983), the conditions complained of related to police custody. This was a habeas corpus case brought on the grounds that conditions in police cells were such as to render detention unlawful. The judge accepted in principle that poor conditions could make detention unlawful, but was unable to find any clear standards by which the particular conditions complained of could be so regarded. A more detailed discussion occurs in Gostin and Staunton (1985), and Treverton-Jones (1989: 26–7).

almost all accountability is *post hoc*, with initial decisions being scrutinized after the event; and the strongest form of accountability within the structure is to tighten the reins of bureaucratic control by insisting that certain classes of decision are referred to a central authority rather than being decided locally. A second result, already noted in Chapter 6, is that many governors must rely heavily on their subordinate officials for day-to-day management expertise. The law may be a framework within which management takes place, but it is hardly ever a good framework for the actual practice of management.

An almost diametrically opposite pattern has emerged in England, where the courts have no say in the making of law, and are often—with a few notable exceptions—content to allow managers room to manage. A specifically managemental ethos is therefore more obvious, with a correspondingly greater use of 'audit' procedures such as inspection.

Some factors are, of course, common to both countries. Courts generally cannot be seized of an issue unless it is referred to them, and must act lawfully in determining whether or not they have competence to hear the case. Such provisions constitute boundaries which ensure that however strong the powers of the courts within their own sphere of competence, that sphere cannot be indefinitely expanded to cover all prison issues.[10] Yet differences also exist, again conditioned by broader legal and socio-legal factors. The courts in England, hearing cases for judicial review, rely on the abilities of the lawyers representing the parties to elicit or present the information and legal arguments on which the case is decided; if the lawyers are not themselves knowledgeable about the ways of prisons the hearing can founder or become sidetracked. It was, for example, frequently alleged that when solicitors were admitted to represent inmates in disciplinary hearings, the procedure slowed down because the solicitors had no background knowledge of prisons to guide them; and when judicial reviews of disciplinary

---

[10] Boundary disputes can and do occur, and typically it is the courts which have the competence to resolve them. They may, however, drag on for several years, e.g. the length of time it took for the English courts to accept jurisdiction to review certain aspects of inmate disciplinary hearings. However, it should also be noted that some possibility exists for inmates to 'shop around' for a court which will accept jurisdiction, as in the US, where lawyers tried repeatedly, and ultimately successfully, to find judges prepared to hear prison litigation. Both these issues are dealt with more fully in Ch. 10.

decisions became more common, many of the lawyers asking for review had only the vaguest notion of the realities that underlay those decisions. In Germany, however, the opposite seemed to be true. For one thing, the court procedures differed in that they were less accusatorial and more inquisitorial; judges in the prisoner complaints courts could call for evidence that they felt crucial even if neither side in the case offered it. For another, the judges involved in prisoner complaints quickly became expert in the routine ways in which prisons and inmates generated such complaints (and, indeed, only a very small proportion of complaints were upheld). That many of these judges might at some previous point in their career have been prison governors is presumably a relevant factor in their understandings of the situations.

After this digression I should return to my original contention, that accountability rests on the images of control and independence. It seems to me that we have to understand the issue of control within prisons as intimately related to the questions of control over prisons, that is, to questions about how prisons are made to account for what they do. General government policies concerning prisons structure the ways in which governments seek assurances that the prison system is 'under control', and, in turn, affect the way that control behind the walls is exercised. Yet those assurances must come from credible sources. Part of the argument about accountability thus rests on the conditions under which particular mechanisms of accountability can be depicted publicly as reputable, credible entities. One precondition for this tends to be the independence of that machinery from government and government policies; hence the symbolic importance of the courts, despite the relative infrequency of judicial intervention in prison life. But that credibility cannot, ultimately, be derived from formal specifications of powers and duties. It rests on the agencies' practical effectiveness within the prison environment, and, for reasons explained below, this has to be repeatedly negotiated.

Micro-accountability

Within prison organizations, as we have seen in this and previous chapters, it is not sufficient to model accountability solely in terms of *bureaucratic-rational* and directive relationships. While a good part of prison life is indeed structured in ways that such concepts can apprehend, it is also necessary to remember that accountability inheres in the nature of relationships which are maintained within and across institutions, over time, and which have informal as well as formal characteristics. In consequence, much of the activity generally described as the practice of accountability might be better understood as *symbolic and exhortatory*. By this I want to indicate that accountability is something that has to be accomplished, and, more importantly, has to be seen to be accomplished. The relationships involved must, if the legitimacy of the system is to be maintained, be depicted as open and co-operative ones in which all parties are knowledgeable about the doings and intentions of others. Those who are held accountable are then seen as moving steadily into line with the current thinking of those who can hold them to account; those who can hold to account are dynamic innovators, encouraging and resourcing 'good practice'. The executive is responsive to the exhortations of the policy-makers; the policy-makers are constantly seeking improvement and upgrading of executive performance, in part at least by finding praiseworthy innovations which can be more widely implemented. The implication is that the key factor in the nature of such relationships is the provision, receipt, and interpretation of information.

But if information is the currency in which any form of accountability is rendered, there is scope for the parties concerned to operate, in practice, in ways that can be described as an *informational or communicative game*. It is worth spending some time considering how such 'games' can work. In this game the players seek to use the given structures of accountability for their own purposes by selectively demanding, providing, hearing, or overlooking pieces of information; or, indeed, by agreeing to accept common definitions of the situation, even if these might be 'convenient fictions' which can be accepted only for limited purposes.

By using the term 'game', I do not mean to imply that there are 'losers' and 'winners'. It is true that the game may be one in which information is one of several unequally distributed resources which can be used strategically. But the idea of a game also encompasses the ideas that there are shared understandings about what moves it is possible to make within a system of broadly-accepted frameworks and conventions; that actions are rule-guided rather than rule-controlled; that the aims of the players, if different, may not be mutually exclusive or even competitive; and that players can only be 'driven off the board' in very exceptional circumstances.[11] In the particular circumstances of accountability, the nature of the game is that, providing the moves that are made are consistent with the rules that apply to different areas of the board, the end result is the provision of grounds for claiming that the prison system is a legitimate and fair administration which is subject to proper governmental controls.[12]

If we accept that a game-playing analogy has some purchase on the way that accountability works, the question then arises: what sorts of games might there be? Two dichotomous variables are likely to be important: first, an unequal distribution of control over information versus an equal distribution of control, and second, a zero-sum game within which there is no collusion or co-operation between parties (and some parties win at the expense of others), versus a collusive or co-operative game in which some participants co-operate in depicting their relationship to others.

Descriptions of the 'accountability relationships' between various parties can thus be cast in terms of the four sorts of relationship identified in Table 7.1. First, there would be a 'pure' rational-bureaucratic model with an unequal balance of power and no collusion between participants. Second, there would be a more equal and co-operative form of accountability which I have described as

[11] One (American) example would be the attacks on the Occupational Safety and Health Administration in the early years of the Reagan administration. Even so, the agency was not wound up but wound down, with a budget cut of 25% in 1981/2 (Calavita 1983). Calavita argues that the setting up of the administration was always intended as a symbolic act, and its winding-down occurred because it had become a means for organized labour to obtain real and incremental gains in workplace safety. The winding-down thus heralded an intention, subsequently carried through on other fronts, to put organized labour on the defensive.

[12] In practice, the difference between the games labelled as 'game 1' and 'game 2' in the diagram is primarily one of the degree to which the two principal players are prepared to accept the need for mutually agreed rules (and to see the interests of the other party as legitimate). It is of course possible that any one player in a 3-(or more) party game is playing different games with different other players. It is also possible that these games are played more or less cynically. While the table suggests that the 2-party game is zero-sum, so that those who are held accountable may not wish to co-operate with those holding them to account, this is not necessarily the case. The table simply indicates the possibility in 'game 1' that there can be non-cooperation and resistance to the demands of accountability, or that at least it can become a 'trading relationship'. Similarly, while both parties may be committed to the definition of their relationship as one in which accounts are rendered and received, it may also be that two parties try to depict their relationship as one in which there is accountability even though in practice neither believe this to be the case—though it may suit their purposes if others believe it to be so.

Table 7.1. *Dimensions of Accountability as a Communicative Game*

	Zero-sum game (no collusion between parties)	Non-zero-sum game (e.g. co-operation between two parties to depict relationship to a third)
Unequal control (one party can demand information; the other must provide it)	Pure rational-bureaucratic model	Game 2: benefits both parties to depict the flow of information as constituting accountability; substantial agreement on the rules of the game
Equal control (one party must negotiate access to information from the other in order to know what are the 'right questions to ask'; the other can provide partial or mis-leading information)	Game 1: information provided and received strategically in order to maximize or minimize exposure to criticism from the other; minimal consensus on rules of the game	Symbolic accountability: two parties depict their co-operation as constituting accountability

'symbolic'. These two forms would be diametrically opposed to each other. In the remaining two boxes there would be forms of accountability which are harder to characterize, and which have simply been labelled as games '1' and '2'.

These four sets of relationships are, of course, abstract postulates; it need not be assumed that all four can be found in the real world. Moreover, any analysis of actual practices must recognize that such games have more than two players, and have informal as well as formal aspects to them. Analysis must therefore deal with issues such as the kinds of informal relationships that must exist in order for the formal system to work effectively, how those involved in any particular relationship depict that relationship to third parties, and, most importantly, the orientations of the participants to the way in which the 'board' is structured. The result is that descriptions of relationships can end up, like a Russian doll, as a series of 'nested' games.

Take first the bureaucratic-rational relationships. These exist, in theory at least, between inmates and staff; between subordinates and superiors at all levels from basic-grade officer up to governor; and in a more attenuated form at the higher levels, including interfaces between prison systems and their ministers. I say 'more attenuated' because at the higher levels of the organization the concept of giving orders is not nearly so clear-cut. Explicit orders are usually quite rare. Instructions as to courses of action may be fairly general, and capable of being implemented in a variety of ways. And those giving instructions are also expected take reasonable note of information provided by their subordinates which would make the instructions unrealistic. The sharpest example of this would be where, in the Woolf Report discussion of the Strangeways riot (Home Office 1991), the comment was made that an official did not have sufficient information to make the decision he did, and that his subordinate—who did have the information, and who suggested a course of action which would have required authorization from the official—should have been entitled to have had his suggestion considered more carefully.[13]

One key feature of bureaucratic-rational relationships is that they encompass situations where individuals who are responsible for things getting done delegate the doing to subordinates. This is as true in a headquarters office that processes, say, prisoner petitions or staff payrolls as it is within the prison walls. Another feature might be that the kinds of goals or aims which such relationships envisage tend to be pre-ordained; senior staff tell subordinates to move inmates from here to there, office staff deal with most matters according to preformulated rules, and so on. But staff in prison systems, like most bureaucracies, also know that, parallel to the bureaucratic-rational relationship there is another kind of relationship, which I have described as an informational game ('game 1'). Provided that orders are fulfilled, senior staff are not usually expected to enquire too closely as to how this was done; and subordinate staff, while not lying to their superiors, may be economical with information as to how things were done. In practice, as virtually every prison researcher from Sykes onwards has noted, getting things done in prison involves occasional tradeoffs, rule-bending, creative application of regulations, and agreements not to share information (because for a superior to know something 'officially' would mean he would have to do something about it).[14] The end result, most of the time at least, is that the formal definitions of relationships do not constitute a full and accurate

[13] This incident is discussed in greater detail in Ch. 11.
[14] See the earlier discussion in Ch. 4.

account of the relationships between superiors and subordinates—even though, if outside parties enquire into the relationship, both sides would wish to depict it as a bureaucratic-rational one (i.e. both sides would engage in what I have described as 'game 2'). In short, there is one kind of formal relationship, another kind of informal relationship which makes the formal structure 'work', and a collusion between both parties to stress the formal rather than the informal relationship to outsiders who wish to know how the system operates.

This kind of analysis does not, however, describe the relationships at the top end of a bureaucracy, let us say between prison governors and regional or headquarters staff. For one thing, many practical decisions are made on a more participative basis, even though authority exists for an individual to make an order and the formal document authorizing a course of action is issued under the signature of a single official. For another, the kinds of decisions made at this level can be a good deal more like agreements to 'change direction' while leaving the precise details to be worked out elsewhere. Several examples of participative decision-making—of different types—are possible. If an inmate is being difficult in one establishment it may be thought worth transferring him. The authority for the transfer is usually vested in a designated official. However, it would almost always be the case that the governor of the destination prison would be telephoned and consulted; it may be the kind of consultation in which the governor feels he cannot refuse, or in which he is told that this is the only option and he should make arrangements to receive the inmate despite his objections. But he will not—or at least not usually—simply have the inmate dumped on him. Another kind of participation would be where a number of governors are brought together for the purpose of working out how the regimes in their respective prisons can be brought into parity (or, conversely, ought systematically to differ from each other). This might be the case where the allocation of long-term inmates is based on the supposed differences (or similarities) between prisons, or where inmates are routinely sent from one establishment to another at a certain stage in their sentence. And a third example would be that of working parties set up to advise on specific problems.

These kinds of situations can be described as ones in which, although specific individuals have defined powers, there is a routine expectation that they will only be used after consultation; in which a lack of proper consultation may be criticized as an instance of poor management; and in which it may be a defence, if a situation turns out poorly, that the process of consultation and the advice offered effectively constrained the choices available to the decision-maker. Here we have, then, a model where rational-bureaucratic accountability provides space for the kinds of consultation I have described as 'symbolic accountability'. None the less, it remains open to participants, depending on their own position and interests, to play either of the two games I have described, both in pursuing their own interests and in depicting their relationships to others—depictions which may sometimes become necessary in rather extreme situations, such as accounting for escapes or disturbances.

By way of an example of the complexities one may find in 'game-playing' outside the rational-bureaucratic sphere, let us briefly consider the case of parliamentary questions.

In England, as elsewhere, members of parliament and parliamentary committees can and do ask detailed questions about the way prison systems run. Those questions are answered; but whether they are the right questions is another matter. Bureaucracies do not usually provide answers where they are not asked for, and parliamentarians must perforce rely on correspondence from inmates, advice from academics and others, and 'moles' in the civil service, to know what questions they should be asking. Scope exists for the playing of communicative games as to how questions are interpreted and how detailed is the information provided. A member of parliament might, for example, ask an apparently straightforward question, such as how many inmates had access to integral sanitation in a given prison on a given date.[15] Theoretically at least, a figure should exist which would, having regard to all considerations, be considered satisfactory. That figure might be satisfactory in that it is much higher than it was two or three years previously, thus indicating that 'progress is being made'.

In real life, of course, the point of asking such a question is strategic. The questioner knows the answer will highlight some defect or problem within the prison service. Such questions are

---

[15] Questions of this kind have been asked in the House of Commons in recent years on a range of issues, including inmate suicides, pregnant female inmates, prison sanitation, the prison population, and prison overcrowding.

often asked, not for the answer to be measured against an agreed level of satisfactory performance, but to highlight a problem and to place on the political agenda a criterion—for example, progress towards integral sanitation—by which service performance can be judged. That there may be arguments as to whether this criterion is a real reflection of adequate service provision need not detain us for the moment. Whether a satisfactory degree of accountability can be said to exist may be illustrated equally simply. If the relevant minister is unable to produce an answer to such a question, that inability can be taken as evidence of a serious lack of management oversight in an area of political concern. In other words, the system is not sufficiently accountable. A converse situation, of course, also exists. The inability of questioners to formulate questions that highlight problems might indicate a lack of accountability, in so far as it suggests that those to whom the service is responsible do not have sufficient grasp of the situation to be able to understand whether the service is operating satisfactorily or not.

The upshot of the discussion thus far is not, or at least is not meant to be, the conclusion that everybody involved, all the time, is simultaneously pretending to be equal partners in situations where one is formally superior to the other; and that each understands that the other is calculatingly economical with the truth, while both collaborate to deny to third parties that this is case. It does suggest, however, that three dimensions of accountability matter crucially. First, whether those producing accounts are subordinate to those to whom they account, or whether each has an informational 'power base' which renders them in some way equal. Second, whether the relationship is conceived of as one in which accountability is an outcome of what the participants do—one party 'certifies' that the other operates correctly—or whether the relationship itself is regarded as a collaborative enterprise in which the process of discussion, exchange of information, and creation of shared understandings is defined as accountability. I also want to suggest that between any two parties, these relationships can be nested one inside the other, so that there can be a mutual recognition that the 'real' business of accountability is different from its formal 'appearance' in laws, organizational charts and so forth. Third, I have also indicated that any one party may be involved in a fairly complex set of relationships where, for example, in theory A is subordinate to B for one purpose (while in practice the relationship is more equal); A and B are equals in theory for another purpose (but in practice B is more powerful); and both A and B wish to depict both their relationships as one of broad equality to outsiders. Analysis of accountability at this stage is not a matter of putting a label on a relationship. In part it consists of looking at the ways in which relationships resemble, or diverge from, the formal statements of how they should work; in part it is a question of recognizing that those relationships themselves form part of the situation being brought to account; in part it is also a matter of looking at how the practicalities of the situation lie in such a way that this or that type of relationship makes sense to those involved; and only then is it a question of looking at what practical consequences might result.

One consequence of the approach I have outlined is that the idea of accountability being 'accountability for something' and 'accountability to someone' begins to seem too simplistic. The relationships through which someone is accountable for something are themselves part of that for which accountability is sought; the depiction of a relationship as being one in which issues are raised, reported on, discussed, and agreed (in particular in relation to policy issues) is often in and of itself a depiction of accountability.

Linking Macro- and Micro-accountability

I suggested earlier that a number of links exist between macro- and micro-accountability, of which the most important is probably that the nature of control *in* the prisons is structured by the nature of control *over* the prisons. But the discussion above of micro-accountability suggests that some other links are also important.

First, in so far as accountability can be modelled as a game, it is not a game in which the players have complete freedom to negotiate their roles or the rules. They must attend to the external reality of the situation; that is, to broader frameworks which structure their roles and the expectations on them. The external constraints on action are the obvious ones. A prison inspectorate, for example, has a clear remit to conduct inspections; and governors in the field cannot refuse to comply with inspectors' requests. A court can deal with cases put before it, but cannot proactively find cases it wants to hear. While minimal compliance with formal structures, and communicative games such as 'being economical with the truth', are possible, outright refusal to recognize formal structures is rare because of the sanctions that can be invoked. Only in cases where no real sanctions can be applied—such as, in England, ministers refusing to appear before select committees—does one sometimes see determined noncompliance in processes of accountability. Second, in so far as accountability can be modelled as a game, there remains the question of what the object of the game is. In Monopoly, players have a relatively simple objective: to accumulate resources and simultaneously impoverish competitors so as to drive them off the board. Real-life games, by contrast, are rarely zero-sum; rules guide rather than control moves; players may not have mutually exclusive goals; and each player may have a number of conflicting goals. Moreover, few, if any, of the players are likely to have explicit and long-range objectives. These two points may lead us to the following thoughts about the way that accountability operates within and in relation to prisons.

One of the few groups required to produce long-range goals is the prison administrations themselves. However, planning usually involves only a relatively small group of staff, and how such plans are created is variable. I earlier discussed the English and Dutch arrangements for institutional plans created by governors, which were then not infrequently overruled by systemic demands or budgetary problems. In France, governors indicated that they were not involved in long-range planning procedures, but were simply told what plans headquarters had for them. And, in addition, such plans have to reflect (or secure) the political wills of governments, which tend to think in broad categories such as 'law and order', 'curbing the power of staff unions', and so on. In relation to accountability, the key point is that planners are typically shy of committing themselves to specific inmate rights or prison regime standards, not least because they wish to maintain operational flexibility and do not want to commit themselves to delivering services for which costs can neither be predicted nor appropriated. This may explain why improvements in prison conditions tend to be more common at the level of actual provision, unaccompanied by entitlements to those provisions. However, different countries do start from rather different baselines in these regards. Some examples cited in previous chapters illustrate this point; the provision of electrical sockets in cells stands as examples of an area in which the norm in one country is regarded as a frivolous waste in another, unlikely to receive any serious departmental support. Similarly, the idea of extending inmate rights in disciplinary proceedings, without being required to do so by the courts, however laudable in humanitarian terms, would generally be seen as attempting to fix something that is not yet broken.

Other formal actors in the process of accountability tend to be more passive and reactive. On the whole their job is one of inspection and *post-hoc* evaluation rather than of forward planning. The courts, for example, tend to be involved in determinations as to whether particular actions by prison officials have been lawful. It is possible to identify series of cases in which thinking as to what constitutes 'lawfulness' changes over time, and prison administrations thus have to revise their practices. But such cases are rare, and courts cannot take the initiative. Even supposing that the judiciary is uniformly committed to a particular set of prison reforms, it must wait for the cases to be brought before its views can be expressed. This, incidentally, is the significance of some outside pressure groups. In the US, the American Civil Liberties Union has been active in identifying problems and bringing them before the courts as test cases; and it has occasionally been able to 'shop around', pursuing one area in different cases before a variety of judges in order to obtain, perhaps after a number of cases, a judgement which could be used as a precedent to enforce change.[16] In England the National Council for Civil Liberties announced its intention, in the late 1980s, to follow a similar strategy. The logic of this is simply that if the courts cannot proactively seek out cases, then one role of outside groups is systematically to identify worthwhile cases and take them to the courts. Finally, so far as inspections are concerned, it tends to be the case that inspectors are primarily involved in comparing actual practice with departmental expectations and regulations. The case of Stephen Tumim, appointed Chief Inspector of Prisons in England in 1987, is unusual in that he announced from the very beginning that he had a long-range goal—to eradicate the practice of 'slopping out'.

Concluding Thoughts

It is attractive to view accountability as something like a net, which can be thrown over the prisons, and in which different agen-

---

[16] See Morgan and Bronstein (1985), and Jacobs (1980).

cies, in scrutinizing events from differing points of view, ultimately ensure that every aspect of prisons is subject to oversight. Unfortunately this is very far from being the case. As previous chapters have pointed out, many of the mechanisms we now regard as part of the 'machinery of accountability' grew up at different times for different purposes, and have little logical or, indeed, practical connection with one another.

It is clearly possible to view the process of accountability in game-like terms. On this view, different parties have available to them varying types of resources, and the aims of the game are partly to do with control over the prisons, and partly also with presentational matters. In the latter, different parties singly or collectively seek to depict themselves and their relationships with others as credible means of accountability, and as providing credible information on the basis of which it is possible to make decisions.

At the same time, we must recognize a number of things that accountability is not. It is not, typically, relevant to long-term planning and development. The reverse is more often true; plans may involve changes to forms of accountability. Nor is accountability often an arena for proactive considerations; it is largely an area in which there is reaction to developments proposed from elsewhere in the system.

These comments, which draw partly from the materials presented thus far and partly from an attempt to find usable theoretical frameworks, seem to suggest that we should not be too optimistic about the practical effects of different forms of accountability; our expectations should be tailored to recognize the political frameworks within which such mechanisms operate. This is indeed the case, as the following chapters, on inmate rights, inspection, and prison discipline, will show.

8

Rights, Complaints, and Conditions

IT makes sense to link inmate rights, grievance procedures, and prison conditions in a single discussion, not because there are clear theoretical linkages to be made, but because it is possible to illustrate the divisions that have arisen between them, and which, on the whole, render grievance procedures and rights effective only in the most trivial of instances. Conditions are primarily the result of government willingness to make provision for prisoners, and governments are not often disposed to make rods for their own backs by recognizing rights which will have fiscal implications that they are not prepared to meet. Complaints procedures are by and large geared to the remedy of matters affecting individuals, so that major and system-wide issues about prison conditions are virtually automatically defined as outside their scope. So far as internal procedures are concerned, rights can rarely be relied upon because relatively few clear-cut rights exist in a form which enables them to be enforced by inmates. In short, neither complaints procedures nor the current level of inmate rights can be seen as important avenues for improving prison conditions, though for rather different reasons.

The following section makes some general observations about the nature of rights so far recognized, the nature of grievance procedures, and prison conditions. The subsequent sections deal more descriptively with rights and grievance procedures in the contexts of the four countries, and discuss the ways in which rights and complaints mechanisms can be used by inmates.

Some General Observations

Richardson (1985: 19) asserts that 'individual rights are increasingly being recognized as an essential safeguard against the power of the state whatever its ideological stamp'. This statement can be taken in two ways: first, to indicate that the the formation of rights has become an increasingly common way of defining the relationship between state and citizen, and second, to suggest that the formation of rights is an effective way of circumscribing the power of government. Both views are true, but only up to a point. Richardson is right in that the recognition of rights is one way in which the state expressly limits its administrative power over inmates, and correlatively, one way in which individual inmates may require the state to treat them (or not treat them) in particular ways. The crucial issue is, therefore, the substantive one: which rights are recognized? It is fair to say that there have been some advances on this front in recent years. However, the discussion below will indicate that Richardson may be over-optimistic in implying that inmate rights are likely, in practice, to be an effective safeguard against abuses.

The idea of 'rights' has been a highly contentious one in legal literature, and has been taken to mean a wide variety of things. Rather than re-open any legal debates as to the meaning of the term, I shall simply start with the proposition that, in most countries at any rate, inmates are no longer seen as 'slaves of the state' who have lost all civil rights, but as citizens whose rights are selectively abrogated or limited.[1] However, the other side of this coin is that prison administrations can make important decisions about their inmates which have few parallels outside the prison situation; security categorization would be one example. This suggests that

---

[1] There are doctrines which limit the extent to which prisoners lose their civil rights. In England, Lord Widgery's landmark judgement in *Raymond* v. *Honey* (1982) stated that prisoners retain all rights except those removed expressly or by necessary implication by the fact of imprisonment. One often finds reference to this principle in the literature on the other three countries, and it is frequently repeated in United Nations publications—partly as a statement of universal principle and partly, one presumes, as a reminder to member states. None the less, it remains ambiguous, because it does not address the question of *which* rights should be expressly denied to inmates and which are abrogated 'by necessary implication'. There is, e.g., no particular reason why inmates should be denied the right to vote, or have restrictions placed on their sexual relations. Further discussion of the legal debates on rights can be found in Richardson (1985).

inmates ought to have 'special' rights which address matters specific to the *additional* powers that the state holds over them by virtue of their imprisonment.

In practical terms, while some countries provide more rights than others, an analysis of the rights that have been recognized reminds one strongly of Hall-Williams's (1975) rather sardonic description of the English situation—namely, that inmates have the right to be photographed, fingerprinted, categorized, and to have a file kept on them. Such rights clearly have a point, even if only in extreme situations; in America, prisoners have asked the courts to require prison authorities to introduce categorization systems on the grounds that, otherwise, all kinds of prisoners are herded together and they are at risk of attacks from other inmates.

Perhaps the single most significant right for inmates is that of complaint. Complaints procedures are primarily intended as a means for inmates to seek a review of decisions or situations which affect them individually. They are an avenue for grievances about the quality of food, conditions in cells, and the like; for a wide range of decisions or actions affecting the individual—the inmate was not allowed to change his library books, his wages were incorrectly calculated—and potentially for serious wrongdoing by staff, such as verbal and physical assaults on prisoners.

It is, however, important to understand what the right to complain is *unlikely* to achieve. Issues affecting the inmate population as a whole are not often treated as serious topics for complaint. Where there is prison overcrowding, short-term local remedies do not usually exist, and providing one inmate with a single cell implies additional crowding elsewhere. Complaints about overcrowding are thus likely to be dismissed unless the complainant can demonstrate why he personally should be held in less crowded conditions. Moreover, complaints about facilities are related to national and cultural norms. Inmates in England or Germany may complain that they are not allowed to have televisions in cells.[2] Yet if the prison regulations do not provide for cell television, they will get short shrift. Complaints must assert that something provided for in the rules or the law was not done, or that something was done contrary to such provisions, or that staff interpreted a rule or

---

[2] In France and the Netherlands inmates are allowed television in cells, and the Woolf report recommended their introduction in England, though as a privilege rather than a right (Home Office 1991: para. 14.31).

law wrongly. Even in such cases, complaints procedures may provide limited avenues for remedy. In England, the prison rules are subsidiary legislation. While complaints about staff and management breaches of the rules may be handled within the prison administration, the courts have not always been willing to insist on compliance with the rules.

In addition, inmates are, in practice, unlikely to have definite rights which have a bearing on their complaint. In this sense, complaints procedures tend to be more about the proper application of administrative rules, and to some extent about fairness and propriety, than about the claiming of entitlements. In most jurisdictions certain areas are reserved to the administration and cannot be the subject of complaints. Security categorization, and allocation to a particular establishment, would be examples. Moreover, while inmates possess fairly clear rights in relation to matters such as physical assault, these are usually provided more clearly through the law than through administrative regulations (which would normally place a duty on officers not to use undue force, etc.). Though complaints about assaults and staff misbehaviour can be handled through complaints procedures, there may be special hurdles to be cleared because of the potentially serious consequences for the officers involved.

There are also practical constraints on providing remedies even where the administration is sympathetic. Complaints about visits and recreation, or the price and quality of goods in the prison shop, may be regarded by administrators as well-founded. But in so far as they have their roots in staff shift patterns, contracts with suppliers, and so forth, they are not often easily or quickly corrected.

Finally, it is worth bearing in mind two overarching factors which have tended to constrain the development of inmate rights, the improvement of prison conditions, and thus also the possibilities for inmate complaints.

First, prison conditions are ultimately the result of government policies and expenditure, not just at one point in time, but over many years. It is widely accepted that tough law and order policies mean larger prison populations and ultimately more prison building. Upgrading inmate facilities, however, may fall foul of 'less-eligibility' arguments. Why spend additional money on prisoners which can be used instead on provision for the elderly, health care, education, and so forth? In England and France, though there are differences as to the details, many of the deprivations inmates might wish to complain about are in large part a function of recent political history. Even though large sums of money have been spent, they have not always been spent in ways that improve the quality of inmate life. Prison conditions have perforce deteriorated, and will take some years to correct even where the political willingness is there to tackle them.[3] In so far as conditions are, in the last instance, determined by political considerations, those considerations limit the practical importance of rights and grievance procedures.

Second, and despite shifts in penal policy towards more retributive philosophies, the influence of treatment-based rhetoric should not lightly be dismissed. It is, for example, still deeply embedded in the text of the French CPP and the German StVollzG. Staff, if they espouse any penal philosophy, often still refer to it even if for no better reason than its symbolic value in terms of professional recognition. The rhetoric of treatment suggests a focus on individualization, highly structured (though flexible) regimes, and staff discretion. It also implies that inmates should have correspondingly few absolute rights. We should be cynical about the extent to which treatment actually occurs, if only because of the history of underfunding or misapplication of expenditure. Yet this rhetoric remains influential and can be mobilized to resist arguments for increased inmate rights.

Inmate Rights

In general, inmate rights can be characterized as either substantive or procedural rights. Substantive rights are those which give an entitlement to a specific thing or service, while procedural rights give entitlements in relation to a decision-making process; for example, the right to have legal representation in disciplinary hear-

---

[3] See e.g. English discussions on prison sanitation. Stern (1987: 165) quotes from a Public Accounts Committee hearing from the previous year in which Sir Brian Cubbon, then head of the civil service, was questioned on the provision of integral sanitation in prisons and acknowledged that a very large proportion of inmates would have to continue 'slopping out' for many years. Even the proposal of the Chief Inspector of Prisons (1989) for a crash programme of toilet installation envisaged that once the political will for the programme had been established, it would still take some seven years for integral sanitation to be installed in all prison cells.

ings, or to have explicitly stated factors considered when decisions are made in relation to a particular inmate. Rights, of either kind, can be contrasted with privileges. The distinction between a right and a privilege is that, for example, a right to X hours of visits would be a minimal expectation that the prisoner could have enforced by some means if it were denied, while a privilege of X hours of visits per week would be a normal part of the regime (maybe only for certain groups of prisoners) which could be altered or cancelled by administrative decision, without any means available to the prisoner to compel the prison authorities to provide it.

On the basis of these distinctions, the current situation is one of weak and very basic rights, leaving prisons with wide discretion (Germany); of developments in procedural rather than substantive rights (England and the Netherlands); of developments in 'moral' rather than 'legal' rights (in the Netherlands), and as developments not in rights but in privileges (in France). In short, all four prison systems have been cautious, if not conservative, in their attitudes towards inmate rights, though they have been cautious in different ways and to different extents. The following sections elaborate on these assertions.[4]

Germany

Formal inmate rights are probably most developed in Germany, though the StVollzG establishes them indirectly by placing duties on governors from which correlative rights can be deduced.[5]

Some duties, and their corresponding rights, are not qualified in any way. These absolute rights include separation from other prisoners during reception procedures (s. 5/1), provision of information about rights and duties (s. 5/2), medical examination on reception (and a duty to allow oneself to be examined; s. 5/3), and the right to have a treatment plan made for one's sentence and to be

---

[4] Descriptions of the roles of various officials and broader discussions of various laws may be found in Ch. 2; explanations of acronyms can be found in the Glossary. So far as possible I have tried to avoid the duplication of descriptive material presented in earlier chapters.

[5] The StVollzG deals with adult prisoners only, though analogous legal provisions also exist for young prisoners. N.B. German law also establishes a number of clear duties on prisoners, including e.g. the duty to allow oneself and one's cell to be searched.

consulted about it (s. 7).[6] Most rights are, however, qualified in various ways. Inmates who apply for placement in an open prison or on a work programme outside the institution (ss. 10, 11) have a procedural right to be *considered* for placement according to criteria and guidelines set out in the relevant sections of the law. They have no substantive right to a placement. Equally, inmates have the right to be considered for twenty-one days' home leave a year, though not a right to home leave as such (s. 13). In other cases, the StVollzG states rights which are then qualified in terms of maximum entitlements and the conditions under which the right can be withdrawn. For example, inmates may send and receive letters and parcels, and make and receive telephone calls (ss. 28 et seq.), subject to certain maxima (to receive three parcels per year, and the contents only to be consumables or tobacco). These rights may also be withdrawn on vaguely specified grounds, such as when 'the security and order of the institution are immediately threatened' (s. 33/3).

These are of course legal provisions. In addition, some limits on rights, and specifications of decision-making procedures, are contained in administrative regulations (Verwaltungsvorschriften, or VV) and house rules. The VV are regarded as purely administrative, and thus a breach of the VV will not be regarded as an admissible complaint in the court-based grievance procedure (which in Germany is the major avenue for complaints). If, however, a right provided in the StVollzG is breached as a result of implementing provisions in the VV, judges are prepared to declare that the latter are incompatible with the law and must be changed (Landfried 1982).

Prior to 1977, neither legislation nor case-law existed on prisoners' rights in Germany. A joint set of administrative regulations was in existence, but legal challenges to them, or based on them, were not possible. The passing of the StVollzG remedied the situation, but, as Feest (1982) notes, it was 'decisively weakened' in its parliamentary progress when the Länder, which finance the prisons, demanded amendments which delayed, put off indefinitely, or reduced the scope of provisions with expenditure implications. Yet despite such financial concerns—the original commitment progressively to increase prisoners' wages was never implemented—court

---

[6] A complete list of German inmate rights is given in tabulated form in Kaiser, Kerner and Schöch (1983: 120–25), in the context of a detailed analysis of the provisions of the StVollzG.

decisions on other areas of rights began to flow quite quickly. One of the first cases taken to the higher level in the procedure, the Oberlandsgericht (OLG) in 1977, for example, concerned the provisions of the VV in relation to home leave (Urlaub). The decision was that the VV's restriction of home leave to those in the last eighteen months of their sentence was contrary to the new StVollzG; home leave thus became available, potentially, to all inmates.

Seeking to exercise one's rights in the German system is not, in principle, difficult. Complaints may, and in some states must, be made to the prison governor and the Land prison administrations. However, the ready availability of the courts means that many complaints are taken fairly quickly to the relevant Strafvollstreckungskammer. This simply requires a statement of the grievance in writing and addressed to the court for the district in which the prison lies. Most of the courts' business is conducted on paper, and personal appearances before them are rare. This ease of access is, however, accompanied by two drawbacks. First, complaints must allege a breach of the law rather than the administrative regulations. Second, although legal representation before the court is permissible, legal aid is not available. The Baden-Württemberg Ministry of Justice, for example, estimates that only about 3 per cent of prisoners have legal representation for their complaints.[7]

In practical terms, the requirement that a complaint must allege a breach of the law means that the threshold at which a complaint becomes admissible is already rather high. Although the legal provisions are more tightly drawn than in, say, England, the specific rights provided are relatively weak and the chances of a complaint being upheld remain correspondingly slim. Kaiser, Kerner, and Schöch (1983: 217) cite three studies between 1976 and 1981 which indicate overall success rates (combining both first instance and appeals to the higher court, the OLG) of between 2 and 3 per cent. Feest (1988) suggests that at the OLG level, prisoners' complaints are upheld in only a little more than 5 per cent of cases.[8]

---

[7] This figure was provided to me in an interview with Baden-Württemberg prison headquarters officials. Since Land and national statistics on the numbers of complaints taken to the Strafvollstreckungskammern are not compiled, and exact figures not kept for the proportions of cases legally represented, we have to rely on Ministry estimates and figures generated from research samples.

[8] Nationally, about 1,400 cases were lodged with the OLGs by prisoners in 1986 (Feest 1988).

Nor is this the end of the story. Though what appears to have been a legislative oversight, the StVollzG provides no mechanism for the court to compel prison authorities to give effect to its decisions. This—and here I rely on anecdotes from German colleagues—appears to be a hangover from Prussian attitudes towards administration, where it would be assumed that any declaration of a court would be complied with as a matter of course. In practice this is usually the case. However, it is quite clear that a prison administration which was prepared to make a stand on a particular issue could defy the court with impunity—though the political fallout would mean that such a move would only be taken with, by its lights, very good reason.

France

The CPP places many of the same duties on French prisons as does the StVollzG on German ones. There are, for example, requirements to group prisoners by penal category, age, state of health, and personality (Art. 718); to leave correspondence with defence lawyers unopened (D69); to inform prisoners of the house rules and of their treatment plan (D94; applies to long-term establishments only); and to provide sufficient productive work to occupy the normal working day (D100). In areas such as these is it possible to take the view that the duty of the prison implies a correlative right on the part of the prisoner. Yet the CPP leaves untouched many areas of prison life, which are thus regulated by administrative direction, staff discretion, and the grant of privileges. One area totally controlled by administrative regulations and, within those, the discretion of the governor, is cell searches.

Several broad principles seem to affect the way in which prisoners' rights are conceived in France. One is that any decision affecting the length of sentence actually served should be a judicial decision. The Juge de l'Application des Peines has control of most matters pertaining to the actual time of imprisonment—home leave, remission, and some parole decisions.[9] But this is not to say that the decision-making process is removed from the practical aspects of prison management. Remission, for example, is not

[9] Parole decision-making is split between the JAPs and the Ministry of Justice. The JAP is sole decision-maker for those sentenced to less than 3 years; in other cases he or she makes recommendations to the Ministry, which, however, has sole responsibility for making the parole decision.

automatically given and then 'clawed back' as a disciplinary measure as it is in England and Wales. The maximum remission possible is 25 per cent: the JAP reviews each case periodically and decides how much remission to award. This decision is explicitly based on the prisoner's record: Art. 721 requires the inmate to show 'sufficient proof of good conduct' to warrant remission, and his or her disciplinary record will be one of the factors considered.[10]

At an institutional level, the general experience of inmates on the importance of legal provisions was summed up by one French inmate, who commented, in answering a question on complaints, that[11]

Mostly the director [i.e. governor] accords with requests because they are provided for by the law. If the criteria for decision are not based in law, it can be arbitrary, it depends on the prisoner . . . The directors do abide by the law but that's it.

But how far the rights implied in the CPP can be used effectively still remains problematic. For example, the requirement in D100 that sufficient productive work should be available to occupy an inmate's normal working day, in principle should provide a correlative right to work. The preceding article, D99, states that remand prisoners and some other special categories of prisoners may request work. Yet if a convicted prisoner is not allocated work, or a remand prisoner requests work but it is denied, it is doubtful that they could compel the prison to find them work. Indeed, the *Rapport Générale sur l'Exercice 1986* noted about 12,500 persons demanding work but unemployed in the first half of 1987—one quarter of the population.[12]

So far as the courts are concerned, the principle enunciated in the Law of 16–20 August 1790 still applies: 'judges in the civil

[10] In addition, for remand prisoners, decisions concerning whether visits are allowed, and whether the inmate is allowed to work, are the responsibility of the juge d'instruction (examining magistrate).

[11] Interviewed by me in French (my transl.).

[12] Out of a prison population of 49,904, some 31,483 were without work. Of the latter, 14,709 did not request work and education classes had an enrolment of 4,253. Thus the 'unemployed' population of inmates eligible for and demanding work, but for whom there was none, amounted to 12,521 inmates (see the *Rapport Générale 1986*, p. 160, Fig. 1). In this context 'work' includes prison workshops, concessions run by outside contractors, persons working outside the prison under 'semi-liberty' arrangements etc.

courts may not, under pain of forfeiture of their offices, concern themselves in any manner whatsoever with the operation of the administration, nor shall they call administrators to account before them in respect of the exercise of their official functions' (Neville Brown and Garner 1983; their translation). In practice, as indicated in Chapter 2, lawsuits by inmates must be addressed to an administrative tribunal in which the judicial function is taken by civil servants, and thence to the Cours d'État. Complaints by inmates to this court would usually have to allege that the administration had acted beyond its powers (excès de pouvoir).

Cases decided at the level of the Cours d'État have not often favoured the inmates, and have, moreover, been dealt with rather slowly. One example is the case of Alain Caillol, which concerned his allocation to a Quartier de Plus Grande Sécurité, a unit roughly equivalent to the English 'control units' of the 1970s, though designed for remand prisoners (similar units, known as Quartiers de Haute Sécurité, existed for convicted prisoners). Caillol's challenge to the legality of the QPGSs began in 1980. After argument about whether the case was administrative or judicial, it was heard by the Cours d'État in 1984, two years after the QPGS/QHS system was discontinued largely because of widespread protests by prisoners (Fize 1984). The judgement was that the decision to send the prisoner to the unit constituted an internal measure not open to challenge on the grounds of 'excès de pouvoir' (that is, the administration was acting within its powers).

An illustration of the complexity of the French court system is provided by the case of Manterola (discussed in Couvrat 1985), which makes the point that merely because a decision is made by a judge, it is not necessarily a judicial decision. Manterola was allowed parole (libération conditionelle) by a JAP in 1983. The Procureur de la République, who has the power to seek from the appeal courts an order quashing this, did so and had his request granted. Manterola then sought an order quashing the Procureur's order, not from the Cours d'État, but from the Cour de Cassation, the highest civil court.[13] The order was made, Manterola obtained his parole, and the

---

[13] Neville Brown and Garner (1983: 36) point out that the dividing line between 'civil' and 'administrative' law in France is rather different to that in England; 'in England, one thinks naturally of the tribunals associated with with social security as typically "administrative". In France, most of the tribunals dealing with such questions are regarded as adjudicating matters of private, not public, law, and are therefore subordinated to the supreme civil court, the Cour de Cassation'. However, the

case law was thereby extended, since it was the first time this particular court had accepted that it had competence in that area.

But the major point that emerged from the *Manterola* case was this. Article 733–1 of the CPP, concerning, *inter alia*, parole decisions, enables the Procureur to challenge a JAP's decision to allow parole; and the decision in the case turned on the issue of whether the prisoner could challenge the Procureur's decision. This left unclear the question of how or whether the decision of a JAP could be challenged by an inmate. A revised Article 733–1, introduced in 1986, gave little hope for progress. Although it clarified many of the procedures and powers of the various actors, it appeared to envisage that appeals against the acts of the JAP would emanate only from a Procureur, and would be based on a violation of the legal requirements of the decision-making process. Inmates thus appeared to be left only with the recourse of an exceptional appeal founded on a claim of 'excès de pouvoir' and put in motion by the Minister of Justice.

If there has been little enough development in inmate rights in France, there has been, since the 1970s, a fairly large increase in the privileges available. Conditions in 1974 were undoubtedly bad; Favard (1987) cites La Santé prison, in Paris, as an example of an institution in which many areas of prison life were no better than, or even significantly worse than, those 100 years previously. In 1974, Dorlhac de Borne was appointed as a special ministerial adviser to make recommendations on prison reform.[14] A number of important concessions were subsequently given to inmates, and regimes became more diversified. In particular, the differences in the frequency with which one can make phone calls, the availability of home leave, and so on are now quite marked when comparisons are made between centres de détention, maisons centrales— also for long-term prisoners—and maisons d'arrêt. Further reforms were introduced in 1983/4 and 1985, with the latter year seeing the introduction of television in cells.

acts of a JAP are regarded as acts of 'administration judiciaire', a complex designation which appears to have given rise to legal dispute in France, but which in the civil code at least seems not to be subject to any appeal, except in exceptional cases for excès de pouvoir (Couvrat 1985: 138).

[14] In fact the recommendations were made before, but implemented after, a series of some 80 major disturbances and 9 prison riots. The timing and the consequences of the disturbances are discussed in greater length in Ch. 11. For further details, see also Dorlhac de Borne (1984) and Favard (1987).

However, and despite the use of amnesties and 'grâces' to release large numbers of prisoners at points where overcrowding reaches crisis points, the prison population has continued to rise. This must have led to a deterioration in regimes in the maisons d'arrêt, the remand and short-sentence prisons, purely by virtue of the pressure placed on existing facilities, many of which prison officials freely describe as decrepit ('vétuste'). Indeed, it can be argued that, given the overcrowding, the implementation of some of the privileges granted in 1983/4 and 1985 may have been robbed of their effectiveness or even contributed to the worsening of conditions. For example, permission for inmates to wear their own clothes was in some places not matched by facilities for laundering clothes, so that at the time of my fieldwork, in 1986–8, one still often saw two or more inmates in a cell surrounded by drying clothes.

England

In England, two views on prisoners' rights are widely quoted, but neither is very useful. The first is the *Raymond* v. *Honey* (1982) dictum that the prisoner retains all those rights not expressly or by implication removed by imprisonment. Unfortunately, this begs the question of what these rights are; and it fails to address any special rights prisoners may gain by virtue of imprisonment. The second view, which has historically been held within the Home Office and by certain judges, has been that inmates are for most purposes subject to administrative discretion which admits of privileges rather than rights. This view is given some substance by the 1952 Prison Act, which gives the Home Secretary wide discretion to promulgate Prison Rules, which in turn lay out privileges which inmates may obtain but to which they are not entitled.

The second view, however, ignores the ability of prisoners to rely on common-law rights concerning negligence, the duty of care, or the rules of natural justice.[15] These have proven important because, on the whole, where inmates have challenged the validity of actions which breached the Prison Rules, the English courts have refused to accept that the Prison Rules are binding on the Home Secretary, since he makes them (see for example *Williams No. 2*, 1981). In addition, one landmark case (*Anderson*, 1984)

---

[15] See e.g. *Ellis* v. *Home Office* (1953), which dealt with the 'duty of care' owed by prison governors to inmates.

established the right not to be discriminated against on racial grounds, though the discrimination was so well-documented as to make the decision easy. Yet a number of important areas—among them access to medical services in prisons—remain thus far untransformed by litigation.

Cases based on the tenets of natural justice, in particular, have seen some marked successes, mostly in relation to prison disciplinary hearings.[16] Over the course of the 1980s, inmates established important procedural rights in this area, of which probably the most significant was the right to be legally represented in hearings before the Boards of Visitors. This set in train the reforms which resulted, in 1992, in Boards of Visitors losing their disciplinary powers.[17]

It is worth noting, albeit in parentheses, that a number of more recent pieces of legislation and policy have had fundamental effects in relation to certain categories of prisoner. The Criminal Justice Act 1982, for example, altered the sentences available for young offenders and the organization of penal establishments for youths. Young offenders previously given indeterminate sentences (in Borstals) now receive determinate sentences in Youth Custody Centres. Inasmuch as this represented an end to the administrative determination of sentence length, it could be seen as an improvement in the rights of these inmates.[18] On the other hand, and

[16] A 'core' of prisoners' rights in relation to disciplinary hearings was progressively established from the 1970s. In *St Germain* it was established that there was a right to judicial review of Boards of Visitors' adjudications; subsequent cases led to the tightening up of specific aspects of procedure, evidence etc. In *Tarrant*, it was determined that Boards of Visitors had had a discretion to allow a prisoner legal representation and, in certain circumstances, should grant it. Subsequently, in *Leech* (1988), and *Prevot* (1988), adjudications by governors were deemed reviewable by the courts. Richardson (1985: 59–60) argues that 'prisoners' general rights have prevailed only when they have related to issues of particular interest to the judiciary, while effective special rights have been recognized only where the challenged decision is seen as quasi-judicial and the courts are able to apply a body of familiar principle'.

[17] Despite these advances, few inmates seemed to stand on their rights. Of 3,765 cases heard by Boards in 1986, legal representation or the assistance of other individuals was asked for in only 194 cases (personal communication with the Prison Department).

[18] This also meant that the new YCC inmates became eligible for remission, with the date of release calculated from the point at which they were taken into custody. Since a Borstal sentence had been regarded as a period of training, with the period determined—within legislative limits—by the inmate's performance, not only was there no remission, but the sentence was only regarded as starting on entry to the Borstal; thus giving no recognition to time on remand or time awaiting allocation to

following changes in parole procedures in 1987, inmates who liti-
gated against the Home Secretary on the grounds that the rules
breached the tenets of natural justice were unsuccessful; it took
several more years and a case at the level of the European Court of
Human Rights to establish even basic rights in the area of parole,
and even then only for a special category of inmates (*Thynne,
Wilson and Gunnell*, 1990).[19]

It should be mentioned, finally, that attitudes towards inmate
rights have begun to change. The Woolf Report (Home Office
1991) argues trenchantly that considerations of justice to inmates
should require many current 'privileges' to be seen as normal
expectations and entitlements—in effect, qualified rights (paras.
14.28–39)—and others to form a structured system of incentives
(paras. 14.40–8). It seems that the issue of rights will be central to
the development of the English prison system for some years to
come.

The Netherlands

The Dutch approach to inmate rights is the most remarkable of the
four systems. The other three countries have been tardy about recognizing inmate rights; and those that have been recognized have
tended to be procedural rather than substantive. They are rights to
have one's case considered, to make submissions, to be informed,
and so on, rather than entitlements to specific goods and services.
Moreover, even these rights have not been absolute; the administrations may legitimately abrogate certain rights in broadly-defined
circumstances. Thus, in relation to almost any area where effective

a Borstal. However, in other respects there were no major changes in young prisoners' rights, although many of the new YCCs adapted or modified existing Borstal
regimes, primarily in order to accommodate youths arriving direct from the courts
and with differing sentence lengths. A more detailed description of the Youth
Custody system is in Burney (1985).

[19] The European Court 'found that current United Kingdom release procedures
for discretionary life sentence prisoners, those convicted of offences other than murder, were unlawful under Article 5(4) of the European Convention on Human
Rights. It was held that a discretionary life sentence consisted of two elements: a
period considered necessary for retribution and deterrence, and a period when a
prisoner's mental instability and dangerousness could be monitored and a decision
for release taken, depending on the risk of danger to the public. The factors of
mental stability and dangerousness might alter with time and new issues of lawfulness of detention thus arise. In view of this, the lawfulness of continued detention
should be decided by an independent court at reasonable intervals' (Eratt and
Neudek, forthcoming).

grievance procedures depend crucially on the development of inmate rights, those rights are minimal, or burdened with heavy qualifications. In the Netherlands, it remains true that inmate rights are not full legal rights, but everyone treats them as though they are. Much of the effectiveness of the system relies on early and informal mediation; but this too has been made possible by the development of a formal complaints system which acknowledges inmate rights even though they are not underwritten by law.

Kelk (1983) notes that in the passage of the Dutch Prison Act of 1951, there was no discussion of prisoners' rights. None the less it gives prisoners, *inter alia*, rights to spend half an hour per day in the open air (s. 11); to wear their own clothes (s. 18); to receive toiletries and other personal hygiene items (s. 20); to purchase items weekly from the canteen (s. 24); to smoke during exercise (s. 25); and to borrow items from the prison library once a week (s. 50). In addition it requires inmates to be given information concerning their detention (that is, to see certain parts of their prison file), to keep various personal items, to correspond with others, and to have visitors. Some rights can be characterized as 'freedoms from' rather than 'freedoms to'. For example, remand prisoners are not obliged to work—though if they do not, they get no pay. The right which has attracted most discussion in recent years, however, is the legislative stricture that there should be only one inmate per cell. In view of the recent pressure on prison places, this has been challenged in parliament on several occasions, though not yet overturned.

None the less, the strict legal status of such rights must be open to doubt, since many of them are laid out, not in the head law, but in subsidiary legislation. Moreover, they are not absolute rights.

Three further observations should suffice to explain the Dutch approach. First, while the prison regulations in general and the house rules for institutions are littered with sections which begin 'The prisoner has the right in the institution to . . .', all rights—with one exception, the right to complain—are qualified by the general provision that:

If security or order in the prison make it absolutely necessary in the reasonable opinion of the governor, he is empowered in the case of a particular prisoner to deviate from the rights the prisoner may derive from the provisions of the Prison Regulations concerning the degree of social intercourse, the division of the stay, medical, spiritual, social and material welfare and education and relaxation. The governor shall notify the prisoner of his decision and the reasons for it in writing, pointing out the possibility of entering a complaint. (Ministerie van Justitie, 1980: their translation)

Rights can be withdrawn in the case of individual inmates (s. 15), or prison-wide (s. 16) for reasons of security, though it is possible to lodge complaints against such restrictions in the same way that any other complaint may be made. The overall effect is that what the Dutch describe as 'rights' are more in the nature of normal, but strong, expectations as to what inmates may do or have, and are qualified by a blanket security consideration. In addition to this proviso, some rights are also qualified in particular situations. For example, an inmate has the right to smoke, with the exception that smoking in a punishment cell is not permitted. While under punishment, an inmate may smoke at specified times only, such as when he or she is unlocked to collect his or her meal.

Second, the Dutch recognize what one might call the 'unwritten etcetera clause' in regulations and laws. No regulation says what it means in so many words, since its interpretation will vary in the light of specific situations. Rather than set off down the path of trying to specify rights ever more precisely, the Dutch adopt the pragmatic position that prisoners should have one primary right, that of complaint to an independent body. Rights are thus created more or less *ad hoc*, as the case for them is argued and won, and are not necessarily applicable to all prisoners or under all circumstances.

Third, the notion of rights that has developed is often more in tune with the idea of 'moral' than 'legal' rights. For example, when the regulations concerning parole (Voorwaadelijke Invreiheitstelling, or VI) were changed so that VI became, in effect, a system of remission, prison staff, governors, and headquarters staff began to refer to it as a right. Their reason for doing so seemed to be that the balance had shifted from the need for a consideration and a positive recommendation, to automatic release unless a negative recommendation was submitted for consideration.

These points, while indicating the direction which rights have taken in the Netherlands, do not explain how inmates set about obtaining them. To understand this it is necessary to explain both the court structure and the arrangements for hearing complaints. Though the latter is more commonly used, it is worth briefly explaining the former. Ybema and Wessel (1978) describe two forms of judicial review: a review of the legality of administrative acts, and an 'evaluation of an administrative act in judging an action in tort against the administration'. However, they warn that 'since a general system of redress of administrative wrongs as existing in France is lacking, it may easily occur that there is no redress available at all'. The most immediate form of court action is the kort geding, an action in the district court to obtain a provisional judgement on administrative acts which are imminent, in order to prevent them taking place until an appeal of the provisional judgement in cassation is made by a higher court. This is clearly of considerable practical importance to inmates, and a few such cases are heard each year.

However, the most common resort for inmates is not to the courts, but to the complaints machinery. Yet the extent to which the complaints boards within the institutions, the beklagscommissies, seek to interpret and apply a formal body of rules remains in doubt. The beklagscommissies are more geared towards mediation than are the procedures in the other three countries discussed here, and their general approach seems to be, where possible, to resolve issues quickly and by negotiation. More formal determinative procedures seem to exist at the level of appeal, the Centrale Raad van Advies, whose decision is final. The Centrale Raad is in fact widely described as a 'court', seen as producing 'jurisprudence', and its decisions are considered to form (non-binding) precedents for the handling of future cases. Its decisions are thus seen, on the one hand, as adjudicating claims in terms of rights, and on the other, as clarifying or establishing inmate rights.[20]

Requests, Applications, Complaints: the Practical Impact of Grievance Procedures

The discussion above indicates that although differences exist between the four countries as to the strength and nature of rights, it would be fair to state generally that rights tend to be limited and often cannot be relied upon by inmates in pursuing grievances. Nor, necessarily, should they be; there exists a whole series of

---

[20] None of this prejudices inmate rights to initiate litigation or to approach the ombudsman, though such complaints must allege breaches of law in the one case and maladministration in the other.

areas of prison life where the specification of rights would make little practical sense. But, that said, I would like now to concentrate on the internal grievance procedures, what they are used for, how often, and to what effect.

The first question is one of definitions. In all four countries, and virtually without exception, inmates would begin not by making a complaint, but a request; the issue would only be seen as a complaint once the request had been denied, and the complainant would then allege that the denial was improper.[21] In some situations, requests pose no problems because what is being asked for is a standard provision; in others, interpersonal factors may affect the outcome, as was indicated by the French inmate quoted above, and the material cited in Chapter 4.

However, the point of making a request is to have it granted at the earliest possible moment. Inmates would thus do what they could to maximize their chances of success at this first stage. They would first identify the correct individual to address; requests for extra phone calls, for example, were best directed to social welfare staff. And they could try to have the request dealt with at one remove, by asking educational or social work/probation staff to argue their case, for example, rather than directly approaching staff or governors. But, as I indicated in Chapter 4, the prevailing feeling across all four countries was that many requests—and indeed many other aspects of their treatment—were dealt with either arbitrarily, or on the basis of the inmate's reputation. And in the Netherlands, it was said by inmates that officers were unwilling to accept responsibility for either granting or denying requests, would often refer the matter to their team leader or a team meeting, and that in the end the matter would be shuffled among the staff until it 'went away'. Finally, there was a general perception, across all countries, that a negative decision on a request would be almost automatically upheld in all the further stages of the complaints machinery; the inmate could expect that challenging a negative decision would involve a major battle with the administration unless a more subtle solution presented itself.

---

[21] This clearly would not be the case if the complaint alleged, for example, that officers had verbally or physically abused the inmate. But in general, complaints concern requests that have been denied.

Procedures for Making Complaints

The administrative procedures for handling complaints can be described fairly easily. In all countries they rest essentially on making a representation first to the governor, and then up the hierarchy to ministry level. After that, internal procedures are exhausted and one must write directly to the minister, or an ombudsman (in every country except Germany, which has none), or to one's lawyer or the courts.

The German system is quite simple. One could in principle take a complaint directly to the prison complaints court, which requires simply the writing of a letter—though in practice it would be unusual not to raise the matter first with the governor. Only in a few of the Länder is there a requirement to ventilate the complaint with the central administration before proceeding to the court. This is, on the face of it, contrary to the ECHR decisions in the *Golder* (1975) and *Silver* (1983) cases, where such a practice was declared inconsistent with the European Convention.

In France, inmates can present requests or complaints to the governor (Art. D259, CPP). In practice this is done in writing, and governors may—though it is a matter for their discretion—hold an interview with the prisoner if the latter states sufficient reasons. Article D260 outlines a procedure in which governors' decisions may be appealed to the regional director, and decisions of the latter to the central administration. Procedures are none the less complicated by the fact that the governor is not the competent authority to decide a wide range of matters. As indicated earlier, many decisions fall within the sphere of competence of the JAP, or, for remand inmates, with the juge d'instruction. It is thus not unusual for inmates to write to the governor, only to be called to interview to be told that they should address themselves to someone else.

The Dutch grievance system is perhaps the most 'complainant-friendly' in terms of access and procedures (though not necessarily in terms of decisions). In principle complaints may concern only disciplinary measures, refusals to allow correspondence or visits, and measures denied which could be construed under the prison regulations as a right. In practice the last head, due to the phrasing of the regulations, covers almost all aspects of regimes (a separate body hears complaints about transfers between establishments, a matter often held in other jurisdictions as wholly within the discretion of the administration and not open to complaint). The three-person Beklagcommissie, which hears inmate grievances, has powers both to substitute its own decision for that of the governor in any particular instance, and to award compensation in cash or in the form of, for example, extra visits. The system thus provides a reasonably pragmatic form of remedy for inmates whose complaints are upheld.

The English system was changed in the management reorganization in 1989, and is again under review at the time of writing, following the recommendations of the Woolf Report. There are in effect two systems, one of applications to the governor followed by 'petitions' dealt with at area or headquarters level, and a second, of application to the Boards of Visitors. The overall impact of the changes since 1989 has been to redefine the Boards' role as one of monitoring internal grievance procedures within establishments, and reviewing and seeking to resolve complaints referred to them.[22] Some structural constraints on their role were removed in 1989. Prison Rule 47(12), which created a disciplinary offence of making a 'false and malicious' allegation against prison staff, was abolished. It had been criticized as a deterrent to complainants, at Board of Visitor level as elsewhere, since many complaints are implicitly, if not explicitly, complaints against staff. Ironically, the revised procedures created their own problem, which the Woolf Report then had to address. Inmates may have 'confidential access' by sealed letter to the governor, senior officials, and the Boards of Visitors. The recipient of a sealed letter typically has the discretion to decide who needs to be informed about it. But where a complaint is made about staff, the Staff Manual states that a copy should be given to the staff member complained about. Woolf notes that while the handling of confidential complaints frequently requires their substance to be disclosed to officials, the automatic provision of such a copy compromises the purpose of confidential access.

Woolf made three further recommendations, of which one has already been acted upon and the other two are in process. The first two address the issue of the inmates' perceptions of Boards as independent. First, he reiterated a point made by many previous

---

[22] Further details are in HM Prison Service (1989), and commentaries on the changes appear in Hirst (1990) and Shaw (1990).

commentators, that boards should no longer conduct inmate disciplinary hearings, since they compromise their independence (as indicated above, this recommendation was implemented in April 1992). Second, he argued that there should be a further shift in their role, towards 'advising and assisting' prisoners about grievances rather than 'hearing' grievances and, in effect, appearing to adjudicate upon them (without, however, possessing any significant executive powers). In particular, rather than making recommendations adverse to the inmate, Boards should simply inform him or her that they would take no action in the case (Home Office 1991: paras. 14.340–41). Third, Woolf recommended a further independent avenue, a Complaints Adjudicator—in effect a prisons ombudsman. The post would be filled by an experienced lawyer, appointed by the Home Secretary, and reporting to parliament in the same way that the inspectorate now does. The role would have two major components in relation to grievances (and some others in relation to disciplinary matters). One would be to seek negotiated solutions to complaints; the other, to make recommendations to governors, Area Managers, Headquarters and the Director General, which would have to be responded to within a set time limit (Home Office 1990: paras. 14.350–62). A consultation document was issued in March 1992, with the general intention of setting up such an office by the end of that year (HM Prison Service 1992).

Resort to Grievance Procedures

Only partial data exist for complaints in Germany. The figures for one Land, Baden-Württemberg, are illustrative, though perhaps higher than other Länder, because it is one of the administrations which requires complaints to be taken to the ministry prior to ventilation through the courts. In 1986 there were about 1,550 complaints of all kinds to the Ministry from a prison population of about 7,500. Unfortunately no statistics on outcomes were kept.

It is fair to say, though (here I rely on interviews with officials in another Land, Hessen), that even if the direct success rate of complaints is low, there is an indirect effect in that complaints may reveal situations which ministry officials regard as undesirable. Even though the complaint itself may be refused—partly, as one official frankly stated, as a matter of 'face'—the fact of its having been made may alert senior officials who will then pursue the issue raised independently of the complaint itself. This is an interesting possibility, though it must be somewhat haphazard. To anticipate the discussion of inspection in the next chapter, arrangements for inspection vary by Land. Baden-Württemberg locates responsibility for inspection with one individual in the administration, who does not always see files on complaints. Hessen divides the responsibility for inspection between all the senior officials, who deal with one or two institutions each and who would normally be alerted to apparent problems by the official dealing with complaints.

In France, the numbers of complaints received centrally by the administration has been monitored since 1984; there were 10,504 in 1984 and 10,926 in 1985. Of these, the majority (54 per cent in 1984, 55 per cent in 1985) related to transfers, and of these again the majority came from convicted prisoners asking to be moved from maisons d'arrêt to training prisons. Over one-third of all such requests came from only four prisons—Rouen, Fleury-Mérogis, Marseille and Fresnes (Ministère de la Justice 1986).

In the Netherlands, the post-1977 complaints procedure has seen large yearly increases in the numbers of grievances lodged, and a substantial proportion of those grievances are found grounded and remedial action taken. But the complaints system has been criticized despite its apparent openness. De Jonge (in Kelk and de Jonge 1982) points out that in several instances, the prison administration has seen fit to change house rules rather than conditions, where the two have been at variance and the latter criticized by CvTs.

Statistics are not the best tools for evaluating complaints systems, but the Dutch figures are particularly interesting.[23] In general, about 15 to 18 per cent of complaints are found grounded by the commissies; about 25 per cent of all first-level decisions are appealed to the Centrale Raad van Advies, either by governors or inmates. The governors have a success rate at appeal of about 65 per cent, and inmates about 25 per cent. In terms of the final outcomes, and despite the level of successful appeals by governors, about 10 to 12 per cent of inmates' complaints are upheld—a much higher proportion than in the other three countries, and a

[23] The figures quoted are my calculations based on information from the Dutch Ministry of Justice. More detailed figures on types of complaints are published in the Dutch Prison Department annual reports, though without any interpretative commentary or calculation of complaint rates.

figure which, moreover, excludes the extent of informal mediation. The effect of mediation cannot be lightly dismissed, since one Dutch study discovered not only that a large proportion of grievances seem to be dealt with at an informal level with members of the commission acting as mediators, but that the members of the complaints commission often saw their principal role as a mediating one (Mante, personal communication).

Since there are no formal barriers to complaints (the CvTs decide for themselves if a complaint is admissible) this may be regarded as quite a high proportion. But the majority concern restrictions imposed on prisoners on the grounds of security and order. In so far as the complaints system can also be seen as generating rights, then, they are tied to very specific situations and have no implications beyond the particular case.[24]

My interviews with Dutch prison governors did not uncover any significant unease about the high level of inmate complaints found grounded. Undoubtedly they presented a 'front', part of which involved the projection of their image as professionals able to handle any situation. In some respects they had little cause for concern. Many of the grievances concerned comparatively trivial matters. As one officer observed—albeit with a certain amount of exaggeration—the idea of complaints had become so entrenched among inmates that they would insist on making a formal complaint if they thought their coffee was not hot enough. In other cases, where the complaint concerned security or control issues, even if the complaint was found grounded, the measure complained of would already have taken place and management objectives been attained, and the issue would be one of compensation. Moreover, the governor would have the possibility of appeal to the national commission. In addition, the management structure is one in which most actions taken with regard to a prisoner will have been the subject of staff discussions and consultations, while there seemed to be no adverse consequences for a governor (such as reprimands from the central administration) if he or she was found in the wrong. As against this, however, one member of a complaint commission I interviewed, a well-known academic, suggested that governors were often more worried by complaints than they pretended to be, because they expected the commissies to take a strict

[24] Though see App. 2 for mention of two exceptional CRvA decisions with system-wide implications.

legalistic view of the matters complained of. His comment on this was that the commission he sat on saw part of its task as assessing the action that gave rise to the complaint, not in purely legalistic terms, but in the wider context of the emotional factors surrounding it; in other words, the commission sought to reduce the governors' apprehensions and increase their willingness to share problems by adopting a non-legalistic, 'partnership' approach.

In England, as indicated above, the complaints procedures have undergone radical surgery in the last few years. Pre-1989 empirical studies cannot now be wholly relied upon in discussing the current situation. None the less they indicate that, historically, one of the main issues has been delays in handling complaints, an area which appears to have been substantially improved in England by the introduction of time limits for replies. A second issue was that of low success rates. In Ditchfield and Austin's (1986) study of petitions dealt with by regional or headquarters offices—which of course predated the reforms—inmates' requests were granted or complaints upheld in only 6 out of 120 cases studied.

It has often been said that Boards of Visitors lack in practice the independence that they have in theory (see e.g. Maguire and Vagg 1984; Maguire 1985; Vagg 1985). Certainly neither the boards nor the arrangements for petitions have been used to considering rights, rather than, for example, privileges, custom, or local expediency. Prisoners have never regarded the complaints procedures as having much to do with rights (Austin and Ditchfield 1985), and Boards of Visitors very often accepted governors' views on the validity of complaints, especially if 'security' or 'order' considerations could be cited (Maguire and Vagg 1984). Woolf's recommendations will very likely give them that independence, though at the price of removing most of their powers. However, since independence needs to be grasped as well as given, it remains to be seen how well these reforms, if implemented, will work.

Making Complaints, Using Rights

In practical terms, it seems quite clear that there is a certain point beyond which an inmate attempting to contest his treatment in prison, whether he or she invokes concepts of rights or not, is going to have a lengthy battle with the administration. The case may be decided long after the measure complained about was terminated: in the case of Caillol, mentioned above, some two years after the regime he was fighting had been abandoned. In some cases the court decision may come some time after the inmate was released. However, the 'certain point' after which the battle will become long-term is not synonymous with the division between making grievances and demanding rights, because different countries treat the idea of rights in subtly different ways. They are most closely woven into the fabric of the complaints system in the Netherlands; recognized in the 'judicialization' of complaints from a relatively low level in Germany; somewhat recognized through the CPP in France, though with many areas left untouched; and in England, following some fifteen years of litigation, are most explicit in the area of prison discipline (though most of that litigation deals with procedures since discontinued).

In the courtroom, where the litigant must demonstrate that his treatment was wrong in law, the likelihood of success is small in every jurisdiction. But it is low, too, in relation to complaints made internally. It is often argued by prison officials that low rates simply reflect the willingness of inmates to make formal complaints about trivial matters. Complaints cost nothing, while every gain, however minor, is still a gain. Yet the suspicion still lingers that low success rates reflect administrative conservatism, even where independent complaints mechanisms are concerned.

One further issue has to do with the form of the decision. Where grievance procedures deal with specific complaints and determine that they are 'founded' or 'unfounded', they are usually supporting either the complainant or the administration. There is correspondingly little room for halfway decisions which acknowledge that part of the inmate's case is justified, or that the facts of the case would admit a complaint other than the one actually made. There has thus been interest in the possibilities for mediation.[25] But, outside of the Dutch system, and the potential for mediation that lies in the post-1992 role of the English boards of visitors, there is little enough scope for it and a paucity of hard evidence as to how it works in practice. Mediation depends crucially on reaching an interpretation of events and an agreement that is acceptable to

[25] See e.g. Wright (1985), which argues that many events dealt with through the *disciplinary* procedures could better be seen as conflicts to be resolved though mediation, and suggests a model loosely based on a number of American procedures, e.g. that developed within the California Youth Authority: not, he explains, intended to replace existing procedures but to relieve the pressure on them.

both sides. However, there is no 'equality of arms' in prison; governors and staff do not often embrace the idea of meeting inmates on neutral ground, or accept that inmates and staff should have equal status within any procedure. Questions of 'face', and of the maintenance of their authority after a mediation session, militate against such ideas. The complaints procedures in the Netherlands depends crucially on the empowerment of inmates through the recognition of a wide range of (qualified) inmate rights; and despite governors' concerns, the situation is still one in which the authority of the governor is maintained—first because he or she is not brought directly into discussion with the inmate, and second because the system is *post hoc*, with management aims often already achieved and the issue being the cost, in terms of compensation, of having achieved those aims.

In sum, it is difficult to see any direct relationship between improvements in rights and improvements in conditions. Many rights have not affected prison conditions generally, since they apply only in specific situations or deal with the treatment of individual inmates, with (except in the Netherlands) only a limited machinery for considering how classes of inmates may be similarly affected.

In some spheres, the development of rights has had adverse rather than positive effects on prison conditions largely because facilities have not been developed to cope with them. In both France and the Netherlands, inmates now have the right to wear their own clothes, yet laundry arrangements have not kept pace; inmates are frequently required to wash and dry clothes in their cells. The right is robbed of its attractiveness by the consequences of standing on it.

Beyond this, there seems to have been a general tendency to maintain the gap between what is available as a right and what is available as a privilege. Both France and the Netherlands have recently begun to allow prisoners to have television in their cells, though this is not a right. Germany and the Netherlands now have, in some prisons, facilities which amount to conjugal visits; there are no laws or regulations—yet—which overtly recognize the point of having a visiting room with a double bed in it, nor which govern access to such visits. Increased privileges and improved facilities for inmates, where they have occurred, have actually opened up new areas in which prison staff routinely exercise discretion but where the idea of accountability is perhaps too coarse or too unwieldy to take hold. This can be illustrated in the following ways. The facility for prisoners to make telephone calls (in the Netherlands) is now considered a right. However, the implementation of the right varies from prison to prison. In some places a coin-operated phone is available to prisoners. In others, particularly closed prisons, the inmates must book a call in advance and are not guaranteed a specific time (or necessarily day) for the call, which can then form the basis for informal negotiations about other matters, such as prisoner behaviour. A second example can be found in France; the right to wear one's own clothes then raises the possibility that inmates sometimes swop clothes or sell them to each other. But this is also an extension of the possibility for governors' control. The right sets up a new type of request that can be made by prisoners, a new potential disciplinary offence, and thus two new areas in which governors can exercise discretion and use it as a lever, however small, to increase or reduce pressure on prisoners.

American Experiences: A Note

Thus far, litigation in European countries has not developed along American lines, where court cases are very frequent and three significant steps have resulted in the courts wielding considerable power in relation to prison conditions. First, the courts allow 'class actions', in which a number of similar cases are dealt with as a single case affecting a class of inmates. Second, courts are willing to consider the totality of prison conditions, rather than specific regime issues, as a proper subject for complaint. In consequence the combined effect of a number of separate factors may be judged to constitute unconstitutional treatment. Third, the courts have used two mechanisms, consent decrees and the establishment of special masters, to intervene in prison administration and ensure that movement towards acceptable conditions takes place.[26]

The US Constitution contains a number of provisions which

---

[26] American prison litigation is reviewed in Jacobs (1980), and Morgan and Bronstein (1985), and these two sources inform much of the following paragraph. However, since the mid-1980s, the climate of opinion in American courts has shifted away from a reforming stance and fewer cases have been decided in the inmates' favour.

prisoners have been able to use in litigation. The First Amendment allows freedom of religion, freedom of speech, and the freedom of the press. It has been used to remove arbitrary censorship of inmate communications. The Fourth Amendment deals with security from unwarranted search and seizure, and thus provides some protection against improper searches and confiscation of personal items. The Fifth and Fourteenth Amendments require due process in the deprivation of life, liberty, and property, and have been used to establish proper administrative procedures for discipline, parole, and segregation.[27] The Eighth Amendment prohibits cruel and unusual punishment, and has been relied upon in cases alleging that particular practices, conditions, or totalities of conditions can be so described.

The slew of court cases brought in the 1970s created significant improvements in prison conditions nationwide, though it has to be said that conditions in certain states were truly appalling. In many prisons they still remain unacceptable, not least in terms of overcrowding, which in 1990 was 22 per cent above design capacity in state facilities, and 46 per cent over design capacity in the federal system (Stephan 1992: 7).[28] The Supreme Court became progressively less sympathetic to inmate litigation after about 1976, and some recent cases have rolled back the gains of the early 1970s by raising the threshold at which treatment may be considered cruel and unusual. In *Albers* v. *Whitley* (1986), a case concerning injuries sustained during a disturbance even though the inmate was not directly involved in the incident, it was decided that prison officials should have wide discretion in the maintenance of order, and that the case could only succeed if unnecessary and wanton, or malicious and sadistic, pain was involved. In *Wilson* v. *Seiter* (1991) a case complaining of a number of conditions amounting to cruel and unusual punishment was decided on the basis that the inmate must prove the deliberate indifference of prison officials to those conditions (for a discussion of both these cases, see Alexander 1992).

[27] The 5th Amendment applies to federal prisoners and the 14th to state prisoners.

[28] Alternative figures, based on 'rated capacity', show a much lesser degree of overcrowding. However an alternative measure, of ft^2 per inmate, indicates that state prisons provided on average 56 ft^2 per inmate and federal institutions, 44 ft^2; in maximum-security institutions at state level, inmates remained locked up for an average 18.7 hours per day, and at federal level, for 19.5 hours (Stephan 1992: ns. to table 9, and table 14).

None the less, in 1990, some 22 per cent of US correctional institutions were under court orders or had consent decrees to limit their population, while 20 per cent were under such orders or decrees for other reasons. Such court orders or consent decrees affected institutions in all but seven states (Jankowski 1992: 47; Stephan 1992: 7).[29] Despite their recent retrenchment, it is clear that the courts remain, in the US, the single most important source of pressure for improvement in prison conditions.

Alternative Forms of Complaint: A Note

Given the freedom of communication that most inmates now enjoy—one of their few unambiguous rights—the possibilities for complaint now extend beyond the mechanisms I have discussed above, and encompass letters to members of parliament to take up a complaint with the relevant ministry, and requests to MPs to refer allegations of maladministration to the ombudsman (except in Germany, which does not have such an official).

In England, the ombudsman, officially titled the Parliamentary Commissioner for Administration, appears to deal with only seven to eight inmate cases each year.[30] This is partly the result of MPs declining to pass complaints to the PCA, and partly a question of the PCA's remit, which is restricted to matters involving maladministration. The end result has been that while many PCA decisions have established that internal regulations are illiberal or unjustifiable, they were applied without evidence of maladministration. This was, for example, the finding in the 1971 case of *Knechtl* discussed in Chapter 2, and is notable in that the case in maladministration failed; it was, rather unusually, a parliamentary Select Committee that subsequently resurrected the case and forced the Home Office to liberalize its regulations on inmate access to medical advice. And in some other cases, although evidence of

[29] Of 1,207 institutions, 264 had court orders or consent decrees pertaining to population limits and 242 pertaining to other conditions of imprisonment. Some institutions were affected by both kinds of orders or decrees. States not affected by court orders or consent decrees were Maine, Vermont, North Dakota, Alabama, Oklahoma, Oregon, and Wyoming. The federal prison system was not subject to any such provisions.

[30] Cited in Penal Policy File 38.6, *Howard Journal*, 29/3 (Aug. 1990). The original source is a response by the National Association for the Care and Resettlement of Offenders to the 1989 government proposals, cited earlier, to change inmate grievance procedures.

maladministration was found and the Department criticized, the conclusion was that the extent of maladministration did not amount to unfair treatment.[31]

The Dutch ombudsman, also, deals with only a handful of prison-related cases each year. Usually, between about one and five investigations each year are carried out. An examination of cases decided in the period 1983–7 reveals a variety of different areas of complaint by inmates, some of which involved matters such as mode of arrest, sentencing, or parole refusals rather than internal prison affairs. Findings of maladministration are relatively frequent; of the thirteen cases during 1983–7 on which I have detailed information, nine resulted in findings that the administration did not act 'decently' in relation to part or all of the matters complained of.[32] However, seeking a specific remedy is more problematic. There appears to be no automatic machinery for compensation or restitution.[33]

Conclusions

To what extent, then, has the recognition of inmate rights led to improved prison conditions or improved treatment of prisoners? The observations above support four sets of conclusions.

First, prison conditions are largely the result of a historical

[31] For a fuller account, see Birkinshaw (1985b: 127–40).

[32] These 13 cases were: 82.03017 (decision 26 Aug. 1983), a complaint about the handling of an earlier complaint to a governor about a member of staff; 83.01739 (19 Nov. 1985), concerning fire safety precautions in a particular prison; 83.01817 (8 Oct. 1986), concerning a refusal of home leave; 84.00721 (1 Nov. 1984), concerning poor handling of a plea for clemency; 84.01364 (9 Jan. 1987), concerning refusal of home leave; 84.01745 (5 Dec. 1985), an inmate given incorrect information about his release; 84.02148 (19 Dec. 1986), a complaint about prices in the prison canteen; 84.02178 (7 Nov. 1986), concerning the lack of a reason for the refusal of the complainant's plea for clemency; 85.00125 (11 Oct. 1985), concerning a 'walking sentence' who did not receive his notice to start his prison sentence and was consequently arrested; 85.00712 (26 July 1987), a complaint about the period of time spent in an HvB awaiting transfer to a mental hospital; 85.01961 (29 May 1987), concerning a refusal to allow the inmate to study outside the prison; 86.00101 (31 July 1987), concerning refusal to allow an independent medical examination in an outside hospital; and 86.01357 (15 May 1987), complaining that the administration did not help the inmate to continue his business while in prison. I am grateful to Ralph Vossen for his summary translations of the ombudsman's reports on these cases.

[33] I have not discussed the French ombudsman's work because his official annual reports make no mention of prison-related cases. It is thus not clear how much, if any, of his workload is generated by inmate complaints.

accretion of policies and funding. England and France have both experienced severe overpopulation and the deterioration of regimes in many prisons, though the deteriorations have not been borne equally across the whole prison system in either country. The Netherlands has also experienced budgetary cuts resulting in some regime cutbacks, the major effect being that inmates now spend less time at work during the week and longer in their cells at weekends. Cognizant of these factors, governments have fought shy of introducing inmate rights which will become a rod for their own backs, particularly in financial terms. As a result prisoners in England, France, and the Netherlands have relatively few legally based and enforceable rights, and although, in the Netherlands, it seems common to describe many features of prison life as conferring rights, they are largely moral rather than legal rights and are conditional in that they can be withdrawn. In Germany prisoners have rights pertaining to many more areas of prison life, though many of these are conditional, and/or stated in very basic terms which leave the administration much room for manœuvre.

Second, in so far as inmates have definite and enforceable rights they have been the outcome of lengthy political and legal battles which have often taken years to resolve. Prison reforms in England have largely centred around the removal of irksome restrictions (for example on correspondence) and the extension of early release. In France they have gone further in the removal of restrictions on inmates, and have in addition permitted new privileges such as televisions in cells. But a differential between the regimes of the different kinds of prison has been maintained. In the Netherlands, the single major reform was the creation of a new complaints procedure, and individual inmates have benefited to some extent. But the major substantive improvement in prison conditions has been the introduction of televisions in cells. Beyond this, inmates can be said to have benefited from the *failure* of reform proposals such as the termination of the 'one prisoner per cell' law. In Germany the whole system was reformed at a stroke in 1977—albeit after a decade of discussion—with relatively little development thereafter. However, with the exception of the Dutch creation of a new complaints system, these reforms have generally been improvements in privileges rather than the recognition of rights.

Notwithstanding the comments above, there have been real improvements in prisoners' rights, though they have not been as strong as might have been supposed. Even where rights have been declared, they have opened up new possibilities for discretionary control. In consequence, the key area of debate remains that of how administrative discretion is employed, and the operation of complaints procedures against discretionary decisions. The inevitable conclusion is that however powerful is the image of reform through the introduction of rights, prison reform is, in practice, more a matter of political willingness, and there is a greater willingness to introduce privileges than rights, not least because the former do not entail enforceable claims against the administration. The main practical advantage of introducing enforceable rights is therefore the upward pressure that rights exert on privileges. However, the Woolf Report in England has come down firmly in favour of the development of inmate rights, though it is careful not to raise hopes too high, and uses the language of 'legitimate expectations' and 'justice in prisons'. The Report has clearly effected a major shift in the political willingness to consider and extend inmate rights. How far this will be translated into practice remains to be seen.

Third, rights are not always easy to stand on. Insisting on them may result in unanticipated or undesirable consequences. Prior to 1992, for example, an English inmate who insisted on legal representation in a Board of Visitors' disciplinary hearing might find himself in segregation until his lawyer, the Board, and the Treasury Solicitor representing the Board could agree a date for the hearing. In France, insisting on the right to wear one's own clothes results, in many prisons, in a cell permanently full of dirty or wet washing.

Fourth, complaints procedures are inherently limited in their capacity to effect change. They are geared to individual matters, usually of a fairly trivial nature, in which there has been some derogation from normal conditions or procedures. They are much less well suited to handling issues which are in effect 'class actions', affecting large numbers of inmates or with massive resource implications. They typically address matters such as the way that administrative rules are applied, not the substance of those rules or the policies which lie behind them. It is not surprising that in all countries the majority of inmate complaints are found ungrounded.

Inmates, understandably, react to this situation in two main ways. Since a request denied can become the occasion for a complaint, and the success rate of complaints is small, they will try to ensure that their request is made in such a way that its chance of success is maximized. Ironically, this tends to mean that inmates do not stand on the few rights they have, but adopt a more 'reasonable', persuasive, stance. The problem they face in this regard is the unpredictability of decisions. In consequence, they tend to describe decision-making procedures as arbitrary. In this they may be right to some degree, though there is often a reason for the decision which is not revealed to or known by the inmate. This perceived arbitrariness leads them to regard the complaints procedure in a very jaundiced light. It has to be said that most inmates, most of the time, find little cause to make formal complaints; and many, as we saw in Chapter 4, are much more concerned in effecting changes in practical conditions than in rights—the 'more videos, not more rights' argument. Yet the situation they find themselves in is one where formal complaints are only rarely effective in dealing with the kinds of grievances they have.

What is surprising about this situation is how many complaints are made, for example in the Netherlands. It seems that the main reasons for this lie in the fact that there is always the possibility that a 'no' may be turned into a 'yes'. However rarely that happens, any 'yes' is an advance of some kind. Because such psychological factors seem to operate, some officials state frankly that the decision may go against the inmate for reasons of 'face'. In so far as complaints have any long-term or general effect, then, it is likely to emerge through their disclosure of problems which are taken up and addressed through other means. This, however, presumes that those who deal with the complaints also have the authority to set such action in train.

Prison administrations in all countries are creatures of government. Even if there is a commitment to reform, the resources have to be found and the commitment translated into politically acceptable plans. One irony in this situation is that, in England and France at any rate, the most politically acceptable plan became, in the late 1980s and early 1990s, that of taking prisons out of the direct management of the prison administration—that is, privatization. The promise that this holds out is that privately-managed facilities may be able to offer fuller regimes and better conditions than state ones. But I shall reserve discussion of this topic for Chapter 13, which considers private prisons in greater detail.

9

Inspectors, Inspections, and Standards

THE idea of prison inspection is not new. In 1777 John Howard, in his review of English and European prison conditions, proposed that:

To every prison there should be an inspector appointed; either by his colleagues in the magistracy, or by Parliament . . . He should speak with every prisoner; hear all complaints; and immediately correct what he finds manifestly wrong . . . A good gaoler will be pleased with this scrutiny: it will do him honour, and confirm him in his station: in the case of a less worthy gaoler, the examination is more needful, in order to his being reprimanded; and, if he is incorrigible, he should be discharged. (Howard 1929: 36–7, orig. pub. 1777)

Howard made his proposal at a time when the majority of European prisons were under local rather than national control, and were not infrequently run for profit, with individuals paying the authorities for the prison 'contract', and charging fees to the inmates themselves. The concept of inspection, though intended to curb venality and abuses of power, was in principle no more radical than the inspection of any goods purchased by a municipality. The authority appointing the gaoler would have the right to inspect his work and, should it be found unacceptable, the contract could be terminated.

In contemporary circumstances the purposes of inspection may yet return to such a task, as privatization of prisons continues. But since the advent of national prison systems (and the creation of inspectors) the role of inspection has been less straightforward. Inspectors have also been planners, investigators, and architects, and have usually been employed by the agency whose work they inspect, or, if not, then by another part of government. In the four countries I studied, inspectors had a wide range of tasks. They were involved in the assessment of prison regimes on the basis of criteria such as humanity, effectiveness, efficiency, propriety, and 'value for money'. They were sent to conduct investigations into riots, escapes, and suspicions or complaints about staff misconduct. They conducted studies of, and proposed policy changes in relation to, matters as diverse as security, grievance procedures, and aid to prisoners following release. They made recommendations on plans for new prison building and renovations, and on technical matters such as changes to staffing levels or staff attendance schemes. In short, they were multi-purpose advisers and troubleshooters as well as inspectors. However it is also worth bearing in mind that these prison inspectors were not the only inspectors of prisons. In all countries, inspections for particular purposes, ranging from medical and hygiene concerns to fire precautions or workplace safety, were conducted by a range of other state inspectorates, though they did not necessarily possess the same statutory powers in relation to prisons that they did in other settings.

This brief outline of the inspectors' tasks fits well with Rhodes's (1982) analysis of inspection. He distinguishes two major and five minor functions for inspection. The two major functions are, first, to ensure compliance with statutory requirements (enforcement), and second, to secure, maintain or improve standards of performance (efficiency). It is hardly surprising that where one arm of government is inspecting another, the second rather than the first of these is stressed. Of the five minor functions, the most relevant are internal management inspection and accident or incident investigations (the others are revenue collection, checking specifications on supplies or products, and quasi-judicial determinations of matters ranging from staff disciplinary infractions to appeals of departmental decisions). The powers of inspectors vary according to their task: enforcement inspectors generally have strong legal powers, while efficiency inspectorates usually have powers only to obtain access to relevant materials and persons. In addition to this brief typology, Rhodes also points out that inspectorates may be given a number of conflicting tasks, such as advising governors on good practice but also reporting to headquarters on what they have found. As he notes, in relation to the English schools inspectorate, 'no indication is given of the purpose of such inspections . . . everything turns on how this obligation is interpreted in practice' (1982: 104).

The Need for Inspection

It is doubtful that anyone would dispute that there is a role and a need for prison inspections. Yet, as the brief outline above indicates, there might be widely different conceptions of what kinds of inspection are necessary and why. The issue of 'need' is important, however, because the availability of other forms of prison oversight can change the nature of the inspector's role. For example, the provision of strong inmate rights and effective grievance procedures may mean that inmate treatment, in terms of its practical implementation rather than treatment policies, need be less of an inspection priority. Where few inmate rights are recognized and inmates are less empowered to challenge their treatment, inspections would need to consider issues such as staff abuse of power and inmate treatment more carefully. And there is, in practice, a kind of correspondence between inmate rights and the inspection role, though it probably did not come about as a result of this particular train of thought.

It is true that inmate rights are not very strong in any of the countries I studied, but they are none the less stronger in Germany and the Netherlands, and to a limited extent in France, than in England. The reverse is the case in relation to inspections. The strongest—by this I mean best-resourced and most professionalized—inspection office is the English. France and the Netherlands have smaller and less well-resourced in-house inspection teams. In Germany, there is some variation across Länder as to specific arrangements, but they include, for example, the delegation of inspection duties among senior officials (each responsible for a small number of prisons and/or system functions), and the use of small in-house inspection departments with very restricted remits (for example, dealing only with security).

One way of explaining this situation might be to refer back to legal traditions. In the countries with stronger Romano-Germanic legal foundations—Germany and France—the law codifies matters such as prison treatment in ways that are supposed to provide a clear model for administrative action (though, as noted in previous chapters, whether it does is a moot point). The implication is that most legal controls on administrative action are reflected in the bureaucratic structure and to some extent the professional ethos of administrators. In common-law jurisdictions, such as England, the law has a much lesser role in formulating general principles of specifying the nature of relationships, and often settles cases using interpretations of broad principles such as 'natural justice'. The Netherlands sits in the middle of this spectrum, drawing something from each tradition. The difference between these two traditions of law should not be over-exaggerated. For example, as mentioned in Chapter 6, German complaints procedures expressly require an administrative decision to be made before a legal evaluation of it is possible. Yet the broad distinction between Romano-Germanic and common law does seem to include a division of world views. In the former, law is seen as the source of obligations and rights which define relationships, for instance between officials and prisoners. In the latter, in the absence of clearly specified relationships, an inspectorate has a strong role to play in considering issues such as humane and proper treatment.

The Inspectors

The English inspectorate is a comparatively large and complex organization. In addition to the Chief and Deputy Chief Inspectors, and the administrative support staff, there are two three-person inspection teams. Each team comprises a prison governor and a senior prison officer, both seconded from the prison service, and a civil servant with no necessary prior knowledge of prisons. In addition, a number of specialist consultants are appointed; at any one time there are usually two or three such persons working within the inspectorate, and in the past they have included a doctor, a buildings consultant, and a lawyer. Since 1983, a social scientist has also been either employed as a researcher by the inspectorate, or attached to it from the Home Office research unit.

Prior to 1981 there existed an in-house inspection office within the Prison Department. While the work of that team was never criticized (in fact it was never open to public criticism because its reports were never published) there had been a fierce debate about the value of having an inspectorate organizationally divorced from the service it was inspecting, and open to public scrutiny. The new, independent, inspectorate came into being on 1 January 1981. Its independent status arises out of its funding from general Home Office monies rather than the prisons vote; its reporting to the Home Secretary rather than the head of the prison administration;

and its legal obligation to provide annual reports to Parliament. In practice it also publishes its reports on inspections of individual prisons, and its recommendations (however, it only recently ceased to exclude comments on security in its published reports).

The inspectorate's terms of reference were first published in April 1981, stated more fully in its first annual report, and placed on a legislative footing in the 1982 Criminal Justice Act. In essence its mandate is to report on the conditions of prison establishments, staff and inmate morale, and the treatment of inmates and the quality of regimes; and the key factors it addresses are humanity, propriety, and value for money.

The French and Dutch inspectorates are smaller, in-house offices with a somewhat different set of duties. The French inspectorate comprises one inspector, two assistants, and secretarial support, yet deals with a prison system slightly larger than that of England. The inspector, like most senior civil servants, has legal training while his assistants are seconded prison governors. Their inspections are correspondingly smaller-scale affairs (though they conduct a large number of them) and are often geared to the investigation of specific problems which have come to the attention of the administration. The Dutch inspectorate comprised, at the time of my fieldwork, one inspector and two consultants—though at that time there were plans to enlarge the office by appointing three further inspectors. It was organizationally part of the prison service and reported to its director. The nature of this inspector's work was very varied. Although he did conduct inspections of establishments, much of his work revolved around advising on proposals for staffing, planning for new establishments, and investigating specific complaints (including disciplinary reports by governors) against prison staff. In neither of these countries were reports of inspections, or of other aspects of the inspectors' work, published.

The difficulty in talking about inspections in Germany is that inspection arrangements varied by Land; inspection was often a part of line management; and officials designated as inspectors often had a very restricted role. In Hessen, the two senior officials designated as inspectors were concerned almost entirely with security.[1] In Baden-Württemberg, each senior official had specific

---

[1] Their inspection checklist included space for comments on: the state of watch-towers; the functioning of alarms, metal detectors, cameras and other hardware; the security arrangements for storing guns and ammunition; the storage and issue of

responsibility to inspect one or two establishments, plus one or two system-wide functions such as workshops, disciplinary hearings, social-educational work with inmates, or security. Every official of 'referent' grade thus became an inspector of some part of the system.

Since 1989, an additional inspection procedure has operated within Europe. This is the European Committee for the Prevention of Torture and Inhuman or Degrading Punishment, which operates under the auspices of the Council of Europe and takes its mandate from the similarly-named European Convention. However, because of its pan-European nature, this body is discussed further in Chapter 12.

Inspection Strategies

The pre-1981 English inspectorate operated a rolling programme of inspection visits that resulted in every establishment being visited about once every five years. The independent inspectorate continued this model, though it was periodically taken away from its programme for special purposes. At the time of my fieldwork, each inspection team, in principle, operated along the following lines. Decisions about which establishments to visit were drawn up annually and agreed with the department—which may itself propose, on the basis of its own concerns, that particular establishments be included. The programme was, however, kept highly confidential, so that governors would only get a few weeks' notice of the inspection. This was unavoidable because he or she would be asked to prepare briefing documents for the team.

The three-person team, plus the Chief Inspector, his Deputy, support staff, and specialist consultants would then arrive, receive a further oral briefing, and meet the prison management and staff unions. They would then split up, with each team member attaching himself to particular prison staff in order to have guided tours of the establishment. The inspection lasted for five days, though not all the team members would be present for the whole period. My interviews with inspectorate staff indicated that the Chief Inspector would normally only be present on the first day, and the

keys; and many similar items. Measures or evaluations of treatment programmes or regimes, or even of basic items e.g. food and inmate clothing, were conspicuously absent.

consultants may only be present for that part of the work that concerns them. Although the inspectorate in its early days, created some checklists of areas to study, it appeared that these were treated as aides-memoires rather than guidelines. The whole process was intended to be open-ended and receptive to whatever emerged from conversations with staff and prisoners.

In the remainder of the inspection, the team would look particularly at issues or functions most often regarded as crucial to the good management of a prison. Records of inmate complaints, the use of special cells (i.e. isolation or segregation cells), and the use of handcuffs and other restraint equipment would always be checked. In addition, there would be discussions with the board of visitors and groups of inmates. On the fifth day, there would be an initial verbal report to the governor. On arriving back at headquarters, the team would then begin drafting its report; and a report on a medium to large prison would often contain 2 or 3 major recommendations and from 50 to 100 minor ones.

Following the inspection, the team was debriefed and recommendations sorted into four groups, to be forwarded to different persons: the governor, the area manager, the Director General, and the Home Secretary. One team member compiled a draft report, subsequently modified in internal meetings. Once the Chief Inspector agreed the final draft, the manuscript was transmitted to the Home Secretary and the report simultaneously processed for printing.[2] However, it was not distributed until the Home Secretary's response was available for inclusion. The principle behind this seems to be that it would be a major obstacle to the Home Secretary placing pressure on the Chief Inspector to revise his report. In practice the Home Secretary's comments in response usually ran to one page or less, and simply indicated what actions had been taken, or were planned, or could not be taken (usually for budgetary reasons). One assumes that the Home Secretary's comments are produced after consultation with the Prison Department. But whatever the reason, the fact of the matter is that many reports are issued up to a year—in some cases over a year—

---

[2] At the time of my fieldwork in 1988, 2 versions of the report were prepared, one containing an appendix on security issues, distributed within government; the other, with this appendix excised, which was published. Shortly after Judge Tumim became Chief Inspector this practice ceased and the security appendix was included in the published version.

after the inspection. One additional factor, though it is probably not too significant given the delays in the system, is that in past years Chief Inspectors seem to have taken the view that releasing several reports simultaneously would heighten the public profile of the Inspectorate.[3]

Two other strategies coexist with this rolling programme of inspection and publication. First, the Chief Inspector will not usually remain with the team for the whole week. He will often conduct smaller, one-day or half-day, visits to establishments near the one in which the team is working, and some of these will be completely unannounced. For some years, the reports on such visits were not published. However, in the last year or so such visits have been the subject of published reports. Second, the inspectorate conducts system-wide reviews of particular problems areas, and these 'thematic reviews' have come to be as important as the large-scale institutional inspections. Past reports have discussed matters as varied as prison suicides, sanitation, security categorization and inmate grievance procedures. Some have been extremely critical of prison department processes, and have, like the review of grievance procedures published in 1987, resulted in Departmental Working Parties and ultimately in major procedural changes.

The French inspectorate divides its work into five categories. 'Missions d'enquête' are inquiries following escapes or incidents. 'Missions de contrôle' are usually routine establishment inspections, which usually involve only one inspector operating in a prison for one or two days. There are about 70 such inspections each year. Although the term 'mission d'enquête' is used to designate inspections following major disturbances, around one third of the control inspections are in connection with incidents (Ministère de la Justice 1986; 1987).

These two activities comprise the majority of the inspectorate's

---

[3] The actual method of publication is usual for many types of government papers, and documents by pressure groups and academics. In essence they are sent to persons or organizations which have standing orders for them, or who ask directly for them (the Inspectorate and the Home Office Library both provide copies). However, they are not available in bookshops; are not published by HMSO and so are not in its catalogue; and carry no 'Command number', since they are not (unlike the annual reports) presented to Parliament. None the less they are usually summarized in the quality press and reported in detail in the Howard Journal, an academic journal dealing mainly with penal affairs. The annual reports, by contrast, are required to be laid before Parliament, carry a command number, are included in HMSO catalogues, and can be purchased from government bookstores.

work. The other three types of inspection are described as 'missions d'observation', 'missions d'étude', and 'missions du milieu ouvert'. The 'missions d'observation' are principally reviews of institutional security. 'Missions d'étude' are studies of particular system-wide functions, usually related to policy reviews or organizational restructuring. 'Missions du milieu ouvert' relate to the prison administration's responsibilities for probation.

Unlike the English system, French arrangements for inspection give a large role to other state inspectorates concerned with matters of broader social-policy import.[4] The Inspection Générale des Affaires Sociales, with responsibilities for health and sanitation, conducts prison inspections just as it inspects other facilities, and it has studied matters such as medical facilities and the issue of prescriptions in prisons.

In the Netherlands, the inspector spends about half his time in the establishments, yet does not conduct inspections as such. He is mainly involved in the investigation or review of issues such as staff malpractice, policy changes or renovation and development plans for individual prisons.[5] His visits to establishments are therefore normally concerned only with the matter that demands attention at that time. Some of these issues can be quite technical—for example, complaints about the level of airborne asbestos. In such cases, the inspector can draw on other government departments for expertise. In matters concerning buildings (including this particular example of asbestos), he can call on the state building agency, the Rijksgebouwdienst, to assist him. Although this may result in the Rijksgebouwdienst, or other agencies, having to evaluate their own past performance, the inspector asserted in my interview with him that this caused no problems; such agencies are large and compartmentalized, with different sections routinely expected to check on the work of others. In addition, one area which is often regarded as the most difficult to inspect, medicine, falls within the remit of

---

[4] In England, other inspectorates, such as the Health and Safety Executive, are able to visit prisons but usually only on the invitation of the governor. Their statutory authority to require changes or to prosecute breaches of health, safety, or other legislation would not normally extend to prisons just as they would not extend to other Crown properties. The major exception is currently in the area of kitchen hygiene. However, the Chief Inspector of Prisons liaises with other inspectorates in order to tap their technical expertise.

[5] The design and building of the the prison estate is, however, ultimately the responsibility of the state building agency.

another Ministry of Justice inspectorate, which has competence to deal not only with prisons but also the state-run TBS mental institutions (though not the privately-operated TBS establishments).[6]

One point common to all the inspectorates was their reliance on collecting impressions and information at first hand. Although prison managements have been expanding their demands on establishments for plans, evaluations, and statistics, these documentary and statistical sources were never seen as having more than ephemeral importance. All the inspectors concurred that inspection must rely primarily upon physically visiting establishments, talking to staff and inmates, and listening to views and complaints. This clearly has implications for proposals to introduce explicit regime standards, and these are discussed below.

Effectiveness: the Case of the English Inspectorate

The previous sections have made clear that inspectorates *per se* exist in only three of the four countries I am considering, and inspections as such are regularly made in only two—England and France. Since the French inspectorate, like the Dutch, is an in-house body, and few details of its work are published, any consideration of the effectiveness of the inspection process is driven back to consider, first and foremost, the English case.

The comments that follow from each establishment inspection are largely to do with physical conditions and regimes. The inspectorate does not pull its punches, and many such comments are damning. A 1992 report of Canterbury prison described it as 'dirty, cramped, dispiriting and inadequate', while the two unused wings at Parkhurst, which had been left empty since a 1979 riot, were described as a 'scandal' in terms of wasted resources (HM Chief Inspector of Prisons 1992a; 1992b). A 1990 report on Leeds referred to 'completely inadequate resources' for some regime functions, while the report on Brixton stated that it was 'corrupting and depressing' and that inmates faced an 'austere, miserable and wholly negative experience' (HM Chief Inspector of Prisons 1990a; 1990b). Although inspectors found some praiseworthy matters in

---

[6] The TBS institutions are custodial facilities for mentally abnormal offenders, though the definition of mental abnormality used in considering allocation there is lower than one finds in England. These institutions were formerly designated by the abbreviation 'TBR'.

some institutions, phrases such as 'wholly unacceptable' are repeatedly applied to a wide range of provisions.

Such comments, in both the establishment and thematic reports,
may broadly be grouped into two kinds of criticisms. First, there
are criticisms that either specific administrative procedures or generally accepted standards of good management are not being
applied. Second, there are recommendations that particular facilities—libraries, medical facilities, and so forth—be introduced or
upgraded in specific prisons, or that procedures should be
redesigned to operate more effectively (for example, the procedures
to identify potential suicides). What the inspectorate has *not* so far
done, at least in its own reports, is to argue that prisoner entitlements to facilities should be recognized and improved. In so far as
the Chief Inspector has supported such argument, it is in the context of Part II of the Woolf Report, which he co-authored.

The Report was commissioned to investigate the riot, in April
1990, at Strangeways prison, Manchester, and the subsequent wave
of major disturbances in the English prison system. Part II deals
with issues above and beyond the riots themselves. It takes the
view that inmate dissatisfaction with a wide range of regime matters was a predisposing factor in the riots, and conducts a thorough review of regimes. Its recommendations—many of which are
discussed elsewhere in this book—will, if implemented, radically
change the nature of English prisons. Finally, the Report comes
down in favour of a code of (non-enforceable) minimum regime
standards, arguing that they will help identify regime aims and the
resources necessary to meet them.

If the inspectorate is, as indicated above, very critical of the
prison system, what has resulted from its criticisms? First, and
importantly, the inspectorate has unequivocally established a reputation for independence. The point of instituting an independent
inspectorate was to bring prison issues more fully into the public
gaze and to provide an authoritative assessment of prison conditions. The inspectorate has fulfilled these roles rather well. Its
reports form a catalogue of chronic problems affecting the prison
system, and, with the passage of time, the repetition of some criticisms must have proved an embarrassment to the Home Secretary.[7]

However, the inspectorate has no executive role. In the terms of

---

[7] See e.g. the Chief Inspector's annual report for 1988 (1989b) and the report on
Risley Remand Centre (HM Chief Inspector of Prisons 1989c).

the discussion in Chapter 7, and Table 7.1, its relationship with the Prison Department sits somewhere between symbolic accountability and 'game 1', where the prisons co-operate with the Inspectorate's requests but would not offer information not asked for, yet the extent of their co-operation is sufficient to justify the claim that accountability for prison conditions does exist. The relationship between the Inspectorate and the Home Office, on the other hand, sits somewhere between symbolic accountability and 'game 2', namely that the accountability of prisons for their conditions is predicated on the flow of *information* rather than any action which follows from Inspectorate reports.

It is for the Home Secretary to deal with the problems or to institute the solutions outlined in the reports, and on this level, results been more patchy. It is true that certain reports have gener-ated policy changes, as indicated with the report on grievance pro-cedures discussed above. Yet the Home Secretary's response to many problems has been dismissive. A criticism of laundry facilities in one prison drew the response that the governor should bid for departmental monies for washing machines (report on Canterbury, 1992a). A recommendation that a young-offender institution should be retained, despite its small size, was prefigured by a statement that it would anyway be closed (report on Eastwood Park, 1991; the statement on closure was made in December 1990). Recommendations on broadening the range of food available to inmates were accepted in principle but implementation would be delayed (report on Leeds, 1990a). The particularly damning 1990b report on Brixton simply drew the comment that conditions would improve as new London prisons came on-stream and overcrowding reduced. The report on Haslar (1990c) expressed concern about its use as a holding centre for detained immigrants; this was ignored.

Several reports have now indicated that problems identified in inspections some years previously have not been resolved. Under these circumstances, the Inspectorate seems to have adopted the pragmatic position that it must choose the issues on which it should make a stand, and try to push along those matters in which it detects a willingness on the part of the administration to change. The report on inmate grievance procedures is a case in point. There had been complaints for some years about the process from inmates, but that in itself would not have been enough to generate a willingness to change. However, the impending re-organization, with regional directorships being replaced by area managers, necessarily meant some re-drawing of the procedural aspects of the complaints system; the 'Fresh Start' programme had opened up a debate about how improvements in prison regimes could be produced; and the role and future of boards of visitors had been opened up for discussion by the controversy surrounding their disciplinary role. It was, therefore, the right time to consider this particular area, and the injection of the Inspectorate report at that time proved highly effective.

One final problem for the inspectorate is that of delay. This has arisen from two sources. First, it has at various times been pulled away from its programme of inspections to conduct other inquiries. This is understandable; inquiries into disturbances form an important aspect of its work. But the impact of such unexpected workloads has been considerable. Morgan (1985) notes that in 1983 the inspectorate spent more than six weeks in Northern Ireland investigating two prison escapes. The latter part of 1986 and the first four months of 1987 were almost wholly taken up with inquiries into the widespread rioting in English prisons in April and May 1986. A return to 'routine' work lasted until April 1990, when the English prison system again experienced a series of major riots. The Chief Inspector subsequently became involved in co-authoring the second part of the Woolf Report. Second, there has been delay in the production of the Home Secretary's responses to the reports, which has in turn delayed publication since the responses are inserted into the reports. The latter problem became so serious in 1990 that the Chief Inspector took the step of issuing a press release complaining about the delay.

A comparative assessment of the effectiveness of prison inspectors is, as I have indicated above, impossible to make. But what is clear is that the idea of 'effectiveness' differs substantially between England and the other three countries, principally because a major component in the English case is the concept of publicly available authoritative statements about the nature of the prison system and its shortcomings. The English inspectorate has unequivocally established its credibility on this score. A second issue in effectiveness would be that of agendas. Where inspectors are in-house advisors or administrators, their remit is largely a reflection of senior officials' preoccupations. They have little control over their own agenda. The English inspectorate, although it has had to take into account the policy interests of the prison department, has been free to construct its own agenda and has had limited success in pushing reforms forward.

Arguments For and Against Standards

The idea of standards is not new. The United Nations Standard Minimum Rules for the Treatment of Prisoners refer, in their Introduction, to a set of rules endorsed by the League of Nations in 1934. The UNSMRs themselves were adopted by the First UN Congress on the Prevention of Crime and the Treatment of Offenders in 1955, and have since been amended several times (see United Nations 1984). They were intended to 'set out what is generally accepted as being good principle and practice' (para. 1), though they were robbed of detail by the necessity to bear in mind the diversity of economic, cultural and climatic conditions worldwide. None the less, they firmly establish the idea that minimum standards have a part to play in creating acceptable prison conditions and inmate treatment programmes.

More forceful arguments for the necessity for standards emerged in England and America in the 1980s, and the arguments about standards have almost entirely been an Anglo-American affair; there is as yet little discussion of standards in continental Europe. The English debate began as the result of, as Gostin and Staunton put it,

a consensus among official bodies, prison employees and prison reform organizations that prison establishments are overcrowded and fall below acceptable physical standards. If the prison system is to avoid the charge that physical and mental degradation has become a *de facto* goal of the system, and one that is routinely achieved, certain basic levels of dignity, privacy and diversity of experience must be built into that system . . . The consequence of a lack of such standards, or a failure to implement them, is quite simply that prisoners will be subjected to the risk of physical and mental harm. (Gostin and Staunton 1985: 91)

The key argument was therefore that standards would provide an objective (or at least agreed) means of determining whether or not prison conditions are minimally adequate.

Yet minimum standards remain a controversial topic, not least among prison inspectors themselves. Among the inspectors I interviewed in all four countries, there was scepticism about the value of explicit regime standards. They were perceived as an attempt to quantify quality. Although basic levels of provision (one inmate per cell, integral sanitation and so on) might be specifiable, many important regime factors were intangibles, matters that could be more easily observed or 'felt' than quantified. While many commentators have held out regime standards as a means of improving prison conditions, inspectors generally were not optimistic about the prospect of drafting workable standards, and claimed that their own expertise provided adequate if not quantifiable yardsticks. In short, they saw their professionalism as an alternative to, and as more sophisticated than, explicit regime standards. The only significant exception to this view is that both the current and previous Chief Inspectors of Prisons in England (Stephen Tumim and Sir James Hennessy) have been receptive to the idea, though they have not been able to carry their staff with them.

The primary objection that the inspectors made in my interviews with them was that exact and measurable equivalences could not be made between institutions, or, in some respects, within them. While an *equity* of treatment, based on professional judgement, could be achieved, an *equality* of treatment based on standards would be impractical. One illustration of this argument was that north-facing cells would get less sun than south-facing ones (assuming the windows were large enough to catch the sun anyway). With a single system of central heating throughout the cell block, some cells would necessarily be cooler than others. Such variations, within limits, had to be accepted as normal and natural. In terms of treatment, equity but not equality could be established in areas such as recreational facilities and workshops, access to which varied across establishments and across inmate groups.[8] And, as several inspectors pointed out, how could one set explicit standards covering matters such as staff conduct towards inmates?[9]

[8] However, arguments about equality versus equity are sometimes settled in favour of equality. In the Netherlands, plans to introduce televisions into cells had led to arguments about whether they should be made available on a rolling programme when the institutions had been wired for them, or only made available once all establishments had made provision for installation. The latter view ultimately prevailed.

[9] There are several answers to this. One lies in the UN code of professional conduct for law enforcement personnel. Another can be found in areas such as race relations training, codes of conduct prohibiting language derogatory of ethnicity, religion, and so forth. The inspectors I interviewed, while accepting such points, replied first that prison staff on the whole come from a working-class background

Faced with such objections, there was clearly a need to 'sell' the idea of standards to inspectors, administrators, or both.[10] Casale's (1984, 1985) 'sales pitch' is primarily that a code of standards would improve life not only for inmates, but for staff who work in prisons; and that the information systems which would need to be developed to monitor standards would also have some contribution to make to effective management. Dunbar (1985) provides a more detailed discussion of how standards might be given, first, an acceptable gloss, and second, a managemental usefulness. He suggests that one might change the terminology, and adopts the term 'benchmark' rather than standard. He then argues that while the administration requires performance indicators which would tell the administration how the system is carrying out its objectives, the inspectorate needs benchmarks which would help it determine 'whether the prison system is doing what it should' (1985: 75). The performance indicators and benchmarks would clearly be linked. Indicators imply the measurement of system functions in ways that enable benchmarks of acceptable performance to be set, while the benchmarks would be standards based on those system functions and objectives, and very likely would be monitored, in part, using the performance indicators. The implication is, of course, that the benchmarks must be practical; there is no point in setting them at unattainable levels, or setting benchmarks in areas where measurement is not possible.

The advantage of this kind of approach is that it argues for standards, not on the basis of minimum levels of provision for every inmate, but as a planning tool. As indicated in Chapter 6, governors on the whole regarded regime monitoring as potentially useful in formulating development plans and contracts with area

and share, prior to training at least, much the same attitudes as that class as a whole; and second, that the problem was not so much one of deliberate discrimination or insults as of officers being offhand with inmates or not taking complaints seriously because they were preoccupied with other concerns. See, however, the comments in Ch. 4 on arbitrariness in staff decision-making.

[10] It is worth noting that some of the inmates interviewed were sceptical about standards, largely because they perceived many of the problems that standards are supposed to address as emanating from their co-prisoners. As one English inmate pointed out, there is little point in having hygiene standards when many of the problems come about because inmates do not clean their cells properly. How true this is is a matter of conjecture, since the facilities for cleaning cells and maintaining personal hygiene are not always very good; if better facilities were available inmates may be motivated to make better use of them.

managers, but irrelevant in day-to-day management. Dunbar does not argue with this view. He is not trying to sell standards to governors, but to administrators, and to the latter, the argument is that performance indicators and linked benchmarks could assist in system-wide planning.

The disadvantage of Dunbar's sales pitch, however, is that talk of performance indicators and benchmarks moves away from the idea that certain levels of provision should be routinely available to every inmate. Instead, it enables measures to be set on a statistical basis—for example the percentage of inmates who have had access to particular facilities in a given time period. It is not, then, quite the same thing as a standard minimum provision or entitlement for each individual prisoner.

The Substance of Standards

Arguments for and against standards are one thing; the practical matter of drafting workable definitions of minimally acceptable conditions is quite another. How have such standards been drawn up? Casale's (1984) proposals simply drew on existing authorities, such as the European Standard Minimum Rules, UK Health and Safety Executive standards, UK legislation relating to shops, offices and factories, and the American Correctional Association standards (discussed below). They covered space and occupancy, lighting, heating, and ventilation, noise, cell contents, personal and general hygiene, food, exercise and recreation, and safety.

Casale makes a number of points, both general and specific, about standards in her text. First, her draft code is not comprehensive, but deals only with areas where a possibility of objective, quantifiable measures exist, and where relevant reference materials or statutory provisions can be drawn on. Second, she rightly indicates that any code of standards is only going to be as good as the mechanisms which enable it to be followed, and these include not only monitoring arrangements but also the legal standing of the code itself. She distinguishes between *minimum* standards and *norms*; and between 'desirable guidelines for judging past and present performance and shaping future prison management', 'minimum standards based on a natural law concept of the rights of the prisoner to certain basic conditions of life', and a code laid down in law as standards to which inmates are entitled as of right, where breaches could result in compensation. As she points out, the substantive content of all three might be identical; but the way in which they might be used, and their practical effects on prisons, would undoubtedly be rather different. And third, she indicates that because they are intended to be minimum standards applicable to all detention situations, no special standards are created for particular groups of inmates—women, youths—where the desirable level of provision may in fact be higher than her proposal for a minimum (as indeed it may be for all inmates).

Casale based much of her work on the American Correctional Association standards, and these are worth brief discussion. In America, the creation of standards was one response to the increasing numbers of court cases alleging breaches of American constitutional rights. The ACA, through its Commission on Accreditation for Corrections, is the main organization in the field.[11] It currently publishes nineteen sets of standards, some for correctional facilities and others for programmes such as food service and health care. The ACA is a private rather than state-operated body, and obtains funds by selling accreditation. Institutions pay to submit to inspection and monitoring procedures, and if they meet the relevant standards they receive a certificate which is valid, subject to various conditions, for three years. The certificate confirms institutional compliance with the standards, and the time-limited period means that institutions must buy into a rolling programme of inspections. Prison systems have been keen to submit to accreditation and to pay for it, for the simple reason that it is widely regarded as a good defence against inmate lawsuits based on prison conditions. Moreover, some authorities—Casale (1984) cites the New York City Board of Correction—have adopted or adapted the standards and, in effect, given them the force of law by passing legislation requiring prison authorities to comply with them.

The ACA/CAC standards, while lengthy, are not necessarily very detailed. The current (3rd edition, 1990) *Standards for Adult Correctional Institutions* comprise 463 specific standards dealing with administration and management, physical plant, institutional operations and services, and inmate programmes. The 21 standards on institutional services, for example, are then subdivided into more specific topics: reception and orientation, classification, food

[11] However, a number of other US systems are in operation, including the US Department of Justice Federal Standards for Prisons and Jails (1980).

service, sanitation and hygiene, health care, social services, and release. Most, however, are rather basic. Standard 3–4274, dealing with reception and orientation, reads in full: 'There is a program for inmates during the reception period'. The explanatory comment accompanying the standard gives some idea of what such a programme might comprise, but this again is very broadly phrased. Even if read in conjunction with other standards, the specifications for reception and orientation simply comprise written statements and verbal briefings in the inmate's language as to institutional rules and routines, testing and interviews to pick up special problems, the availability of normal inmate services such as work, exercise, and religious services, and separation from the general inmate population during the reception process but up to a maximum period of four weeks. Many other standards simply require that written policy and procedure should provide instructions to staff on specific institutional processes, without dictating what the substance of those processes should be. Furthermore, many of the standards' requirements—such as that the reception process should be no longer than four weeks—do not appear to be based on any clearly articulated principles.

The standards that do contain specific and measurable requirements are those one would expect. A single management unit should be no more than 500 inmates (standard 3–4124). Each inmate should have 35 square feet of usable space, and 80 square feet if confined for more than 10 hours per day (standard 3–4128). Lighting in inmate rooms, at desk level, should be 20 foot-candles (standard 3–4139). Noise levels in inmate housing should not exceed 70dBA during daytime and 45dBA at night (standard 3–4143). Other standards, or their explanatory notes, give minimum permissible temperatures for dishwashing and maximum temperatures for cold food storage.

In short, while a great deal of thought has gone into the creation of the standards, and many of them undoubtedly do make demands that, if met, would benefit both staff and inmates, they inevitably leave large areas of inmate life uncharted, and do not address a number of concerns that inmates themselves might be expected to have. Standards on libraries do not require that all inmates have access to the library (other than a law library). Those dealing with recreational programmes refer to National Recreation and Park Association guidelines, and the explanatory comments describe the kinds of facilities that should be in place, but again simply state that there should be written procedures, trained personnel, and regular equipment maintenance. They are silent as to the substance of recreation programmes and the amount of time for which inmates should have access to them.

In England, meanwhile, the Home Office spent most of the 1980s dragging its feet on the standards issue. A call for minimum standards came in the Chief Inspector's annual report for 1982, and the government undertook to develop such a code. Following House of Commons Select Committee prodding, a Prison Department code of standards on new prison building was developed, though it was hardly a ground-breaking document.[12] *Inter alia*, it specified, for new prisons—several of which were being planned at the time—the intended size of most spaces within an establishment of any given type. Cells intended for single occupancy and with integral sanitation were to be 6.8m², yet the design clearly envisaged continued use of double-occupancy cells with integral sanitation in remand establishments; these were to be 9.2m². This suggested that there was no firm commitment to single cells in new establishments. They then specified matters such as the ratio of inmates to WCs, wash basins, showers and so forth in different parts of the establishment, ventilation, heating and lighting standards, and a number of ancillary matters. However there was no explanation of how these standards were drawn up and several diverge radically from the figures suggested by Casale (1984).[13]

Pressure is clearly on, however, for the department to come up with more comprehensive standards. The Woolf Report records six reasons presented by the department for its reticence: minimum standards are already provided in many areas via the new building standards, Prison Rules, Standing Orders and so forth (para.

---

[12] 'Current Recommended Standards for the Design of New Prison Establishments', HM Prison Service, 1984.

[13] e.g. the 'newbuild' figures were 200 lux (SI measure of illumination) for cell tables/desks and 50 lux for night use. Casale suggested illumination of 37.2 lm/ft² for reading and 0.01lm/ft² at night. Lumens per ft², also known as foot-candles, are the imperial measure; the conversion factor is 1 lux = 0.0929lm/ft². Translated into lux, Casale's standards would be roughly 400 lux for reading (as opposed to 200), and 1 lux as opposed to 50 for night-time in-cell illumination. The newbuild standards are thus clearly much less than Casale's for reading, but much greater for night lighting. It should in this connection, however, be noted that the newbuild figures relate to installed capacity, not actual levels of illumination used, and so night-time illumination in practice might not be as bright as the newbuild figures suggest.

12.102); no government would endorse a code unless it is also prepared to provide the resources for its standards to be attained; the range of subject matter is unclear; different standards would need to be drawn up for different kinds of establishment; improvements can be made without a code of standards; and the preparation of a code would require substantial resources (Home Office 1991, para. 12.104). Woolf disposed of the argument that minimum standards already exist by pointing out that they are neither codified nor comprehensive. Nor was he impressed by the other arguments, since he recommended (para. 12.108) that 'the Prison Service should prepare its own Code of standards which sets out what the Prison Service considers should be the standards which it should set itself to achieve'. In addition, he pointed out (para. 12.119) that such standards would anyway have to be developed, in view of the intention to privatize the management of selected remand institutions, so that contractors' compliance with their contracts could be monitored.

Woolf's vision of such a code is fairly detailed. It should be comprehensive, and 'sufficiently explicit as to be meaningful, but not so detailed as to be unrealistic or unhelpfully rigid' (para. 12.108). It would, in its first version at any rate, apply across the board to all persons in custody; and it would be little more that a set of desiderata. However, Woolf recommends that an accreditation system be set up to monitor progress towards the targets set by the code, probably administered by the Chief Inspector of Prisons. In addition, the standards would form part of the contract between area managers and governors, and, indeed, the contract between the prison service and inmates. Although they would not be treated as providing corresponding rights for inmates, the standards and the accreditation (or withdrawal of accreditation) could, even at this early stage, be introduced as evidence in inmate litigation, and might be enforceable by judicial review. The clear aim is that, once it has proven possible for the department to meet the minimum provisions contained in the code, there would be no reason for those standards not to be considered in law as coterminous with inmate rights.

One pertinent question about the standards debate is: why have the English and Americans taken this route, while in other European countries it has not been an issue? The answer seems to lie in the entrenched belief that the CPP in France, and the

StVollzG in Germany, already constitute codes of minimum standards. It is true that in many areas they do specify levels of provision. The CPP asserts that work should be available to fill a normal working day, and that inmates should be kept one to a cell, save for temporary arrangements in view of overcrowding or during the day for reasons connected with work (D84). Yet in practice both provisions are routinely ignored. Not enough work is available, and 'temporary' doubling of cells has existed for many years. The gap between legal provision and actual practice is probably not so wide in most German prison systems, but the key point is the same. The CPP and the StVollzG are the product of a view of law as an abstract specification and regulation of relationships. But in practice they represent, in Casale's terms, little more than 'desirable guidelines' which, in the French case, stand little real chance of being achieved in the near future. As David and Brierly (1985: 83) point out,

An effective public law calls for a highly developed civic sense in both administrators and in the public at large. It is not, in short, workable unless the feeling is widely shared that all government bodies must bow to judicial control. It supposes, in addition, that government officials view the people as citizens rather than as subjects. Past experience has demonstrated how great the difficulties can be when an effort is made to prompt the state to carry through with some act of elementary justice or to renounce some unreasonable project.

Conclusion

The title of 'inspectorate' does not do full justice to the range of work that such bodies undertake. In France and the Netherlands, where they are offices within the administration, they are also advisers to the head of the service and deal with many different kinds of organizational issues. At the same time, however, their work is closely linked to political and managerial exigencies. The independent inspectorate in England came about as a solution to a particular political problem, the need to provide authoritative information about prison conditions to a critical audience made up of members of parliament, active pressure groups, and the media. It therefore 'just happened' that linkages can be seen between models of law (and thus of specifications of prison conditions and inmate rights) and the particular form that the inspectorates take. In England, where the common-law approach has not emphasized inmate rights, the inspectorate can at least be seen as policing prison conditions. In the other countries, where the law specifies the nature of inmate rights a little more clearly, the general view seems to be that inmates can pursue their own complaints about conditions, and inspectorates are thus divorced from the need to conduct regular and stringent reviews of conditions and inmate treatment. The inspectorates are thus more or less technocratic advisers to the head of the administration.

The English inspectorate has developed its public role with great vigour. Its broad remit, separation from the prison administration, and publication of reports, allow it to create its own policy agenda. In recent years this can be seen in its thematic reports and in the Chief Inspector's contribution to the Woolf Report. However, it has no executive role, and can only point to problems, not solve them. In order to prompt action, it must retain a sense of political realism. Thus far it has been most effective in proposing solutions to problems already recognized within the administration, and where there has been a political willingness for change.

So far as standards are concerned, it seems that the most pressure for their introduction has been in the common-law countries (England and America) rather than countries influenced by the Romano-Germanic legal tradition. The most likely explanation is that the concept of a penal code in the Romano-Germanic structure includes the specification of relationships between inmates and the administration which bears some semblance to prison standards. None the less, as I have argued, it is the weakest rather than the strongest version of standards, desiderata which may in principle form the basis of inmate litigation but which in practice may turn out not to be explicit rights. The situation in England, meanwhile, has been one of growing pressure for standards which could be effectively monitored and enforced, against a history of a foot-dragging prison department. How quickly standards will be developed now that Woolf has come down in their favour has yet to be seen.

10

Good Order and Discipline

PRISONS are largely about control; and the most overt forms of control over inmates are disciplinary sanctions, and measures such as segregation which may be applied for managerial purposes. Formal punishments, and 'special' measures such as administrative segregation, are by no means always severe, but at their harshest they can be very oppressive. Their application does not usually follow procedures as full as those of courts, rules concerning evidence are not so precise, and there is sometimes more latitude allowed in the attribution of guilt. The nature of prison discipline is occasionally held to be a matter of 'preserving good order', so that unfairness on the basis of individual treatment may be justified by reference to wider inmate-control issues. The ramifications of disciplinary processes include their influence on remission, parole, security categorization, and location. In addition, as Chapters 4 and 5 made clear, more informal disciplinary processes may exist.

This chapter addresses three issues. First, given the emphasis on control in prisons, how important are formal disciplinary arrangements? Second, what kinds of interrelationships are there between disciplinary procedures and other control measures? Third, since formal disciplinary procedures, in principle, protect inmates from arbitrary punishment, to what extent can they be considered as a mechanism of accountability? The first half of the chapter describes formal disciplinary rules, adjudicatory processes, and punishments. Since some infractions of discipline are also breaches of criminal law, the extent to which the courts become involved in prison discipline is also briefly discussed. In addition, the idea of an 'informal' structure of discipline is considered, along with control-related administrative decisions. Only after these discussions is the material drawn together into a consideration of the questions listed above.

Disciplinary Rules and processes

The specification that only certain acts, or classes of acts, are punishable provides in principle a measure of protection for inmates against the abuse of power and arbitrary use of sanctions. In most jurisdictions, prison rules pertaining to discipline either specify the duties of inmates, with failure to fulfil the duties being punishable; or proscribe particular behaviours and set out the sanctions applicable to inmates who engage in them. However, the degree to which these rules do provide effective protection for inmates against the arbitrary use of disciplinary powers is, in every system, somewhat compromised by requirements to obey direct orders given by staff[1], and 'etcetera clauses' prohibiting actions likely to endanger good order and discipline and not explicitly prohibited elsewhere in the rules. Between them, these two rules provide for a considerable degree of control over the inmate population, which is tempered formally by review procedures intended to detect and quash unfair decisions, and informally by factors such as officers' 'face'—that is, the negative reputation officers can acquire if they report inmate infractions too readily.

Bearing in mind this duality of formal protection from arbitrary control and the operational need to maintain flexibility in the use of discipline, let us look in more detail at the disciplinary rules in the four countries.

In England, the list of disciplinary offences is collected together under Prison Rule 47. In the other three countries, statements as to what actions constitute breaches of discipline are scattered through a number of clauses in the penal codes and prison rules. This in itself might alert us to the relative importance of discipline in England. None the less, the lists of offences in all four countries cover the same general ground. There are requirements that inmates obey officers and obey the 'house rules' specific to the particular establishment, and prohibitions of disorderly, noisy or uncontrolled behaviour.[2] Convicted prisoners, who are required to

[1] Such orders must, however, be lawful.

[2] For these regulations, see the following: for England, Prison Rule 47; for France, the CPP, Troisième Partie, Arts. D241, D242, D243, D244, and D245; for the Netherlands, Beginselenwet Gevangeniswezen 32 and 45, Gevangenismaatregel (GM) 28, 49(1–2) and 100(1–5), and Huishoudelijk Reglement (HR) iii 1; for Germany, the StVollZG, ss. 37, 41, 81, and 82. This does not, however, constitute a complete list of regulations pertaining to discipline in any country.

work, must work satisfactorily. There are some differences as to the items prisoners may possess and the services they can receive, with consequent variation as to what is unauthorized and may thus form the basis of a disciplinary offence. However, most forms of barter or trade between inmates that is not expressly approved via a request to staff are defined as infractions. And, as mentioned above, all sets of rules prohibit actions injurious to good order and discipline not covered by more specific prohibitions.

While the acts defined as offences are largely identical, the disciplinary proceedings are not. The only common features are these. The 'hearing of first instance' is held before the prison governor or deputy.[3] Hearings usually take place only one or two days after the alleged offence, and do not normally take longer than a few minutes. In general, prison disciplinary matters can be seen as primarily 'dispositional' in the sense used by Flanagan (1982). That is, if a formal report of an offence is made, there tends to be pressure to impose a punishment; otherwise the implication is that the officer was wrong in filing the report, with adverse consequences for staff morale. But notwithstanding this orientation, there is a broad recognition, across England, France, and Germany, of some aspects of natural justice. Inmates receive a written statement of the charge prior to the hearing, and may (though many do not) present a defence or mitigation in response, either verbally or in writing.

Differences between the four countries exist in the physical arrangements for the hearing, its formal design, and the involvement of prison staff, social workers, and doctors. English and French prisons almost all possess specially-furnished adjudication rooms, often adjoining the punishment cells. In England the rooms are mostly provided with a chair and table for the inmate to sit and take notes of the proceedings, though the furniture is frequently bolted to the floor. In France, they mimic a courtroom, with inmates standing behind a rail or bar. Dutch prisons are not so equipped; the 'adjudication' usually takes place at the door of the inmate's cell.

---

[3] There are, however, special arrangements for hearings to be conducted by other senior officers in specified circumstances. In esp. France and Germany, these include arrangements for 'satellite' institutions headed by non-governor grades. Following Fresh Start in England, a unified grading system integrated chief officers into the governor grades; and governors were redesignated by function so that the deputy governor became the 'head of custody', who in practice would now normally conduct disciplinary hearings.

The structure of the hearings in England is partly inquisitorial and partly adversarial. It replicates the distinction between prosecuting council and prosecution witnesses, in that a senior officer lays the charge while the 'reporting officer', whose evidence forms the basis for the charge, is treated as a witness. The hearing comprises two sections. In the first, it is for the adjudicator to determine the facts of the matter and to decide upon guilt or innocence.[4] Evidence is formally presented, inmates may speak in their own defence, and they may also cross-question the reporting officer and call witnesses. If the inmate is found guilty, the second part of the hearing comprises evidence in mitigation and a determination of the punishment.[5] In Germany and France the procedure is purely inquisitorial, inmates have fewer formal rights to call or question witnesses, and indeed the member of staff who reported the infraction may not even be present. The procedural aspects of the hearings are relatively simple, with no formal separation between the consideration of guilt and mitigation. However, in France the hearings must, and in Germany they may, be attended by social workers who advise on the inmate's behaviour and personality, and may comment on the suitability of particular punishments.[6]

[4] Unlike the criminal procedure, there must be an investigation of the facts even if the inmate pleads guilty, ensuring that inmates who plead guilty because they are placed under pressure to do so can still be found not guilty. Whether this fully addresses Flanagan's point about the 'dispositional' characteristics of discipline remains arguable.

[5] Prior to 1992, rules also existed which required certain kinds of offences to be remitted to the Boards of Visitors for consideration. Prison Rules 51 and 52 provided that charges of escape, attempted escape, assault or gross personal violence (implying more serious harm) on an officer, gross personal violence to any person other than an officer, mutiny, and incitement to mutiny must all be heard by the boards of visitors. However, following the *Campbell and Fell* European Court case, discussed later, arrangements for 'strengthened' adjudication panels to hear particular kinds of case, and empowered to give greater punishments, were dropped. In addition to these requirements, the governor was able to refer other cases to the board at his discretion, under one of two heads: first, that the charge was serious or repeated, and second, that the governor felt that the punishment he could impose would be insufficient (R51(2)). The determination as to whether or not a case should proceed to the Board was made at an initial formal hearing by the governor.

[6] The other staff who are in attendance are: in France, the 'surveillant chef de détention' ('surveillant chef' is the highest uniformed grade; one of the surveillants chef is given special responsibility for the regime and is known as the 'chef de détention'), and sometimes, in maisons d'arrêt and centres de détention, a member of the social-educational staff; in Germany, sect. 106 of the Strafvollzugsgesetz requires that for particularly serious offences staff who have worked with the prisoner also be present, and that if the inmate is a pregnant or breast-feeding mother, the institutional doctor should also be present.

In England, a staff member who knows the inmate will usually be asked to make a statement, which is formally treated as a plea of mitigation. Although some such statements indicate that the inmate was under stress, or that the infraction was out of character, many appear to state simply that 'this prisoner keeps his cell tidy'—which may be taken to mean 'I don't have any strong impressions about this inmate'.[7] Inmate statements, which are formally treated as either presenting a defence or mitigation, are rather more interesting. Some simply state, for example, 'I don't know what came over me' or 'I have problems, I am aggressive, I am seeing the psychiatrist'. Some are juvenile attempts at a defence, such as 'I didn't say that to the officer, I said it to myself' (inmate charged with calling an officer a 'fucking arsehole'). But a large proportion of the written statements are restatements of grievances which inmates allege led to the offence, while not a few express straightforward opposition to authority. Some illustrations of complaints are:

I have already stated on my last notice of report my refusal to return to work until I'm paid the £2.30 which I believe is owed to me from when I was working in the plastics shop. Sir I look to you today to clear this matter up for me once and for all, not leaving any doubt in our minds as to my belief in saying why I should work a week, spend a few days in the hospital here, come out and find I've lost £2.30 earnings for being sick. I wish to apologise for any trouble or inconvenience I may have caused by my refusal to work until I'm paid what I believe I'm owed for work already done. (charged with refusing an order to work)

I entered the barricade of my own free will to express the way I felt against the regime. Remand prisoners get treated worse than convicted prisoners. Why can't we get medical treatment until we are convicted? Why is exercise compulsory for remands? Just a few of the points I get stroppy about. I couldn't give a toss about the punishment but we took the governor at his word and hope he sticks by it and looks into our requests. (charged under Rule 47(20): 'in any way offends against good order and discipline')

Of the more combative responses, the following are typical:

---

[7] My data, on which this comment is based and from which the examples of inmate statements are taken, comprise a sample of 200 adjudication forms for adjudications in 1986–8; half came from a training prison and half from a local prison. The inmates may write a statement on the back of the charge form or make it orally at the hearing. Since my data were collected from the completed records of adjudications, they cover both written and oral statements.

I never got three warnings I got one. I treat this nicking with the contempt it deserves. Next week you'll be nicking people who don't shit once a week. The YPs [young prisoners] were talking but the screw blanked them out and walked over to me. Why did he nick me when other staff were nearer? And why is his name not on the white sheet? (charged with making noises during a chapel service—officer's name apparently omitted from the charge sheet by clerical oversight)

Governor, how many times do I have to tell you I don't take orders from no one not even you!!! Let alone your wanky pals. NOT GUILTY. (charged with disobeying an order)

Dutch disciplinary report procedures are similar to those elsewhere. There is a standard form on which to report the alleged offence, including a two- or three-paragraph description of the incident. This is passed to the governor, who must—except in specified circumstances—deal with the case within twenty-four hours. The hearing, however, differs markedly from the other three countries'. Since the governor or assistant governor will often conduct the adjudication at the door of the inmate's cell, the process can clearly be rather informal and the likelihood of third parties (officers, social work staff, inmate witnesses etc.) being called is small.[8] If the governor imposes a punishment, the reasons for it are entered onto the standard form, though they may be no more than formulaic. In one case (provided for me by a Dutch colleague) the offence was: 'You allowed your bird to fly in the block, then insulted officers, then refused to come to the observation office with the officer'. The reason for the punishment (seven days confined to cell) was: 'These endangered the good order of the establishment'.

Punishments

Table 10.1 lists the maximum sanctions available to governors in the four countries. In essence, France, Germany and the Netherlands provide for confinement to cell, or to a punishment cell, as the most severe penalty; the maximum periods are forty-five days, four weeks and two weeks respectively. In England, while the power to confine an inmate in a punishment cell is limited to three days, governors may, unlike their continental counterparts, order

[8] I am indebted to Max Kommer, at the WODC, Netherlands Ministry of Justice, for this account.

TABLE 10.1. *Maximum Penalties Available for Disciplinary Infractions, England, France, Germany, and the Netherlands*

England (source: Prison Rules 50, 51, 52)
1. Caution.
2. Forfeiture or postponement of privileges in Rule 4 (e.g. purchase of items in prison), maximum 28 days.
3. Exclusion from associated work, maximum 14 days.
4. Stoppage of earnings, maximum 28 days.
5. Cellular confinement, maximum 3 days.
6. Forfeiture of remission, maximum 28 days.
7. (Unconvicted inmates only) Forfeiture for any period of the right under Rule 41(1) to have articles (e.g. books, newspapers, writing materials).
8. (Unconvicted inmates only) Forfeiture for any period of the right to wear own clothing under Rule 20(1) if guilty of escape or attempted escape.

France (source: CPP D250)
1. Warning.
2. Declassification at work (if offence committed there).
3. Removal of privilege to buy beer, cider, toiletries, etc., to receive items from outside prison, or other privileges, no maximum limit.
4. Removal of privilege of use of individual radio.
5. Removal of privilege of open visits (if offence committed there).
6. Punishment cell, under conditions of Art. D167–9, maximum 45 days.
Payment for damage to materials may also be demanded but is not a sanction.

Germany (source: StVollZG s103(1))
1. Warning.
2. Limitation or deprivation of prison monies and purchases, maximum 3 months.
3. Limitation or deprivation of reading matter, maximum 2 weeks; of radio or TV, maximum 3 months; of both together, maximum 2 weeks.
4. Limitation or deprivation of free time activity or group events, maximum 3 months.
5. Separate accommodation during free time, maximum 4 weeks.
6. Deprivation of daily leave (if applicable), maximum 1 week.
7. Deprivation of work or activity, maximum 4 weeks.
8. Limitation of communication (except urgent matters), maximum 3 months.
9. Punishment cell, maximum 4 weeks.

Netherlands (source: Ministerie van Justitie, n.d.)
1. Confinement to isolation cell, maximum 2 weeks.
2. Confinement to own or other cell, maximum 2 weeks (may be a confinement only for certain hours of the day).
3. Reprimand.

that remission of the sentence (which is automatically granted at the beginning of a sentence) be removed, up to a maximum of twenty-eight days on any one offence.[9]

[9] In England it is accepted that while the court sets the maximum period of imprisonment, the executive may allow early release for good conduct. Remission is automatically granted at the beginning of the sentence and most inmates need serve

There are differences between prison systems in the collection of data, and these make detailed comparisons of punishments actually imposed rather difficult. England has historically placed much emphasis on collecting and publishing disciplinary statistics. German Land systems also collect extensive data, which is required by the federal Ministry of Justice. France began to collect rudimentary data in 1985, and it now appears in annual departmental reports. In the Netherlands, no information is routinely collected centrally and only samples drawn from particular institutions for specific research purposes are available.

Tables 10.2, 10.3, and 10.4 deal respectively with the levels of punishments, their frequency, and the nature of offences. As indicated above, their differing provenance suggests a need for caution in making comparisons. Even so, they still lend support to the following propositions.

Although sanctioning rates clearly differ across prison systems, the variation within systems is very large. Much of this variation is related to different inmate populations, though it is difficult to say whether it is caused by the population makeup, the disciplinary approach, or both. In general, the most frequent use of disciplinary measures is in female and youth establishments, and against young adult males.[10] In England, remand prisons show higher rates of

only one half of the court-ordered sentence, unless a period of remission is withdrawn as a disciplinary punishment. A 21-day withdrawal of remission is in effect similar to a court-ordered sentence of 42 days (on which half remission would be given). A system of parole exists alongside the rules on remission of sentence, though it is not directly affected by decisions on remission. In France, any matter affecting remission is dealt with by the JAP. Remission of up to a quarter of the sentence is possible, and supplementary remission may be given in certain cases, e.g. on the successful completion of education courses. Parole is possible after half the sentence in most cases, determined by the JAP or the ministry, depending on the original sentence length. In Germany remission does not exist, though parole is possible after a third of the sentence, provided a minimum of 2 months is served. In special cases, parole may be limited to half of the sentence. In the Netherlands the parole system was changed in 1986, so that it was granted automatically unless a specific decision was taken to the contrary—turning the procedure in effect from one of parole into one of remission. While governors may make recommendations to block the grant of parole, they have no authority to change release dates as a disciplinary measure. Parole is possible after two-thirds of the sentence, provided 6 months has been served.

[10] Flanagan (1982) reports some American data on this topic. He found that younger (under age 25) and single inmates were likely to be sanctioned more frequently and more severely than older and/or married offenders. However, in a further analysis of the kinds of violations involved, he concluded that the relationship between age and sanction severity was 'primarily a function of differential handling

TABLE 10.2. *Disciplinary Sanctions per 100 Inmates, Germany, by Land, 1986*

	Annual average population	Sanctions per 100 inmates	Annual average population (rank order)	Sanctioning rate (rank order)
Baden-Württemberg	7,406	71.0	3	6
Bayern	9,645	80.0	2	4
Berlin	3,527	31.4	6	9
Bremen	831	26.5	11	11
Hamburg	2,286	82.6	8	3
Hessen	4,683	30.5	5	10
Niedersachsen	5,398	50.9	4	7
Nordrhein-Westfalen	15,014	72.0	1	5
Rheinland-Pfalz	3,189	86.2	7	2
Saarland	940	90.0	10	1
Schleswig-Holstein	1,618	34.5	9	8
Total	54,537	64.8		

TABLE 10.3. *Illustrations of Sanctioning Rates from Different Institutions in the Netherlands*

Institution name and brief details	Source of data	Basis for calculation	Rate per 100 inmates per year
Bankenbosch (mainly short-term, zelfmelders), capacity 156	(EW)	551 incidents in 24 months (Apr. 80– Apr. 82)	156
HvB Den Haag, population 161	(MK)	31 reports in 43 days	163
HvB Breda, population 108	(MK)	15 reports in 18 days	281
Doetinchem hoogbouw (sentenced inmates, 14 days– 6 months), population 90	(JV)	92 reports in 207 days	179
Den Haag prison, population 116	(JV)	(a) 104 reports in 1986	90
		(b) 21 reports in 181 days in 1987	37
Het Schouw, population 116	(JV)	303 reports in 1986	261
De Singel, population 86	(JV)	66 reports in 181 days, 1987	155

Sources: (EW) Erkelens and Van der Worp (1983); (MK) Data collected by Max Kommer, WODC, the Netherlands (NB based on small samples). (JV) Data collected by Vagg.

of order violations and fight/staff violations' (1982: 230). See also Flanagan (1983) for a discussion of the predictive abilities of variables such as age, sex, race, current offence, and sentence length in relation to rule violation.

TABLE 10.4. *Types of Infraction and Punishment in 99 Cases, Het Schouw HvB, The Netherlands, 1986*

Type of infraction	Number of cases	Of which, number confined to cell	Maximum penalty actually imposed
Refusal to obey an order	33	13	Confined to cell 4+ days
Possession of prohibited item	18	10	Confined to cell 6 days
Aggression (to inmate or staff)	16	12	Punishment cell 7 days (NB 11 of the 12 confined in own cell—max. for 4+ days)
Lateness (to work, or returning from home leave etc.)	11	7	Confined to cell 4 days
Absence from proper place	6	2	Confined to cell 1 day
Verbal aggression or insults (to inmate or staff)	5	5	Confined to cell 8 days
Fighting	4	1	Confined to cell 4 days
Non-cooperation in cell search	1	1	Confined to cell 1 day
Attempted escape	1	1	Confined to cell 13 days
Drinking	1	1	Confined to cell 2 days
Damage to cell	1	1	Confined to cell 6 days
Creating disturbance	1	0	Excluded from recreation, 1 evening
(Unknown)	1	1	Confined to cell 1 day

† Available records show punishment served; that imposed could have been higher.

sanctioning than other adult male establishments. In France, too, disciplinary measures are more frequent in the maisons d'arrêt, with remand and short-sentence populations, than in long-term prisons. In the Netherlands, the highest rate is again in the remand facility, though the semi-open prison also uses disciplinary punishments frequently, most often (though the Tables do not illustrate this) for relatively petty drugs offences and abuses of home leave.

   In addition to general factors such as the nature of the population, particular institutional traditions and management objectives have an impact on the rate and levels of punishments. For one thing, different institutions have differential 'opportunity structures' for offending. In England, where most inmates do not have access to cookers, possession of a makeshift cooker (usually an old tin can with a candle inside it) was regarded as an offence. In other countries, small cookers were available either on the landings or in the cells. And in low-security prisons in all countries, drugs proved to be a significant problem. But an additional complication was the availability of the means of punishment. For example, at the Dutch semi-open prison I studied, offences involving drugs or violence would normally lead to a period of cellular confinement (strafcel) followed by a transfer to a closed prison. However, the establishment had only four punishment cells. As the assistant governor explained,

The prison culture is seven days strafcel for threatening personnel severely, five days for using heroin, three days for smoking a joint, sometimes one day for smoking a joint depending on circumstances. If it's a serious threat to personnel they go for two weeks to Maastricht or Veenhuizen [special segregation units]. If I want to punish for 14 days I will send people there—I can't afford the space. We don't give suspended disciplinary punishments as such. It depends on the use of the strafcel. If we need it and it is occupied we will conditionally release those in it. But this is not often used.

The second consequence of such offences, transfer to a closed establishment, also posed problems. The governor pointed out:

The only place they can go is Breda, the only problem is Breda is very occupied. We have to wait till a room is free. We may be able to make a three-way transfer; we send to Breda, Breda to Alkmaar, Alkmaar to here.

The problems posed by such transfers had led the governor to adopt the policy of giving longer punishments within the establishment but not seeking to transfer the inmates.

So far as management objectives are concerned, it is clear that governors can and do affect the level of use of punishment in their prisons. In one German establishment for young adults, for example, the governor had decided to change a disciplinary policy followed by his predecessor. If an inmate refused to work he would simply remain locked up for the day rather than being given a formal punishment. The idea was to rely on boredom rather than punishment as a disincentive. The result was almost immediate, since refusals to work constituted a large proportion of all offences. The rate of disciplinary hearings per 100 prisoners per year fell from 157 prior to his arrival (in 1979) to 55 in 1979, and to between 18 and 42 in all subsequent years.

Although the use of punishment is clearly affected by offending opportunities, institutional traditions, and management resources and objectives, it is not always clear what kinds of behaviour are most frequently punished, or what the threshold is at which a particular kind of behaviour comes to be seen as an offence. Data from England and the Netherlands can be used to illustrate this observation. In both cases, the most common offence was that of disobeying an order. This is a kind of 'etcetera' clause in the disciplinary rules, and probably reflects the nature of control more than it does the frequency of particular behaviours. Whether or not there is a specific rule prohibiting certain behaviour, a prisoner can be directly ordered to do (or desist from) something, and failure to comply is itself an offence. In addition, and despite the frequent use of Dutch prisons as models to be emulated in other countries, they recorded higher rates of aggression and fighting than the English prisons. This may, however, have resulted from a lower tolerance of aggressive behaviour, which would imply more, though less serious, charges. Dutch prisoners, when interviewed, suggested that institutional tolerance of aggression was very low.[11]

Comparisons of punishments actually imposed are problematic. The only clear conclusion is that the majority of punishments imposed were substantially less than the available maxima. None the less, in France, the punishment cell is used in over half of all cases; and the length of such punishments was usually between eight and fifteen days rather than the two or three often found in England.[12]

Along other dimensions, any comparison of punishments has to consider two factors. First, in England, loss of remission is often imposed alongside cellular confinement; this is not the case in any of the other countries, where matters concerning the time actually served are the province of magistrates, judges, or ministries. Second, even in comparing ostensibly similar punishments such as cellular confinement, there are clear differences between countries as to what such a punishment comprises. Cellular confinement is often served in a punishment cell in England, while in the Netherlands inmates are usually confined to their own cells. They

[11] Though see the example of the threat to throw a slop-bucket at a staff member, descr. in Ch. 4. This, however, happened in a cell and was not witnessed by other staff. This may be why it was not charged as an offence.

[12] German data on levels of punishment are, unfortunately, entirely lacking. The files do not appear to be collated into a statistical form.

thus retain personal possessions, including televisions.[13] It would initially appear that punishments are less stringent in the Netherlands than in England. Yet we should also take into account the *relative deprivation* that such punishments entail. In the Netherlands, the reduction in regime implied by cellular confinement is rather high, because inmates typically have access to substantial recreational and other facilities. English inmates typically have access to lower levels of provision, so that the proportionate level of regime deprivation is probably similar in both countries. If the deprivation involved in punishment is related to the regime, the key issue becomes the generally impoverished English prison regimes rather than the punishments themselves.

Scrutinizing Disciplinary Decisions

Routine scrutiny of governors' disciplinary decisions, in England and France, takes the form of requirements to submit reports to the regional or area administration. At the time of my fieldwork in England, all records of disciplinary hearings were forwarded to regional offices. They were scrutinized by an assistant regional director to ensure that no unsafe decisions had been taken. At that time there were some 72,000 records submitted each year. These were checked by only four assistant regional directors, who therefore were expected to deal with some 1,500 records each month. The amount of time spent on each case cannot have been large. Even with the change to fifteen areas, implying an average of some 400 reports per area per month, it is unlikely that more attention would be paid to them, because the areas have commensurately fewer resources.

In France, a similar procedure exists. The regional director's office receives notification of disciplinary punishments for the region. With nine regional directorates and around 42,000–43,000 cases each year, the possibility for detailed scrutiny is very likely greater, though no information is published on the numbers of punishments altered or quashed as a result. The JAPs, informed of disciplinary punishments because of their bearing on remission,

---

[13] It should be noted, however, that almost all televisions are rented from the establishment. Since inmates confined to cell cannot work and therefore earn no money, if the period of confinement to cell is sufficiently long they will be unable to keep up the rental payments and lose the privilege in this way.

have no power to alter punishments. Their role is limited to varying the 'punishment-remission' equation if they feel the punishment
was wrong, and notifying the regional director of cases they feel
are problematic. However, governors and JAPs I interviewed
claimed this was a rare event.

In Germany and the Netherlands there was little, if any, monitoring of disciplinary decisions. The openness of the complaints
mechanisms to grievances concerning discipline was held to justify
this lack. If or when problems seemed to exist with the disciplinary
process they would be investigated in the same way as problems
with any other area of prison management—that is, through an
internal committee or a request for inspectors to consider the matter.

Using the Courts

Certain acts, such as possession of drugs, constitute criminal
offences in most penal codes as well as disciplinary infractions
within the prison rules. The possibility therefore exists for inmates
to be prosecuted and, perhaps, given an additional sentence.
Indeed, none of the four countries regards the use of disciplinary
punishments and court proceedings as double jeopardy, since the
two procedures are held to have different aims.

Despite this, prosecution appears rather rare in practice, even
though the potential for its use remains high. Of the governor-
grades I interviewed, none had ever been involved in the prosecution of an inmate. Nor, apparently, is it easy to mount a
prosecution. Menard and Meurs (1984) illustrate this with a study
of inmate aggression in France. They start by outlining the legal
requirements that follow a serious incident. Article D280 of the
CPP requires any serious incident affecting discipline or security to
be reported to the regional director and the central administration,
and also to civil and procuratorial officials, the Prefet and the
Procureur. Article D40 requires any official who, in the exercise of
his functions, discovers a criminal offence, to report it to the
Procureur. Menard and Meurs note, however, that neither article
defines what would constitute a serious incident, while internal circulars dealing with incidents are mutually inconsistent. On the specific matter of inmate aggression, for example, they employ
different criteria in assessing whether an assault on staff is an inci-

dent in its own right if it is committed in the course of an escape attempt. And even if reported to the prosecuting authorities, not all such cases will be proceeded with. Of 49 aggressions committed in 1982 and studied by Menard and Meurs, only 27 resulted in further sentences of imprisonment.

Nor is it clear that incidents which could result in prosecution are always reported to the relevant authorities. The French prison staff I interviewed pointed to the lengthy sentences available to the courts for criminal offences committed in prison, in particular in relation to drugs. They described prosecution as virtually automatic in drugs cases. Yet none could remember any drugs case being prosecuted, despite the number of drugs possession cases known to them. It seems that the ambiguity in internal circulars as to what constitutes a serious incident allows matters to be dealt with internally, by way of disciplinary proceedings, even where they are also criminal offences.

Similarly, in the Netherlands, where the level of disciplinary punishment available is comparatively low, neither the ministry staff I spoke to (including officials in the legal section) nor the governors interviewed could recall a case ever having been prosecuted. In England, however, even though the absolute numbers were low, and even though most serious cases were, until 1992, heard by Boards of Visitors, the South West Regional Office (since replaced by several Areas) had 22 such prosecutions in 1985 and 27 in 1986 (the regional average daily population was about 7,500 against a national average of about 49,000 at that time). The country with the most severe internal sanctions also had most resort to the courts; yet even so, it seems highly probable that many more cases that could have been prosecuted were not.

The question remains, however; why are prison inmates so rarely prosecuted for crimes committed in prison? One American study (Eichenthal and Jacobs 1991) describes non-prosecution as a 'de facto policy' which may be explained inter alia, through the desire of prison officials to maintain the 'constructed reality' that correctional institutions are insulated from outside scrutiny; a lack of credible witnesses (inmate witnesses are likely to be hostile and in any case are discredited by virtue of their imprisonment); inadequate expertise among prosecutors in handling such cases; the necessity, within limited agency budgets, to make choices about which cases need prosecution most urgently; and not least, the fact that since the offender is already in jail, additional sanctions may be both limited and pointless.

My own discussions with prison officials indicated that the major problems were the length of time involved in the process, which resulted in large part from the need for prosecution to be sanctioned by and arranged through headquarters, investigated by the police or investigating magistrates, and fitted into court schedules. It is also true, as Eichenthal and Jacobs point out, that since the offender is already in prison anyway, prosecutors may see little reason to pursue an additional sentence, except in special cases such as major riots or homicide where an overriding public interest may be at stake. In short, while it may be appropriate for very serious offences, prosecution is simply irrelevant to the internal order of the prison, which is better assured through the use of disciplinary sanctions, perhaps combined with measures such as changes in security categorization or transfer to other establishments or units. This, of course, leads us into the question, considered below, of administrative measures which are 'punishments in effect'.

Punishment in Effect: the Use of Administrative Measures

Inmates can be punished for breaches of discipline. They can also, in addition or as an alternative, be transferred to another establishment or unit, removed from their workshop, given a higher security rating, or have certain privileges withdrawn. In some circumstances they may be placed in administrative, rather than punitive, segregation; although there is usually no difference between the two in terms of the cell they will occupy. Moreover, even if loss of remission is not a formal punishment, the fact of the punishment may be a factor in determining how much remission they will gain. Much the same could be said of home leave, open visits, and parole.

It has also occasionally been asserted that in addition to these disciplinary and administrative consequences of misbehaviour, the dependence of inmates on staff makes them vulnerable to an 'informal' disciplinary system, in which the 'punishments' are along the lines of lost mail, delayed permission for visits, delays in getting to the toilet, and the like. The issue of informal controls has been taken sufficiently seriously in England to warrant discussion in the Departmental Committee on the Prison Disciplinary System (the Prior Committee: Home Office 1985). Evidence presented to the Committee included the assertion that there existed an 'alternative disciplinary system', of which the main components were adminis-trative segregation, transfers, security re-categorization, and removal of privileges. It was argued that increasingly formal disci-plinary hearings, lengthy delays prior to hearings, or large propor-tions of technical acquittals would result in the greater use of this alternative system (Morgan *et al.* 1985; Prison Officers' Association 1984).

These arguments were made in the context of a reconsideration of the disciplinary role of Boards of Visitors—a role which the Committee wished to retain with modifications, but which was subsequently terminated as the result of further legal challenges. But they are propositions which have a general application, and the assumptions that underlie them are interesting. They presup-pose, first, that there is, and should be, a clear distinction between the formal and informal aspects of the treatment of inmates; and second, that there is, and should be, a clear division between disci-pline and other administrative action, because it is somehow easier to use administrative decisions as punishments than it is to invoke formal disciplinary procedures. It is these two propositions that the following paragraphs explore.

In all countries, it is thought necessary to provide the governor with a variety of control measures that can be implemented speed-ily. Several such measures are, in practice, available to governors; though in all jurisdictions a distinction is made between 'discipli-nary' and 'administrative' measures. Since both types of power may *de facto* result in similar outcomes so far as the inmate is con-cerned, the definition of certain powers as 'disciplinary' is largely symbolic, indicating that the measure is being imposed as a punish-ment (whereas other powers may be used for wider purposes). None the less, the powers are granted under different sections of the prison rules, and may be subject to different forms of over-sight. In addition, the stronger the governor's powers, of whatever kind, the greater is the need felt by the central administration to maintain oversight over their use. This can be done in two ways. One is to insist on a set of formal procedures which include safe-guards against the abuse of power, including, for example, the ability of inmates to complain about the decision; the other is to review governors' decisions at a higher level within the administration on a routine basis. This state of affairs can, in principle, result in a situation where oversight is less stringent over administrative rather than disciplinary measures, either because they are considered less severe, or because of the procedures employed to scrutinize them. The reverse situation, while possible, is less likely to exist, since one needs to find an occasion which can be described as a disciplinary infraction before punishment can be imposed.[14]

The fence between disciplinary and other matters is most rigorously sustained in England, with its relatively elaborate disciplinary hearings. The erection of this fence, however, was the cumulative result of inmate litigation coupled with the courts' willingness to address the issues they raised.[15] That willingness was initially lim-

[14] Some measures, e.g. transfer, require the consent of other governors, and regional or headquarters officials, and are less likely to be used as 'alternative' disciplinary sanctions unless the consent is automatically forthcoming—see the later discussion of 'lay-downs' in England, and n. 19.

[15] Much of this is documented elsewhere (e.g. Morgan *et al.* 1985; Zellick 1981; 1982). But a summary of the process would have to highlight the following legal decisions: in *St Germain* (1978), the Court of Appeal determined that the Divisional Court had the power to review boards of visitors' hearings where it was alleged that some procedural defect had occurred. The case was then heard by the Divisional Court (*St Germain No. 2*, 1979). Fitzgerald (1985: 34) notes that while the court accepted that the rules of natural justice applied, it 'declined to assert any general right to call witnesses whatever the circumstances, or to assert any general rule excluding hearsay. But it recognized that in this context fairness would generally require that a prisoner be allowed to call any witness with material evidence to give, and that administrative inconvenience alone could never be a reason for refusal'. The consequence was that inmates had a clear right to seek a judicial decision setting aside the punishment on the grounds that the disciplinary procedure had been conducted improperly. The practical impact of this was minor, since the punishment would have been served by the time the decision was made, and compensation was unlikely. But the cases marked a radical change in attitudes towards prison discipline. In several cases, evidential issues were raised, e.g. *McConkey* (1982) and *King* (1984). The former concerned evidence of 'participation' in drug-taking where cells were occupied by several individuals. The latter dealt with evidence of 'unauthorized possession' of drug-taking equipment, again in a cell occupied by more than one person. Other cases concentrated on the ability of inmate defendants to call witnesses, and indeed to have the fact that witnesses existed called to their attention. In *Tarrant* (1983), it was decided that inmates should be able to obtain legal representation at board of visitors' hearings. This was not intended to be a blanket right. The court stated that inmates should request legal representation from the boards of visitors, and that such requests should be granted according to guidelines set out in the judgement. These guidelines implied that a right did exist, because boards had to be very careful in refusing representation, and a refusal of itself might be the subject of a judicial review (see Morgan and Macfarlane 1985, and Quinn 1985). These cases, taken together, radically

ited to board of visitors' hearings, where the court-like nature of the hearings and the levels of punishment they could impose created some resemblance to magistrates' courts. European Commission and Court cases have taken a more relaxed line on this. In *Campbell and Fell* (1982), the Commission took the view that loss of remission in itself could not be regarded as equivalent to a further sentence of imprisonment (para. 125). It was the length of lost remission at stake—570 days for Campbell—rather than the principle of lost remission as a punishment that went beyond what the Commission could consider reasonable as a disciplinary measure.[16] In the 1976 case of *Kiss*, a loss of 80 days' remission (and a maximum penalty of 180 days for the offence involved) were not considered unreasonable.

The major 'alternative' to a disciplinary punishment in England would be the use of Rule 43, dealing with administrative segregation. Although inmates may request segregation under this rule it remains open to the governor to isolate a prisoner on the grounds of good order and discipline—normally described as 'Rule 43 GOAD'. There have been suggestions that governors use this power as an alternative to punishment, if, for example, they know the inmate committed the offence but cannot prove it. Yet safeguards do exist. Any use of Rule 43 requires that any isolation longer than twenty-four hours must be authorized by the board of visitors (for one month, renewable) or the Home Secretary (in practice, by the administration). Maguire and Vagg (1984) noted that authorization by the boards was often routinely forthcoming, though this has almost certainly changed in more recent years, due to internal discussions among the boards' memberships. The use of

changed the nature of boards of visitors' disciplinary hearings so far as procedural matters—evidence, witnesses, fairness, legal representation—were concerned, and set the stage for the demise of the boards of visitors' disciplinary powers. Yet the importance of this trend cannot be underestimated, in view of cases such as *Leech* and *Prevot*, in which the courts accepted that governors' hearings too could be subject to judicial review.

[16] The Court, however, was somewhat more conservative on this point. Its judgement (10/1982/52/85–86) of 28 June 1984 concluded in Campbell's case that the failure to comply with Article 6 of the European Convention on Human Rights existed only in relation to the fact that the board of visitors did not pronounce its conclusion publicly, and did not allow Campbell legal assistance prior to, or legal representation at, the disciplinary hearing (p. 60). The judgement does state, however, that the Court's view was influenced by developments in the procedures for board of visitor adjudications which had been implemented in the period between Campbell's adjudication and the European Court hearing.

Rule 43 segregation thus remains, on paper, an alternative to disciplinary action, but in practice it cannot be achieved by the governor alone.[17] In other cases, while inmates may complain to the Boards of Visitors, the Boards have no executive powers and would have to refer the issue to the governor for reconsideration, unlike the Dutch situation described below.

In France, the fence between disciplinary and other powers is equally rigorously drawn in principle. The CPP lists types of behaviour which may result in a disciplinary punishment. In addition it requires that punishments may only follow a hearing in which the governor is apprised of the circumstances of the infraction, and may only be used against individuals, and not as 'group punishments'. On the other hand, it defines isolation as an administrative act and not a punishment. Allocation to the punishment cell does not in itself require an inmate to be isolated, though administrative segregation may be ordered alongside the punishment (CPP D249, D250, and D171 respectively). However, the governor must report any use of isolation to the JAP, for discussion at the next meeting of the commission de l'application des peines, the prison doctor, and the regional office (CPP D170). Since these 'internal' controls are similar to those covering disciplinary offences, there is little reason to prefer administrative isolation over formal discipline, especially in view of the lack of strict procedural requirements at disciplinary hearings.

Where the distinction between discipline and punishment does seem to fade, however, is in the handling of remission. Any decision affecting the length of time served in prison falls within the province of the JAP. However, the regulations on remission require the JAPs to make decisions solely on the basis of the inmate's conduct (D250–1); they also provide that remission may be withheld if the inmate has not proven his good conduct (CPP 721). Remission is not calculated automatically, but granted on the basis of periodic reviews. It is therefore straightforward for the JAPs to base their decisions on remission on the number of days inmates spend in the punishment cell. Moreover, the legal framework would seem to encourage this approach. In my French fieldwork, different 'equations' appeared to operate in different establishments. But one

---

[17] See also the later discussion of 'lay-downs', which in England rely on segregation under Rule 43 being honoured by the governor of the prison to which the inmate in transferred.

example—the practice of a JAP in a large maison d'arrêt—was simply to deduct 1.5 days' remission for every day spent in a punishment cell. The distinction between disciplinary and judicial decisions is thus significantly weakened by legal provisions which not only enable, but implicitly require decisions on remission to be linked to decisions on disciplinary punishments.

In Germany, governors have few restrictions on their use of disciplinary powers, but have even more extensive discretion to use isolation for periods of up to three months (longer must be with the consent of the ministry). Moreover, the complaints procedures are the same in both cases; remedy lies with the courts, but only where the relatively permissive StVollzG is breached, and not where there is a misuse of administrative powers. Thus there may be little motivation to use administrative isolation as an alternative to discipline. The federal Ministry of Justice compiles annual rates for the use of disciplinary measures and special administrative measures of all kinds, of which administrative segregation is the most common. A comparison between the German Länder indicates that in fact, all four possibilities given by low or high rates of discipline and special measures exist in practice (Table 10.5). It is difficult, on this data, to sustain the proposal that there is any trade-off between the two areas. It is more likely that there are simply differing practices as to the level of use of both types of decision, perhaps conditioned by factors such as local or institutional traditions.[18]

Of all the four countries, the fence between disciplinary and other administrative measures seems weakest in the Netherlands. Disciplinary processes are informal, the use of isolation as a security measure is governed by similar rules to those covering its disciplinary use, and the maximum period in isolation is the same whether it is for disciplinary or administrative purposes. Isolation for longer periods is possible by asking the ministry to transfer the inmate either to one of the two institutions with special isolation units, or to the 'bunker', the highest-security section of the only high-security prison.[19] As one guard noted, it is possible for

[18] However, a recent study of the Bremen prison system has claimed that within individual states which have historically had low disciplinary rates, it is possible to see administrative segregation being used as an 'alternative disciplinary measure'; see Hoffman (1990), and the discussion of Hoffman's paper in Dünkel (1993).

[19] Since this chapter was written, the Dutch prison system has adopted further segregation procedures. The so-called 'EBI' extra-secure institutions comprise 8

TABLE 10.5. *Distribution of Disciplinary Sanctions and Administrative Security Measures, by Land, Germany, 1986*

Administrative sanctions per 100 prisoners per year	0–50	51–60	61–70	71–80	81+
Disciplinary measures per 100 inmates per year					
41+					HB d=82.59, s=84.08
31–40	BL d=31.41, s=44.66			NW d=72.02, s=40.26	
21–30	HM d=26.47 s=26.59		FRG average d=64.79 s=24.45		SR d=89.89, s=25.96 RP d=86.20, s=24.40
11–20	HS d=30.47, s=18.39 SH d=34.49, s=17.68			BW d=71.01 s=11.21	
0–10	NS d=50.89, s=4.61			BY d=80.04, s=3.35	

Key to Länder: BW Baden Württemberg; BY Bayern; BL Berlin; BM Bremen; HB Hamburg; HS Hessen; NS Neidersachsen; NW Nordrhein-Westfalen; RP Rheinland-Pfalz; SH Schleiswig-Holstein. d = disciplinary measures, s = administrative sanctions.
Source: Land Justice Ministry returns to Federal Justice Ministry, 1986.

inmates to be placed in the punishment cells not as a punishment but for 'observation', in which case the period can be indefinite; but in such cases the local practice is to lighten the conditions by allowing free access to the cell's individual airing yard. Since the complaints systems will hear complaints about any measures taken against inmates, disciplinary or otherwise, and including transfers for control purposes, a single means of remedy covers all possible measures.

units, with a combined total of 48 places, situated within existing prisons but administratively separate from them. In addition, 4 newly-built institutions each contain 2 6-inmate units for escape-risk prisoners, discussed in Kelk (1992). It remains unclear, however, how the opening of these units has affected the dynamics of the 'carousel' mentioned later in this Chapter.

There is, however, one problem. Complaints deal with identifiable decisions taken against individual inmates. Yet 'alternative' disciplinary measures need not be specific and identifiable decisions. As one inmate pointed out, prison regimes may be altered for reasons perceived by inmates as, in effect, a group punishment:

They say that you don't get a punishment for escaping . . . but in October '86 [after an inmate escaped] the regime gets tougher. There is a strict group-separation of side [i.e. wing] A and B. No walking around any more. In the evening-hours no sports. The groups have to use the telephone separately. One group in the cells, the others telephoning, washing, taking a shower. One group evening recreation the others not, so recreation is halved. These changes are probably due to the fact that the staff couldn't control it any more. The prisoners had too much influence. (inmate, Dutch maximum-security prison)

If the balance between administrative and disciplinary measures is such that governors have no particular incentive to bypass the formal disciplinary mechanisms, this leaves open the question of how often control measures are used for purposes other than breaches of discipline, but which are none the less *felt* as punishments. The obvious examples of such practices are 'lay-downs' in England and the 'carousel' in the Netherlands, both of which involve the transfer of inmates who pose control problems.[20] The felt punishments involved in these processes are numerous. Transferred inmates may be separated from personal possessions for a period, will be temporarily unemployed and thus lose their prison wages (and purchasing power in the prison canteen) for

[20] 'Lay-downs' are more formally known as transfers made under CI37/1990. They take two forms. A short-term (up to 1 month) transfer may be made from a dispersal to a local prison as a 'cooling-off' period, and may include segregation while there. A longer term (up to 6 months) transfer is possible if the inmate remains on normal location. It should be emphasized that the circular specifically requires that these transfers not be used as a punishment, *nor may they reasonably be seen to do so*. A record of the reason for the transfer must be placed in the inmate's record. During my own fieldwork, the previous instruction, CI10/1974, was in force, with broadly similar provisions. However, while the CIs cover inmates in dispersal prisons only, regional directors also operated a *de facto* similar procedure for other inmates. In the Netherlands, the 'carousel' referred to the movement of an inmate between 3 or 4 establishments in an attempt to find a regime he would not disrupt. It was alleged by inmates that the administration had a 'hit list' of some 25 inmates who would be liable to go on the carousel if they were disruptive. The administration denied it. Probably both were right: in so far as there was a 'list', it very likely existed only in the heads of senior officials, and not in any documented form.

days or weeks, have to abandon vocational or educational courses they may be undergoing, and may have delays in obtaining visits until they can notify family members of their new location. Although the numbers of inmates involved are not large, the threat of such a transfer is none the less a real one for inmates who are disruptive, and the practices involved in such transfers are often designed as a show of force. As one inmate in a Dutch prison pointed out, when one is targeted for a transfer, no warning is given; one is simply asked to go to the governor's office. On entering the office, the inmate will find a large number of staff waiting for him; the only rational response is to ask, 'Do you want to cuff my hands in front, or behind?'

The case of *Hague* (1990), in England, provided an interesting perspective on the distinction between administrative and disciplinary matters. Hague was a high-security (Category A) inmate who twice attended exercise periods to which he was not entitled. He could, in principle, have been charged with a disciplinary infraction and on his account of the events, would probably have been found not guilty. The deputy governor did not charge him, but used a transfer and segregation procedure available to him under the circular instructions, CI10/74 (since revised). His grounds for doing so were that the inmate was disruptive, but also that to have laid a disciplinary charge would have embarrassed the many staff who saw the inmate going to exercise, should have known that he was not entitled to it, and should have stopped him. This was, it was submitted for Hague, a clear case of administrative discretion being used as an alternative to punishment. A second part of the case concerned the propriety of the governor and the board of visitors, in the prison to which Hague was sent, in effect honouring a decision on segregation that was made in another prison.

The court asserted, significantly, that it did have jurisdiction over the matters of segregation and transfer, though it decided that the particular matters raised by Hague disclosed no unlawful actions. It dealt with the second issue by noting that since segregation under CI10/74 was subject to the same procedures as Rule 43, the governor and the board of visitors had the power to return Hague to the general prison population at any time, and an obligation to consider whether or not to do so; that is, whether to discontinue or uphold the measure imposed in Hague's previous prison. Since they had considered this question, the continuation of the segregation imposed at the original prison was not unlawful. As to the allegation that Hague had been dealt with by an administrative measure when he should have been charged with a disciplinary infraction (and would thus have been able to make a defence to a formal charge), the court had several observations. First, the information before the deputy governor, and on which he made his decision, while it could have justified a disciplinary hearing, also constituted reasonable grounds for the action that was in fact taken. Moreover,

There is also, in our judgement no obligation in a prison Governor to prefer a disciplinary charge as a precondition of acting upon information, which demonstrates commission by a prisoner of an offence against discipline, before he may act upon that information as a ground for placing the prisoner on rule 43 for transfer under CI10/1974. Such a precondition would impose upon the exercise of the Governor's discretion, in his task of maintaining discipline, a fetter which is not expressly imposed by the Prison Rules and which, in our judgement, is not only not implicitly required by those rules, upon their proper construction, but seems to us to be contrary to the fullness of discretion which in such matters those rules intend a prison Governor to have. (unrevised judgement, p. 45)

In short, therefore, English law now accepts jurisdiction over certain administrative acts such as Rule 43, but the situation is that governors retain discretion to use either administrative or disciplinary measures, or both, provided only that they act in ways that can reasonably be justified by the information at their disposal when they make the decision. However, the very acceptance by the court that such a case may be heard is in itself significant, since it provides one further avenue for the review of administrative decisions and thus an additional safeguard against their improper use.

The remaining issue, for this part of the discussion, is the likelihood of informal sanctions applied by the officers themselves rather than by administrative design. Interviews with Dutch and English prison staff confirmed that despite their view that their work involves few decisions, they have wide discretion in practice over many day-to-day aspects of prisoners' lives. This could clearly be exercised in ways designed to punish inmates. At its lowest level, this may take the form of staff telling inmates to 'fuck off' when they asked for something (cited in Maguire and Vagg 1984). In the interviews conducted for this study, inmates talked about staff 'blanking them out'—simply not attending to requests or complaints.

Clearly more sophisticated 'unofficial punishments' can be imagined. Yet beyond the straightforward and interpersonal 'punishments', it was no simple matter to apply unofficial sanctions. The inmates who qualified for such treatment were, almost by definition, the uncooperative, difficult, and ultimately litigious prisoners who would be most likely to complain, to engage in game-playing with complaints in order to manipulate staff, and to seek legal redress. To use informal sanctions, staff said, would often provoke more trouble rather than solve problems. One English inmate, who alleged that he had been subject to such sanctions, simply responded by making virtually all requests, however minor, through the medium of his solicitor. The amount of paperwork this generated was, in his eyes, a reprisal against staff for their informal sanctions against him.

None of this means that informal sanctions are not used. However, the interviews I conducted with English inmates suggested that it did not happen often. It was much more common, they asserted, for staff to be inconsistent or arbitrary in day-to-day matters than it was for them to discriminate systematically against particular inmates.[21]

In Germany and France, the basic-grade staff had less discretion in such matters—and, in France, were often described by inmates as being frightened to make decisions. Inmates and staff agreed that the likelihood of staff imposing informal sanctions, along the lines described in England, was very low. However, in France, the use of the cahiers d'observation, in which staff recorded details of inmate behaviour, did mean that troublesome inmates could be marked out for particular attention by senior staff. And in addition, as one inmate observed,

After the first or second time, punishment is more frequent, they search you out, provoke you; not systematically, but for several inmates it has happened. (inmate, French maison centrale)

In sum, three points can be made. First, it seems that the internal controls and external oversight of isolation and disciplinary measures vary across prison systems. Yet the procedures do not differ sufficiently *within* prison administrations for administrative

[21]   e.gs. of such inconsistency were presented in Ch. 4.

measures to be systematically preferred to disciplinary ones, or vice versa. Second, administrative measures, such as isolation and transfer, exist and can be used against troublesome inmates. They are often felt as punishments and, indeed, examples can be adduced of their being implemented in ways that make their punitive content very clear. But they are not often an *alternative* to disciplinary punishments. They may be used following an infraction, and alongside a disciplinary punishment, as a response to the inmate's perceived disruptiveness. They are, in the last instance, management methods of control used against inmates who pose threats to discipline which are more diffuse, and possibly more grave, than the specific acts listed in disciplinary rules. Third, therefore, the best way to characterize the distinction between formal and informal disciplinary measures may be to suggest that the problem is not the one it is usually taken to be—namely, that of different 'levels' in the system using the disciplinary resources at their disposal to accomplish their immediate aims. This may well happen, but it seems unlikely that it takes place systematically because so much of what happens in prisons is unsystematic and arbitrary. Rather, we should recognize the ambiguities that necessarily arise in a structure that allows both formal disciplinary measures and administrative measures to be used either alone or in combination; in which each kind of measure can reinforce the other; in which administrative measures may be used precisely because the inmates understand them to be 'punishments in effect'; in which circular instructions therefore require administrators to make their decisions in the light of inmate perceptions of administrative motives; and in which inmates, knowing this, can try to engage in 'image management' in order to affect others' perceptions of the inmate's perceptions.

Accountability for Discipline

It should be clear from the descriptions above that governors are constrained in their application of disciplinary and administrative controls on prisoners. What, then, are the problems of accountability for inmate control?

First, the variation in disciplinary rates within prison systems, and even across establishments of the same type, seems to suggest that whatever rules and procedures exist, they can be applied in

very different ways according to factors such as institutional traditions. The idea of system-wide equality of treatment, however
desirable in principle, seems to be unobtainable in practice.
Internal controls on disciplinary decisions are, in ordinary circumstances, weak or non-existent. Certainly they do not suffice to
address this problem. The major avenue of accountability thus lies
to the courts and the complaints systems. Here again, however,
there are problems. In England, matters of procedure may be challenged in the courts, though the actual punishments cannot. In
France, appeal lies not to the courts but to administrative tribunals. Realistically, inmates are better advised to explain to the
JAP why their punishments should not be reflected in a loss of
remission. In Germany, inmates may complain to the courts about
disciplinary matters just as they would about other grievances; but
they must allege a breach of the prison law, and the law in this
area is not detailed. Ironically, notwithstanding the relative deprivation that punishment entails, it is in the country with the weakest sanctions—the Netherlands—that the grievance procedures are
best geared to hearing complaints about discipline. Yet even here,
redress for poor decisions follows the imposition of the punishment
and the achievement of management aims.

Second, there is a tendency for prisons to prefer disciplinary
rather than criminal proceedings in all but the most serious cases.
The advantages of internal discipline are its speed, the generally
lower evidentiary requirements, the straightforward procedure, the
nature of the punishments, and ultimately its ability to support
management. The problem, as the *Campbell and Fell* case showed,
is that the advantages may lead to internal procedures being used
for cases which in European Convention terms should be considered the province of the criminal courts. Nor, as Menard and
Meurs show in their study of inmate aggression, is this purely an
English failing.

Third, governors may impose administrative controls instead of, or
as well as, disciplinary ones. To be sure, their powers in this regard
are not unlimited, nor are they beyond challenge. But for the more
recalcitrant inmates, and with the agreement of the central administration, segregation and other measures can be imposed for extremely
lengthy periods. Indeed, in England, it appears from the *Williams*
(*No. 2*) (1981) case that special measures not provided for in the
Prison Rules may be imposed with the Home Secretary's consent.

Fourth, the controls on discipline and administrative measures are based on the assumption that such measures are applied to individuals. The example of the Dutch prison, given above, suggests that following events such as incidents or escapes, a number of screws in the regime can be turned. While individually minor, they can amount to a change felt by the inmates to be a collective punishment.

Conclusions

This chapter began with three main questions. The first had to do with the importance of formal disciplinary arrangements. Notwithstanding the extent to which they are used, the key point is that placing certain formal controls in the hands of the governor does symbolize his or her leadership role in the institution. The governor is perceived as being more than a functionary in a large bureaucracy. Beyond this, there is probably some truth in the argument that disciplinary measures came to be such an important issue in England because of a historical accident. English courts have shown great interest in disciplinary matters since the 1970s, and have had a strong impact on disciplinary processes. Their interest probably came about largely because England, alone of all the four countries, allows loss of remission to be used as a disciplinary punishment. Until the *Campbell and Fell* European Court decision, boards of visitors could give virtually unlimited loss of remission in serious cases, and powers to withdraw very substantial periods of remission remained within their grasp until 1992.

Second, however, the distinction between the formal disciplinary system and other administrative controls is less strong in practice than it is in theory. For one thing, the 'etcetera' clauses in the disciplinary rules allow a wide variety of acts to be punished, through the simple expedient of issuing a direct order and reporting disobedience. For another, the use of disciplinary sanctions does not preclude the use of administrative controls such as transfer. This observation also answers the third of my initial questions, which related to the ability of a formal disciplinary system to protect inmates from arbitrary punishment. It does offer this protection, in so far as formal charges must be framed, evidence presented, and the case determined by the governor. It does not, in so far as catchall offences exist, and administrative measures can be applied

without any formal hearing. The check on such abuses lies primarily with the grievance procedures and courts; and that is to say,
remembering the conclusions of Chapter 8, that any remedy for
abuse will have to travel a long and rocky road with little likelihood of success.

The third issue included a question about the extent to which
prison governors are accountable to others for their use of disciplinary procedures. It seems clear that in England and France, where
there are internal monitoring mechanisms, they must be fairly cur-
sory; the workload and the number of officials involved would not
suggest anything else. In the other countries, inmates are in essence
given the responsibility to bring governors to account via their
complaints, but examination of these procedures suggests that the
major satisfaction inmates will receive is moral rather than practical; remedies will emerge long after the managerial purpose of
the punishment is accomplished.

Finally, however, we must return to a matter which has been
discussed previously in relation to inmate perceptions (Chapter 4)
and complaints procedures (Chapter 8). One of the major characteristics of the inmate view is that decision-making is arbitrary—so
arbitrary, in fact, that an alternative disciplinary system, even if
staff wanted to operate it, probably could only be sporadically
implemented. The problem here is an ironic one. If there were a
clearly defined alternative disciplinary system, with offences and
punishments laid down in custom and tradition, it is unlikely that
inmates would contest it. Yet many administrative measures are
applied in what inmates perceive as an arbitrary way—and such
measures are also perceived as punishments, in effect if not design.
In consequence, many inmates depict themselves as living in a
world in which arbitrary punishments are routine. This thought
underlines the importance of the Woolf Report's comment (Home
Office 1991, paras. 14.19–20):

There must also be justice in our prisons. The system of justice which has
put a person in prison cannot end at the prison doors. It must accompany
the prisoner into the prison, his cell, and all aspects of life in prison . . .
Justice needs to be built in to the operations of the Prison Service, not
imposed by the Courts.

This must be a priority, not only in England, but across all four of
the countries discussed above.

11

The Politics of Riots and Disturbances

THIS chapter deals with two fairly straightforward propositions. First, the inquiries, reports, and debates that follow major collective disorder in prisons constitute a major form of political and managerial accountability, in the senses coined for these terms in Chapter 1. However, as Chapter 7 indicated, political and managerial accountability can be viewed as a fairly sophisticated set of games played between groups with differing aims and resources. Viewed in this way, inquiries are no more than one of the 'game strategies' that may be employed. Second, neither disturbances nor inquiries, in and of themselves, lead to improved prison conditions or inmate rights. At the very least, they must occur at a time when the political possibility for liberalization already exists.

The Concept of an Incident

Typical examples of incidents would be assaults on staff, fights between groups of inmates, hunger strikes, sit-down protests in exercise yards, arson (usually in the form of burning mattresses in cells), individuals climbing on roofs, barricades erected in cells, escapes, and riots. Clearly some such events are more serious than others. In this context, seriousness has to be considered along several dimensions, including the nature as well as the number of inmates involved; damage or injury; the duration of the incident; and the use of force by police or prison staff to terminate the disturbance. Thus the escape of a single dangerous inmate from a top-security institution, a rooftop protest or barricade lasting several days, a riot lasting only a couple of hours but involving substantial damage, a protest requiring the marshalling of a large number of staff in riot gear, or the killing of a prison officer, would all be treated as major incidents. Although any one of these factors by itself may be sufficient to label an incident as serious, what come to be regarded as major incidents often involve a combination of them.

Incidents may not be a daily occurrence, but neither are they rare. The English Control Review Committee, reporting in 1984 on the management of the long-term prison system, noted ten major incidents in the fifteen years from 1969 to 1983 in the eight dispersal prisons. On an actuarial basis, this amounts to extensive damage to some part of a high-security prison once every twelve years. These incidents involved combinations of: demonstrations by inmates at work, on exercise, or in their cells (for example, smashing cells and throwing debris out of windows); the erection of barricades by inmates; destruction of parts of wings, usually behind the barricades; and rooftop protests, including throwing roofing materials onto staff attempting to regain control. Three continued for at least forty-eight hours, four resulted in massive damage to parts of establishments, and five involved the mobilization of staff and riot equipment on a large scale. In April and May 1986, disturbances took place in more than 40 English prisons and Youth Custody Centres—that is, in about one third of English penal establishments. Some 45 inmates escaped and 800 prison places were lost (Home Office 1987: 1). Two prisons, Northeye and Bristol, were severely damaged. In April 1990, a riot took place at Strangeways prison, Manchester, in which some inmates were in control of parts of the establishment for twenty-four days. During the course of the Strangeways episode, and partly as a result of the transfer of surrendering inmates to other prisons, there were five further serious riots and a number of smaller disturbances.

In France, major riots took place at Toul in 1971, Nancy in 1972, and in 89 prisons nationwide in July and August 1974. Further major disturbances took place in 1977–8 (mainly in the form of hunger strikes), 1983, and in 1987–8 (riots at St Maur and Ensisheim respectively). In the Netherlands, the only major disturbances took place in 1971 and 1974, both in Groningen, though some less violent protests have also taken place in more recent years.

The causes of prison riots and disturbances have been extensively discussed. In a wide-ranging survey of control, Ditchfield (1990) indicates that incidents may be one consequence of ideologi-

cal or managerial changes which undermine informal inmate control structures and adaptations; that age and, to some extent,
offence type and sentence length are associated with infraction
rates, so that mixing long- and short-term inmates, in particular,
can lead to control problems; that increased levels of physical security may do little to promote control in the long term; and that
because prisoner adaptations do not reflect official views of incentives, it remains unclear what incentives may promote control. The
implications are that breakdowns in control, and disturbances, may
emerge at times of managerial change, or where prison populations
are not sufficiently differentiated (the 'toxic mix' theory), or where
the capacity to control inmates is destroyed by inappropriate incentives or deteriorating regimes.

Useem and Kimball (1989) offer a complementary, though
broader, view. They identify the major factors in riots as, first,
administrative breakdown, and second, the erosion of security.
Using case studies of American prison riots, they argue that:

> Prior to all the riots under study, there was a breakdown in administrative
> control and operation of the prison. Prison riots are a product of that
> breakdown and should be thought of as such. Constituent elements of this
> breakdown included scandals; escapes; inconsistent and incoherent rules
> for inmates and guards; fragmentation, multiplication of levels, and insta-
> bility within the correctional chain of command; weak administrators,
> often 'outsiders' to the system; conflict between administration and guards,
> often resulting from strong guards organized in a union or otherwise cohe-
> sive and with good bargaining position relative to the state; public dissent
> among correctional actors; and the disruption of everyday routines for eat-
> ing, work, and recreation. (1989: 218–19)

They then point out that the erosion of security not only makes
disturbances more likely, but also makes them more likely to succeed. Finally, Useem and Kimball broaden the discussion into a
consideration of why such factors may precipitate riots. Their central argument is that changing legal and social doctrines in the US
have called for prison reform while several factors, including conservative prison staff and state fiscal policies, have blocked it.
Prison systems found it increasingly hard to defend their legitimacy
when they were perceived by inmates to be operating in violation
of the constitution and of court orders. In that context, riots
became a kind of direct and collective intervention in the political
process of reforming an institution which had lost its legitimacy.

This kind of analysis has some implications for a discussion of accountability. The signs of administrative breakdown are the very things which mechanisms of accountability ought to be able to detect. Yet if such a breakdown is the result of fiscal constraints by states, one implication might be that riots are one result of a situation in which there is too great a distance between arrangements for operational accountability of the prison, and state fiscal policies which may not take into account the practical implications of spending constraints. In addition, Useem and Kimball's argument suggests that the involvement of the courts in making determinations about prison conditions not only widens the arena in which questions of accountability are played out, but may do so in ways that undermine any willingness of inmates to accept the legitimacy of the prison order.

Incidents and their Aftermath

Incidents and 'major incidents' may have different consequences within prisons in terms of the degree of post-incident reporting and review. It seems fairly obvious that a scuffle between inmates and officers, with no wider ramifications, may attract less subsequent attention than the handling of, say, a roof protest. None the less, the borderline between a 'major' and a 'minor' incident remains blurred. First, administrative criteria and procedures are often used to determine whether an incident is 'major', but in practice they can be operated with some discretion. Decisions to open an 'operations centre' to manage the incident, to suspend daily arrangements or transfer large numbers of inmates, or to assemble large numbers of staff or call on 'outsiders' such as the police, might all be taken as moving the situation into the arena in which senior headquarters staff and perhaps ministers will have to authorize or approve courses of action, and may be held responsible for the consequences.

Second, these kinds of decisions may be made differently not only in different prison systems but in the case of different prisons. As one Dutch prison governor observed, maximum-security prisons have much less room for manœuvre in handling incidents than other establishments. Because of the nature of their population, the administration is much more concerned about them, governors are expected to report emerging problems more promptly, and the headquarters staff are readier both to issue instructions about the handing of incidents and to investigate minor incidents. And third, the role of the media cannot be ignored. Extensive media coverage of an incident exposes senior officials to a greater likelihood of criticism, and while it may not affect the handling of the incident at the time, it may embroil administrators and ministers in controversy after the event. An incident might come to be defined as serious *post hoc*, simply because of the damage limitation exercise that must be carried out subsequently with the press.

Once the dust has settled after a disturbance, a number of things may happen. A major concern will be the answers to questions such as: what caused the disturbance? who, if anyone, should be blamed for allowing it to occur and to escalate? how might it have been handled more effectively? and what lessons can be learned for the future? At the same time there will be some attempt to detect and punish the alleged ringleaders, either through internal disciplinary sanctions or the courts, though this may be complicated by inmate allegations that they were systematically attacked by staff after the event or that personal property was destroyed during post-incident searches. To ignore such claims is tantamount to asking for legal action, or another disturbance, or both. Claims of assault and damage were made after almost every major disturbance in the England in the 1970s and early 1980s, and Favard (1987: 95) notes that, after a riot at Melun in 1973, staff used 'impermissible violence' against inmates after the prison had been re-taken, even while the director of the prison administration was in the establishment. And finally, if the incident appeared to have been associated with prison regime factors—evidence for this may come from inmate grievances, specific complaints or demands aired during an incident, or from sources such as prison inspectorate reports—prison officials may also make (or be put in a political position where they have to make) a commitment to investigate such complaints.

Several forms of inquiry may take place after a major incident, simultaneously or sequentially, by different groups of investigators.

First, a senior official or group of officials (or other distinguished 'insiders' such as retired officials) may be dispatched to the scene in order to find out what happened and why, and to determine whether the actions of any inmates (or staff) should be the subject of internal disciplinary action, or prosecution. For example, where two inmates escaped from Gartree prison by helicopter in

December 1987, a retired Deputy Director was called upon to carry out an investigation. In the Netherlands and France, where the prison inspectorates are internal departments of the prison administration, inspectors are often sent to investigate the disturbance. Reports of this nature are almost always confidential, though one, the 'Schmelk Report', on the French riot at Toul in 1971, was published. Robert Schmelk, appointed to chair an inquiry into the riot, was an ex-director of the French prison administration.

A second option exists in England, where the inspectorate is independent of the prison administration. Where the Prison Inspectorate is called upon, more resources may be available to do the job, and the reports can be wide-ranging and thorough (though one problem for the Inspectorate has been that the investigation of incidents has, on occasion, derailed its own inspection programme) but the results will be published. The English inspectorate has also been called upon to undertake investigations outside England, for example after the mass breakout from the Maze prison (operated by the Northern Ireland prison service) in 1983.

Prison systems tend to consume their own smoke if they can. But sometimes this is not possible, because of the broader political implications of a disturbance, for example. This was the case with the appointment of Lord Justice Woolf to investigate the wave of English prison riots in 1990. The size of the initial disturbance (at Strangeways prison, Manchester) and the number of subsequent disturbances meant that the credibility of prison management policies was on the line. However, the number of reports of this nature is small, with most coming from England, which appears to have developed a tradition of this kind of inquiry, with a comparatively low threshold, beyond which the appointment of an outsider is considered necessary.[1] In the Netherlands, only one report has ever

---

[1] I have not dealt in detail with the procedure of such inquiries, which is a matter of some interest. As Morgan (one of the assessors for the Woolf Inquiry) notes (Morgan 1991): 'Departments are not monoliths. Ministers may have agendas different from senior administrators, and so on. But whether consensually or not, inquiries are the subject of substantial attempts by the department to influence the course of events. The key source of departmental influence is the written evidence it submits. This is invariably extensive, well presented, carefully documented, authoritative, delivered on time and supported by oral or written elaboration if the inquiry deems this necessary. All resources required to ensure the smooth delivery of this material will be devoted to the task.' Departments will, Morgan notes, often create a special unit to 'support' the inquiry and its members will have an inside track to the inquiry's thinking.

been published following a disturbance, though it must be said that the system has only experienced two riots in the last twenty-five years.[2] Both were at Groningen prison, in 1971 and 1974, and it appeared that the root cause of the riots lay in the nature of the prison regime there (though similar regimes existed in other Dutch prisons). The report of the van Hattum Committee in 1977 dealt largely with the reform of prison regimes, therefore, rather than with the disturbances themselves. In France, only one disturbance has resulted in a report—the Schmelk Report, mentioned above—although additional information on disturbances has been published in the memoirs of prison administrators from the 1974–7 period, and in some research studies.

In addition to the few, mainly English, published reports on inquiries, some information about their results may emerge through mechanisms such as parliamentary questions. In Germany, reports on disturbances have never been published as such, though the fact that the prison systems are operated by individual Länder, and that the Land parliaments have asked for and received information on disturbances, may have rendered publication in a strict sense superfluous.

Dealing with Ringleaders

Disturbances rarely happen by accident. Most are initiated by ringleaders of some sort, though the nature of their leadership is clearly variable. One of the post-disturbance tasks is almost always to determine who had been involved in planning it and who, if anyone, had a clear leadership role. Such persons will then be subject to internal disciplinary sanctions, prosecution, special security measures, or all three.

While this is not directly a question of 'accountability', it can become so. For example, after the riot at Hull prison in 1976, the inmates were dispersed to other establishments and alleged ringleaders dealt with by sanctions imposed by the Hull Board of Visitors, members of which divided themselves into teams and toured the country dispensing administrative justice to their ex-inmates.

---

[2] Though other kinds of disturbances have occurred. De Jonge (in Kelk and De Jonge 1982) refers to protests among young prisoners in Rotterdam, in which inmates destroyed cell furniture and broke windows although they did not apparently attempt to break out of their cells.

This procedure was roundly criticized: for example Taylor (1980) contains what is described as a complete transcript of one disciplinary hearing, which must have taken less than a minute. The eventual outcome of this and similar episodes—about ten years and several judicial reviews later—was a major change in the conduct of adjudications which recognized inmates as having a number of procedural rights. This history, and subsequent developments which further improved English inmates' rights in relation to disciplinary proceedings, was discussed in Chapter 10.

Other problems may emerge if several differently directed inquiries are running simultaneously. For example, it appears that several alleged ringleaders of disturbances in the English 1990 wave of unrest were not interviewed by the assessors working on the Woolf inquiry. There was a separate exercise being conducted at the same time, directed towards the production of evidence for use in internal disciplinary cases or prosecutions. Legal advice received by Woolf's team indicated that if the ringleaders were interviewed, the results of those interviews, directed at finding out how the unrest started, could have been demanded for production as evidence in court. This might have harmed the credibility of the inquiry, which was primarily intended to uncover social causes leading to the disturbances, rather than evidence against ringleaders.

In fact the Woolf report suggests that some thought had gone into the creation of the Strangeways disturbance. Notes concerning the riot were circulating in the prison several days prior to the event, and were alleged to have been written by high-security prisoners. Most inmates were expecting some kind of trouble to break out at a specific time and location, namely a chapel service. Some brought along with them hoods and clubs. Two inmates in particular signalled the beginning of the riot, one by taking the microphone from the priest and inciting others, and the second by echoing the incitement after the priest had snatched back the microphone. As inmates surrendered and were removed to other establishments, they became 'carriers' of unrest into those other prisons and some were involved in further disturbances there.

But although someone must, as it were, make the spark that sets off the prairie fire, the Woolf Report was more concerned to know why they should have done so. Indeed, most published reports on prison disturbances have placed little stress on the activities of

allegedly troublesome ringleaders—even though they may be identified and their actions described—but have concentrated instead on
the social factors which may have turned an institution into a riot
waiting to happen. Part of this, of course, is to do with the nature
of an inquiry whose report is destined for publication. For matters
to have reached the stage at which such an inquiry is commissioned, there is usually a great deal of prima-facie evidence for
believing that something is wrong with prison conditions and
administration. The nature and investigation of these social causes
is the subject of the next section.

The Social Causes of Disturbances: England

It has sometimes been observed that, until the early 1980s at least,
major disturbances were largely confined to long-term, high-security prisons. Such prisons were usually comparatively well-
resourced and not overcrowded, though admittedly their high level
of security may have curtailed regimes. With this observation in
mind, it was often suggested that disturbances were probably created by those who had experienced the inconsistencies of regime
between various long-term prisons; who had, or had acquired
while in prison, an anti-authority and sometimes also a politically-
charged mindset; and who had relatively little to lose by taking
their grievances onto the roof.[3]

This view changed in the course of the 1980s in England,
because the 1986 wave of riots included a wide variety of establishments and large numbers of short-term, low-security, or remand
inmates. On the one hand it was pointed out that short-term and
remand prisoners might have even less stake in the prison system
than long-termers; they would soon be released, while a large proportion of remand prisoners receive non-custodial sentences. This
might mean that they would be as prepared as long-termers to create disturbances. On the other hand, it had become quite clear by
the time of the 1986 disturbances that prison conditions in general
had been declining, and that the local and remand prisons were the
worst-resourced in the English prison system. This lent credence to

[3] But not all long-term inmates would be prepared to riot. Many were sufficiently
co-opted into the system to feel that riots and the consequential instability of
regimes would erode any privileges they had; however planners and instigators of
disturbances would be more likely to exist among the long-term population.

the view that prison conditions needed to be explored as an explanation for disturbances. Similar situations had, incidentally,
occurred in the 1970s in both France and the Netherlands, and are
discussed below. In England, reports on both the 1986 and 1990
disturbances attached weight to the background of generally poor
and probably declining regimes, many features of which had previously been roundly criticized by the Chief Inspector of Prisons.

The 1986 Riots and the Prison Inspectorate Report

As the Chief Inspector of Prisons (1987a) noted in his report on the
1986 wave of unrest, conditions in many English prisons had been
poor for quite a number of years, both in the physical state of
prison buildings and in the quality of regimes. Most prisons,
though not all, experienced a significant degree of overcrowding.
This did not only mean prisoners sleeping two or three to a cell
intended for one. It also meant that some inmates were held far
away from home, or in conditions which were not best suited to
contain them. For example, Northeye prison in Sussex, a low-
security closed prison housed in buildings originally designed as an
RAF radar station, contained many inmates from London; a large
proportion of young inmates, often regarded as more volatile than
older prisoners; a substantial group of persons convicted of drugs
offences; and had a high turnover of short-term inmates. All these
factors were held to have reduced the stability of a prison which,
because of its low level of security, relied substantially on inmates
co-operating with staff and maintaining self-discipline. Northeye
was virtually destroyed in the unrest.

At a national level, although the number of staff in post had
increased in the years prior to 1986, so also had the numbers of
inmates they had to deal with and the number of tasks which they
were required to perform. The increase in the number of inmates
was disproportionately an increase in the numbers of remand prisoners, who must be produced regularly at court; it fell to the
prison service to provide the staff to accomplish this. In addition,
staffing levels, as a consequence of agreements between the Prison
Officers' Association and the Prison Department, were set in such a
way as to allow for the working of substantial overtime by officers.
Officers had become dependent on overtime pay to boost their
wages, and the prison system was highly dependent on staff overtime in order to function smoothly. However, industrial action by

prison staff had taken place sporadically since the late 1970s over a range of issues and industrial relations were generally poor.

The Prison Department was thus facing a position in which it was paying for more staff, and continuing to pay for substantial amounts of overtime, but in which rapidly increasing demands on staff time were being made. The overall result was increasing expenditure but declining prison regimes. In order to combat these problems, and to meet Treasury cash limits, the Department arranged for a team of management consultants to review its staffing and shift systems. It was initially thought that something of the order of 10–15 per cent of staff time was being lost through inefficient staffing practices. The whole situation was one in which staff were both resentful that the Department was more concerned about its budgetary constraints than about its employees, and anxious that the intention to reduce dependence on staff overtime would erode either their take-home pay or the numbers of staff on duty. 'Fresh Start', the staffing and salaries package that was subsequently negotiated, is discussed further below in relation to the 1990 riots.

None of the factors mentioned above made riots inevitable. The overcrowding, the poor industrial relations, and the other factors had existed for some time, though admittedly they had worsened in the years immediately prior to 1986. By that year it seems likely that the bulk of the inmates—remand and short-term as well as long-term—were discontented not only because of the poverty of their physical environment and regime, but also because of the inability of the prison service to allocate them to establishments which would have made sense in terms of, for example, the maintenance of easy home contacts. The business of routine, everyday control by staff of prisoners was very likely becoming increasingly difficult, with fewer incentives for inmates to behave, and an increasingly stretched and alienated staff attempting to control them. When staff began industrial action at Gloucester prison in 1986, the spark behind the initial disturbance was inmate fears that the staff dispute would reduce an already poor regime to complete poverty.

It was in the context of these kinds of observations that the 1986 report made the following recommendations. First, a factor leading to industrial action had been the lack of agreed criteria over manning levels, though the Chief Inspector hoped that the Fresh Start programme would reduce this problem. Second, he argued that poor physical conditions had disposed inmates towards protest; that attempts to manage prisons efficiently had been driven by the need to cut costs and had not been used as attempts to improve regimes; and that continuing reliance on prison staff to escort prisoners to court had drained some prisons of adequate numbers of staff in the establishments. Recommendations here included the establishment of a separate escort service, a change of attitude towards the purpose of efficiency exercises, and a critical review not only of conditions in general, but also of ineffective improvements which had been made, as, for example, where workshops had been upgraded but left under-used.

In addition, the Chief Inspector came out in favour of the creation of regime standards, with both target standards (towards which establishments would aim) and minimum standards (below which institutions should not be allowed to fall). Some further observations in the report were concerned with the availability of prison population data, which, it was held, could have alerted the prison administration to mismatches in particular establishments between the level of security and the types of inmates being held; and some problems in relation to contingency plans and headquarters-field communication (the report cites an example of a governor being led to believe that headquarters had a contingency plan to handle the industrial action in his establishment when in fact it did not).

In short, the review of the 1986 riots identified the following as causative factors: poor prison conditions; the fact that inmates knew conditions were bad and might, because of staff industrial action, get worse; the lack of agreed criteria for staffing which contributed to poor conditions and was a factor in the industrial action; and a failure to ensure that inmates were being held in appropriate levels of security, so that 'volatile' prison populations were built up in low-security establishments. Whether these matters could be put right was another question. Fresh Start would, it was hoped, resolve many of the difficulties. Yet in 1990 it was evident that the implementation of Fresh Start had contributed to the riots.

The 1990 Riots and the Woolf Report

As detailed earlier in this chapter, 1990 saw a major riot at Strangeways prison, Manchester, and a series of other major disturbances throughout the English prison system.[4] The result was the Woolf Report, a 600–page document co-authored by a senior judge (Lord Justice Woolf) and the Chief Inspector of Prisons. Although its initial brief was to inquire into the causes of the Strangeways riot and to make recommendations (and it was subsequently expanded to cover the other prisons in which unrest occurred) it was, in fact, a thoroughgoing review of the state of prison regimes system-wide, starting from precepts some way removed from those to which the prison service had been accustomed, and making a number of recommendations intended to improve regimes as a whole. Moreover, it was significant that in addition to the four questions it was commissioned to answer, concerning the causes of the riots, the way they were handled, and preventive steps against future riots (para. 1.1), two other questions were listed in para. 1.3 which Woolf appeared to have asked himself. They were: why did the riots not happen earlier? and why were their consequences not even more serious?

The Woolf Report conclusions were of several kinds. First, there were detailed comments on matters such as how misunderstandings arose between individuals responsible for handling the disturbances. Some appeared to relate to the personalities of those involved. One person is described as having shown 'great strength of character and leadership in being prepared to take a clear and firm decision', which he was unfortunately not in a position to take, since he did not possess relevant information. Although such observations have implications for prison management, they need not be laboured here.

Second, a number of criticisms of prison regimes were offered. Inmates at Strangeways were reported by Woolf as claiming that legitimate means for airing grievances were ineffective.[5] In various other establishments, even though staff-inmate relations and certain aspects of the regimes were praised in some establishments, praise was usually balanced by points of criticism. These covered areas such as overcrowding, visits, food, the availability of work and education, and petty regulations.

[4] It is also worth noting that some prison disturbances in 1990 in Germany were held by German commentators to be a 'knock-on' effect of the English riots, which received world-wide media coverage (Frieder Dünkel, pers. comm.).

[5] In Ch. 8, I argued that the problem may be deeper than this. Grievance procedures are typically ill-placed to deal with complaints concerning a prison regime as a whole, since they are designed to cope with matters affecting only individuals.

Third, security arrangements and contingency planning were not immune from criticism. Woolf's generally liberal views were tempered by the realization that if prisons were to function in the way he envisaged they should, firm and clear security and control were prerequisites. Scattered throughout the report were comments such as that a quick and clear show of force, for example through the visible mobilization of large numbers of officers in riot gear, would probably have been sufficient to dissuade those planning disturbances from carrying them out.

Fourth, a number of comments were made about national policies and their implementations. There was, for example, a lengthy discussion of the implementation of the Fresh Start staffing arrangements (in paras. 13.12–79), which concluded that the targets for improving efficiency in staff use were too ambitious, had not properly been communicated to staff or indeed to governors, and had been too quickly implemented, leading to sudden staff shortages and reductions in regimes.

Fifth, one recommendation made in the 1987 report, but which was not implemented, was forcefully restated. This was a call for minimum regime standards. In addition, Woolf set out a number of principles which should govern the treatment of inmates and the relationship between inmates and staff and which would, by implication, need to be reflected in minimum standards. Three were particularly important, and can be summarized thus:

- Inmates should be able to take some responsibility for their own lives in prison, *and to be accountable for the choices they make*; this is regarded as a fundamental feature of life outside prison which must be replicated inside the walls in order, if nothing else, to avoid psychological deterioration.
- Inmates should be entitled to expect to be treated with justice; arbitrary decision-making, failures to supply reasons for decisions, and so on need to be weeded out.
- Woolf also proposed that the concept of privileges was now outmoded. Many of those things described as privileges and available at the discretion of the governor, he suggested, ought to be a normal expectation: 'It should not be a matter of grace or favour that inmates are allowed to associate, or that they are able to buy books, newspapers, or have radios' (para. 14.32).

The overall conclusions contained twelve recommendations. They implied (though sometimes did not explicitly state) general principles capable of application in a range of ways. In consequence, they were further discussed in more specific sub-recommendations, and in some 204 numbered proposals for implementing them. In looking at the content and scope of these recommendations, it seems that Woolf started with the view that whatever the immediate cause of the riots—in terms of ringleaders and copycat disturbances—the state of prison regimes generally was such that 'any spark could light a prairie fire'. Since in many ways the sparks could be considered almost routine, it seems that Woolf was less interested in explaining their occurrence than in 'fireproofing' the 'prairie'. This required a systematic approach which dealt not only with security and contingency plans, but with standards of treatment for inmates, and clearer staff policies. Thus the recommendations strayed far beyond the immediate causes of riots, and covered such disparate issues as the management of HIV-positive prisoners, staff uniforms, integral sanitation in cells, more hostel places for offenders, industrial relations, and the creation of a new complaints mechanism.

The Years 1986–90: What Happened?

The key factors in both 1986 and 1990 appear to be poor staff-management relationships and industrial action, in a prison system which anyway offered rather poor conditions—factors which Useem and Kimball (1989) subsume under their general label of administrative breakdown. After the 1986 riots, the Chief Inspector placed his hopes in the Fresh Start, the industrial relations package which was intended to create greater efficiency and remove service dependence on staff overtime. But although, in some respects, inmate rights (for example in relation to prison discipline) were improved, both inspectorate reports and academic research (cited in Chapter 3) suggest that regimes continued to deteriorate over 1987–90 even though the level of prison overcrowding eased slightly. Further, the key point raised by Woolf in relation to the 1990 disturbances was that the implementation of Fresh Start actually made much deeper cuts in staff availability than staff or governors had anticipated, and these more than anything else adversely affected regimes, and created the conditions in which inmates began to feel that they had little to lose by taking their complaints onto the roof.

The Social Causes of Disturbances: France and the Netherlands

The Netherlands has seen only two major prison riots since the 1970s. They were treated as a warning of a prison service moving in the wrong direction, though corrective action was limited in scope. France, on the other hand, experienced waves of riots in the 1970s and 1980s which appear to have sprung from much the same situation as the English riots in the late 1980s and early 1990s.

In the Netherlands, the van Hattum committee, reporting in 1977, was not charged with investigating the Groningen riots in 1971 and 1974, but with recommending improvements to prison regimes. None the less, as Constantijn Kelk, a member of the committee, notes (Kelk 1992), many of its recommendations were designed to remove from prison regimes the factors which had been seen as contributing to those riots. One key recommendation was for smaller establishments to be built, or created out of the compartmentalization of existing larger units. With a maximum 120 inmates per prison, they argued, greater staff-inmate contact was possible, security problems became easier to deal with, and tendencies towards the degradation of regimes were easier to counter. Under the circumstances it was unfortunate that several much larger prisons were already too far advanced in their construction to be halted or modified.

The French history of riots and responses to them is much more intensive. We may conveniently start with the riot at Toul in December 1971. Some 475 inmates, almost the whole of the prison population, was involved; one of the two wings of the prison, including workshops and educational facilities, was destroyed, the prison roof was stripped, and inmates remained in control of the prison for about five days. The fact that they gained access to the roof enabled them to communicate their grievances to the media. The commission of inquiry headed by Robert Schmelck published its report on 8 January 1972, less than a month after the riot had ended. It called into question the severe discipline of the establishment, and argued that the riot was the result of 'a general malaise attaching to the defective conditions of detention, a maladjusted and clumsy disciplinary regime, and also the lack of co-ordination between different services in the prison'.[6]

[6] Schmelck Report, p. 7, my trans. The French original is quoted in Favard

In addition to the major riots at Toul in 1971 and Nancy in 1972, there were disturbances in eighty-nine prisons nationwide in July and August 1974.[7] They were primarily protests at the appalling state of French prisons at that time. Reforms followed, at least chronologically, if not as a direct result. One retrogressive reform was the establishment in 1975 of eleven high-security units. Inmate protests against the units took place in 1977–8, and their use was abandoned in 1982, undoubtedly partly as a result of the protests. Further prison disturbances took place in 1983 and 1988, though on these occasions their relationship to reforms was less clear. In fact they they followed, rather than preceded, the announcement of reforms. What lessons can be learned from these events?

In the week following the Toul riot, prisoners destroyed the Charles III prison in Nancy and occupied the roof. The Garde des Sceaux, the minister with responsibility for prisons, had earlier implied that the conditions described by Schmelk at Toul were not replicated elsewhere in the prison system. And he asserted, following the Nancy riot, that conditions there gave the inmates no serious grounds for complaint, that the riot had been led by subversive elements, and that the result of the riot would be to hinder reforms. Reform proposals had in fact been under discussion for some time, but were implemented on the day following the Nancy riot (leading to unfavourable press comments that the administration had given in to the rioters). They consisted of a package including prison leave and remission, devolving parole decisions for inmates sentenced to less than three years from the central administration to the JAPs attached to the prisons, and allowing inmates to obtain and read journals (Favard 1981, 1987).

These reforms were promulgated by decree in September 1972, and by law in December. Clearly the riots did not in any sense

(1981: 178): 'un malaise général tenant aux conditions défectueuses de la détention, un régime disciplinaire inadapté et maladroit, ainsi que le manque de coordination des différents services de la prison'.

[7] The following account is largely based on Dorhlac de Borne (1984) and Favard (1981; 1987). Dorlhac de Borne was appointed in 1974 as Secretary of State for Prison Conditions, a newly created position within the advisory cabinet of the Garde des Sceaux. Favard was an official in the prison administration from 1970 to 1975 and technical adviser to the Garde des Sceaux from 1981 to 1986. I have omitted consideration of some notable 'one-off' riots, and a further wave of riots in 1985, which do not have any particular relationship (in terms of timing) to prison reforms.

lead to the reforms; and indeed the Schmelck report on the Toul riot indicated the need for measures quite different to those actually proposed. None the less, the Toul and Nancy riots together must have underlined to administrators the need for an overhaul of the system, and lent some urgency to the reform project. They, together with continuing unrest in the following year, lent weight to the need for a new Secretary of State for Prison Conditions—a post created in 1974 by Giscard d'Estaing and filled by Hélène Dorlhac de Borne.

Dorlhac de Borne was appointed in May 1974, and had been in post less than two months before a new wave of disturbances rocked the system to its core. The first riot was at Clairvaux, on 19 July; this was a long-term prison in which staff-inmate relations had deteriorated suddenly, with many small disturbances and also fights among groups of inmates. After one such fight on 18 July, two inmates were punished with fifteen days' segregation, and there was some altercation between staff and other inmates as to whether these two should be punished so severely; this appears to have been the spark for the riot. On the following day there was a riot at Nîmes, also a long-term prison although it appeared that some short-termers were serving time there. When the Clairvaux riot was quelled some mutineers were sent to Fleury-Mérogis, just outside Paris, and the largest prison in the system. In due course there were copycat disturbances there and in other establishments.

There seem to have been several contributory causes to these riots. One was that conditions in most prisons were poor; so poor, indeed, that when the media were later invited to film inside several prisons the resulting images appeared to swing public opinion in favour of reform (and this may well have been the effect that was anticipated by the administration in allowing access for film crews). And not only conditions were poor, but also prison pay; the only systematic demand made by inmates seems to have been for increased pay in the prison workshops.

A second factor was that the administration had probably lost its grip on at least some prisons in previous years. Dorlhac de Borne (1984) indicates that in the early days of the 1974 riots it became clear that while prison staff unions differed markedly in their views on the need for prison liberalization, their members uniformly felt that the demands placed on them were frequently unrealistic if not incompatible, and that their training had been inadequate for the tasks with which they were faced. In effect, the staff had lost control even before the inmates took it from them.

A third, and related, point was that the level of overcrowding within the system had caused the separations between short- and longer-term inmates to break down, while judicial delays meant that remand prisoners were spending increasing amounts of time awaiting trial. The prisons were not only broken down physically and operationally, but had also ceased to function as a system.

The need for a reform package of some description had been seen prior to the riots—hence the appointment of Dorlhac de Borne. But discussion of reforms had to continue alongside incoming reports that yet another prison had lost its roof. Thus the riots themselves did not lead to reforms but ran parallel to the reform process, underlining the need for action. None the less, the fact that the reforms had been under discussion during the riots led to resistance in their implementation from prison governors and staff, who claimed that they had been developed under duress and, in essence, met the inmate demands while reducing the power of prison staff (Seyler, n.d.). The reform package was announced on 7 August, two days after the riots had subsided. The 'progressive' system for long-term prisons was abandoned, and regimes were diversified, with those in centres de détention becoming easier than those in maisons centrales. Restrictions on correspondence were partly lifted, inmates in centres de détention were permitted in some circumstances to make telephone calls, the suspension of sentences became possible under certain circumstances, parole possibilities were enlarged, prison leave was introduced, and new rules allowed the grant of additional remission for educational or other rehabilitative progress. These measures were implemented in mid-1975.

There was, however, a quid pro quo. Security in the maisons centrales (for long-termers) was tightened, and eleven special units, 'quartiers de securité renforceé', were built to house the most recalcitrant and subversive inmates. These units were the subject of further organized inmate action, mainly hunger strikes, in subsequent years. Fize (1984) lists hunger strikes against their lack of transparent windows in 1976; more widespread hunger strikes, and not just by those in the units, in 1978; protests by penal pressure groups in the same year; interviews, given secretly and published in major journals, by Jaques Mesrine after he escaped from a similar (but differently named) unit in La Santé prison, in which the psychological cruelty of their regimes was described; and the protest suicide of Taleb Hadjadj in 1980, as adding to the pressure to abandon the units. They were abandoned in 1982, and the decision to do so seems to be the only clear-cut case in which inmate pressure, along with some support from interest groups, was able to change prisons policy.

After the 1975 reforms, prisons policy flip-flopped between the hard and the soft. The 1981 'Security and Liberty' law introduced a 'security period' for those convicted of certain serious offences, the effect of which was to bar them for the first part of their sentence from measures such as semi-open conditions, home leaves, and so forth. The 1983 reforms did away with this, and then went further. The measures applied in centres de détention in 1975 were extended to all sentenced inmates, and a number of other relaxations were introduced, of which two were especially significant; prisoners were no longer required to wear uniforms except under special, security-related, conditions, and open visits were introduced. In this case, however, demonstrations took place in several prisons in protest at the length of time between the announcement of the reforms in December 1982 and their implementation in January 1983; a case of the reform process leading to riots rather than the other way round.[8]

Although additional privileges—including television in cells— were authorized in 1985, the overall prison situation in the 1980s remained bad in three primary ways; overcrowding, poor physical conditions in often old establishments, and a rather rigid structure which created delays and bottlenecks in making proper arrangements, for the transfer of inmates between establishments, for example. In 1987/8 two riots occurred which were held to be significant in what they implied about the management of the prison system as a whole.[9] In November 1987 a major riot took place at St Maur, a modern, purpose-built maison centrale. It was completely destroyed. In April 1988 a hostage incident and riot took place in Ensisheim, a maison centrale housed in an extremely old and physically decrepit structure. The prison was almost completely gutted. On the surface the immediate cause of the

[8] This is not unknown elsewhere. Similar situations have been described for the USA by McCleery (1961), and for Poland by Hołda (1991).
[9] Largely based on reports in Le Monde, 19 and 21 Apr. 1988.

Ensisheim riot appeared to be twofold. First, some of the St Maur rioters had been transferred there, and second, a high level of tension had been created by the discovery and frustration of an escape attempt. But it was not lost on commentators that one of the most modern and one of the oldest maisons centrales had both been completely gutted by riots in a matter of months, suggesting that the underlying problem was the way in which maisons centrales themselves were administered. On this occasion, however, the political tide had turned yet again towards the right: no concessions were made to the demonstrators and, rather than risk copycat riots through transferring inmates to other establishments, they were kept (in very overcrowded conditions) in the few remaining habitable cells in the prison.

In order to draw lessons from this series of events it is necessary to add a few further observations; but the lessons are as follows. First, it is clearly not possible to suggest that riots in French prisons led directly to the improvement of prison regimes. In 1971 prison administrators and politicians were faced with a major riot in a prison known to be 'hard', at a time when they were already considering some changes, not in prison regimes, but in the opportunities for early release. The report on that riot, although acknowledged to have identified the underlying problem correctly, was dismissed as relevant only to that one institution. Reforms of the prison regime which would have been pertinent as a result were either not considered, or discounted. The pre-existing plans were carried out instead. In 1974, the situation was slightly different in that the general condition of the prisons prompted a high-level review which had already begun at the point that inmates showed, often on the roofs of prisons, that they had had enough. The riots might indeed have damaged the possibilities for reform, but the administration was able to use them, via the media, to sway public opinion towards the view that reforms were necessary. They were less able to sway their own staff, who resisted the reforms on the grounds that they appeared to capitulate to the inmate demands during the riots.

Second, however, there is one case of inmate protests—albeit at a sustained level over several years—having affected penal policy. The use of high-security units, introduced in 1977, was abandoned in 1982 following an inmate campaign including hunger strikes, suicides, and messages smuggled to the media over a period of about five years. And third, we have an example, in 1983, of riots occurring because of the gap between the announcement and the implementation of reforms.

Some Lessons

Where there are disturbances, a number of subsequent events are related, sometimes directly and sometimes tenuously, to issues of prisons accountability. At one level there are questions of managerial accountability which are directed at discovering what went wrong. Sometimes there will be political pressure on ministers for answers to this same question—pressure which is inevitably passed on down the chain of command. One way of producing such accounts, used in England on repeated occasions, though rarely in other countries, is to commission an inquiry. Such inquiries, for a range of reasons, tend to concentrate on the social causes of the disturbance. More straightforward detection of perpetrators is often a separate activity. In any event the questions at the back of the inquiry's mind must be: why did a large number of inmates support or condone the actions of (usually) only a small number of perpetrators? and in particular, what aspects of prison conditions might have led to this situation?

Inquiries, internal and unpublished as well as independent and published, can come up with wide-ranging criticisms of prison regimes. The most complete document in this respect is the English Woolf report, following riots in 1990. Woolf identified a number of areas in which regimes were deficient, but a major factor was essentially a combination of industrial relations and the stress placed by management on the need to achieve financial efficiency, a term primarily defined with reference to cash limits and budgets. The ultimate source of this pressure was, as one might expect, the Treasury.

Industrial disputes have been a common factor in both England and France; it seems likely that the impact of industrial disputes on prison regimes, and the fears that they arouse in inmates as to the possible worsening of regimes, tends to heighten tension and may make disturbances more likely. In addition, industrial disputes often reveal the kind of 'Babel of evaluative languages', and 'contested criteria of accountability' noted by Day and Klein (1987). For example, one official of the prison officers' branch of the

French union, Force Ouvrière, argued after the Ensisheim riot that a major consideration had to be the lack of an effective penal policy, a lack that he blamed on the prevalence of legal rather than managerial personnel in the senior reaches of prison administration. While this can be read as a pitch for a better career structure for officers, it may also have some grain of truth in it. For many years the administration had been, in the general tradition of French law, much more concerned about general principles than administrative details, and more interested in judicially-oriented reforms, such as parole, than in reforming the interior of the prisons. Thus the staff and inmates probably shared an idea of what the key issues were, while the administration held a radically different view of the priorities for action. The same might also be said, of course, in the 1986 and 1990 English riots.

Producing recommendations is not, however, the same as having them acted upon. Recommendations from the report on the English 1986 riots were largely shelved, and arguably that was a major contributing factor to the 1990 riots. The main optimistic note the Chief Inspector allowed himself—that Fresh Start would resolve manning problems—turned out to be misplaced, since the implementation of Fresh Start did not so much resolve problems as create them. Why and how recommendations are acted upon, or shelved, is a difficult business to unravel. However, the obvious point is that any recommendation, in order to be implemented, must conform to the wider political and economic views of the government of the day.

One point worth mentioning here is the role of the media. In England the media have attracted various negative comments (in the Woolf Report, for example) for the way in which they reported staff industrial action, which, it was argued, may have contributed to unrest; and later for their reporting of disturbances, which, it was again argued, may have created further unrest. The general attitude of officials towards the press has been a censorious one. But in France it seems that the opposite approach has been used at least once, with reportage of prison conditions being used to foster a change in public attitudes supportive of prison reform.

Ironically enough, reforms can produce riots in some circumstances, probably because of the stress that the reform process can create in prisons. But while it is also fair to say that riots can lead to reforms, they typically do so only in the long term, after sustained campaigns of disturbances (the French experience with high-security units) or as an unanticipated effect (the English experience with disciplinary procedures after the Hull riot). As a final comment, then, although major disturbances result in a great deal of attention being paid to prison conditions, the practical results are often compromised. No senior official or politician wishes to appear to be 'giving in' to rioters' demands; any recommendations from an inquiry will take the usual slow route through the usual bureaucratic and political channels; and in any event, factors external to the prison system—governmental convictions and financial policies—often supervene.

12

The Impact of Europe

IN some respects Europe remains as divided as it has ever been. Yet the perennial arguments both for and against closer co-operation within a 'single Europe' have gradually drawn the European countries into an occasionally incoherent, but none the less important, supranational political entity over the last forty or so years. The role of the Council of Europe has often been overshadowed by groupings such as the EEC. None the less, in prison affairs, the Council of Europe has been an important, if not always well-publicized, player.

There are three major points of interest. First, the Council of Europe has promulgated standards for the treatment of inmates since 1973. The original 1973 Rules were, and the 1987 revised Rules remain, phrased in ways that scrupulously observe the niceties of international agreements. Their influence has been limited and symbolic; persistent breaches have rarely been aggressively criticized. Yet they provide an important frame of background expectations, within which efforts to improve prison conditions can, however slowly, move forward. Second, the machinery of the European Commission and Court of Human Rights, while slow, limited in scope, and cumbersome, has enabled prisoners to pursue litigation beyond the national level. In not a few cases the outcome has been a change in national prison regulations and practices. Third, and more recently, the arrangements for the European Convention for the Prevention of Torture and Inhuman or Degrading Treatment of Punishment included a Committee which has, for obvious reasons, been somewhat interested in prisons. For all practical purposes there is now what amounts to a European mechanism for inspecting prison conditions.

A number of less immediately important developments have also taken place. They range from conventions dealing with issues such as the repatriation of foreign prisoners and reports on European prison matters co-authored by members of several national administrations, to less formal arrangements such as study visits of prison staff between different systems or bilateral discussions on sentencing policies.[1] However, in this chapter I propose to concentrate on what I take to be the three major initiatives—the European Prison Rules, the Commission and Court of Human Rights, and the Torture Convention—and to outline the effects that they have had on prison administrations in recent years.

The European Prison Rules

The European Prison Rules (EPRs) which came into effect in 1987 have a long and distinguished genealogy. In 1934, the Standard Minimum Rules for the Treatment of Prisoners were accepted by the League of Nations which, in the following year, passed a resolution requesting member governments to promulgate them. The United Nations, as successor to the League of Nations, agreed a substantially similar set of Standard Minimum Rules in 1955; and in 1973, the Council of Europe adopted a slightly modified version of that document. The Council requested, in addition, that member states provide quinquennial reports to the Secretary General on progress towards the implementation of these Rules.

Neale (1991: 205) states that the need for a specifically European version of the Rules 'was envisaged as giving them a European emphasis and reflecting the particular need and circumstances of the prison systems of the member states of the Council of Europe'. Although the Council's member states were also members of the United Nations, it is probably fair to say that by the early 1970s, the Council of Europe had come to be seen as the more important venue for discussion of European prison matters.[2] Yet even as the 1973 Rules were promulgated, Neale asserts, there was 'a body of opinion in the Council of Europe that aspired to a new, more creative and

---

[1] There have been e.g. biennial Council of Europe conferences for directors of prison systems since 1971; other conferences and seminars on specialized topics; working groups e.g. on prison management (see Brydensholt *et al.* 1983) and prison regimes (see Council of Europe 1986); and not least, the dissemination of Europe-wide prison statistics and other information through the establishment of the Prison Information Bulletin.

[2] The development of the European Prison Rules is described at length in Neale (1991).

distinctively European model'. A European Select Committee reported in 1980 on a possible successor to the Rules; the European Committee on Crime Problems commissioned the drafting of a new set of rules; and the Committee of Ministers ultimately adopted the new European Prison Rules in 1987 (Council of Europe 1987).

Neale describes the main changes as an updating of the previous Standard Minimum Rules, including a more specific recognition of the aim of rehabilitation; a remedying of omissions, including more specific recognition of fundamental human rights and dignity; and some presentational changes, intended to make the new Rules fall into a more rational sequence.

What does this 'creative and distinctively European model' look like? The preamble provides a more specific and rather stronger statement of purpose than that found in the UNSMRs. It states that the EPRs are intended to establish minimum standards which will 'provide basic criteria against which prison administrations and those responsible for inspecting the conditions and management of prisons can make valid judgements of performance and measure progress towards higher standards' (Preamble, section d); to stimulate policy development and management style and practice; and to promote professionalism among staff. It observes, in addition, that prisons should normally be expected to operate above the levels established in the EPRs, which are unequivocally designed to be minima rather than norms. In these respects, then, the EPRs signal a stronger consensus on what imprisonment should be than can be found in the SMRs. By implication, too, they speak to a context in which states can reasonably be expected to have the resources that the individualized treatment of inmates is likely to require, and in which there is already much common political ground on matters such as the purpose of prison labour, prison leave, and the social and political rights of inmates.[3]

The EPRs themselves are arranged in five parts. Part I sets out six basic principles, dealing with respect for human dignity, the prohibition of discrimination, the (rehabilitative) purposes of

---

[3] See e.g. the Council of Europe Committee of Ministers Resolution (75)25 on prison labour, Resolution R(82)16 on prison leave, and Resolution R(62)2 on electoral, civil and social rights of prisoners, discussed in Reynaud (1986), which also reproduces them as apps. Consensus e.g. on the role of prison management, prison regimes, and prison leave has been achieved, at least in principle, through Council of Europe study groups, working parties, and select committees of experts: see Brydensholt *et al.* (1983) and Council of Europe (1986).

imprisonment, the inspection of institutions, the protection of inmate rights through a judicial authority, and the availability of the Rules themselves to inmates. Part II (rules 7 to 50) addresses prison management, Part III (rules 51 to 63) personnel, and Part IV (rules 64 to 89) the objectives of treatment. Part V (rules 90 to 100) provides supplementary rules applicable to untried, civil, or mentally abnormal inmates.

There are no particular surprises among the rules, which broadly follow the precepts of the Council of Europe Standard Minimum Rules which preceded them. Since they were constructed, even within Europe, to include a wide variety of national practices and social and economic conditions, they are phrased in fairly general terms. The rules often refer to 'adequate arrangements' (in storing personal clothing, for example—rule 22.3—and the treatment needs of prisoners, rule 66); to standards laid down by authorities (for example nutritional standards laid down by health authorities, rule 25.1) which presume that such standards exist; and to somewhat diffuse objectives, such as developing methods of encouraging inmate co-operation with their treatment (rule 69.2). Since the rules do not say what they mean in so many words, they are supplemented by an Explanatory Memorandum which outlines the thinking behind each of the rules, and sometimes also offers guidelines as to how one might set about implementing them. These notes, though helpful in general terms, are none the less peppered with qualifications such as 'where practicable' and 'wherever possible' when it comes to concrete proposals.

At the same time, some rules appear to have been included as aspirational goals or to generate acceptance of a particular standard, while tacitly recognizing that full compliance with such a standard is not likely to occur in the foreseeable future, in some countries at least. Thus while rule 14.1 provides that inmates should normally spend the nights in individual cells, rule 14.2 specifies that shared accommodation should be occupied by inmates who are suitable for association in such conditions. The Explanatory Memorandum which accompanies the rules refers initially to situations where cell-sharing may be beneficial, such as where inmates are considered potential suicide risks, but appears to accept the necessity of cell-sharing in many other circumstances by laying down some broad guidelines for the management and supervision of shared accommodation.

The EPRs have no formal, legal, Council of Europe-wide authority. Insofar as they have any legal status, it derives from their
recognition in the domestic legislation of individual countries. In
England domestic legislation makes no reference to them, and they
are neither legally binding nor enforceable. In France, the rules are
often bound in with copies of the Code de Procédure Pénale issued
to prison staff, but the legal situation is similar. No complaints
mechanism exists to which complaints of breaches of the EPRs can
be addressed. Nor are the EPRs necessarily bound in with the
processes of the European Commission and Court of Human
Rights, both of which deal with alleged breaches of the European
Convention on Human Rights. Even though the Commission now
appears to regard the EPRs as a 'virtual code for the treatment of
prisoners', breaches of the EPRs do not necessarily imply breaches
of the Convention. Moreover, cases brought to the ECHR which
have cited evidence based on the EPR's predecessor, the Council of
Europe Standard Minimum Rules, have generally not been successful (Council of Europe 1987: 79). Prison governors and administrators may know the detail of the EPRs, but basic-grade prison staff
usually do not. Yet even if governors are fully aware of the rules,
they are not relevant on a day-to-day basis, since they do not form
part of the formal regulatory structure of any prison administration.

Given the lack of specificity in many rules, and the practical
impossibility of enforcing them, what—if any—efficacy do they
possess?

First, as a set of rules promulgated by the Council of Ministers
and thus agreed by all the party states, they have a strong moral
force and may be used as a yardstick against which national regulations and practices may be compared. The Council of Europe
(1987: 79) has claimed that their importance 'lies more in their
influence in raising standards of prison administration than in the
field of legal application', and that they have a 'symbolic importance' in 'inspiring prison administrators and prison personnel in
their pursuit of higher standards' (1987: 81). These phrases may
attract the derision of prison governors who spend their time seeking to fit more and more inmates into poorer and poorer conditions. Yet they have some vestigial function, since governments
have committed themselves to reporting every five years on
progress towards implementing the Rules. It would hardly be

surprising if they did not paint their efforts in the best possible light, and the deliberate openness of many rules, despite their amplification in the Explanatory Memorandum, undoubtedly makes it possible to colour such reports several shades lighter than the reality they purport to describe. Yet the fact remains that they must, every five years, go through such a process, and that certain shortcomings cannot be completely ignored whatever complexion is put on them.

Second, even though the EPRs are not directly enforceable, they may be cited in evidence in domestic inmate litigation. In addition, both the ECHR bodies and the European Committee for the Prevention of Torture (discussed below) have made reference to the EPRs in their decisions and reports. The Rules do, therefore, stand as a kind of benchmark against which moral judgements about prison conditions can be made.

Third, member states are continually involved in meetings, conferences, working parties, and the production of reports which involve them in consideration of their own practices in relation to the EPRs. And fourth, there is a resultant pressure for administrations to harmonize national legislation with the EPRs. This is, it must be said, resistible in the short term; but pressure of this sort is continuous, and long-term non-compliance demands repeated justifications and explanations.

None of this is to suggest that the EPRs have a direct or immediate impact on prison conditions. They do not. But they do place an indirect and long-term pressure on governments to improve their prisons. The pressure is not likely to be great, because no government can criticize others with the confidence that would come with absolute compliance. None the less it has the same kind of effect as a dripping tap; there comes a point at which fixing the problem is psychologically easier than continuing to try to ignore it.

The European Commission and Court of Human Rights

The European Convention on Human Rights, which came into force in 1953, was based on the 1948 United Nations Universal Declaration of Human Rights. It was intended to secure human rights within Europe, and the European Commission and Court of Human Rights, which were created under the auspices of the

Convention, thus form an important means of complaint on human rights matters. Although the Convention is applicable in a large number of areas, imprisonment is one of those where human rights considerations are never far from the surface, and it is not surprising that roughly one third of all registered applications have been made by persons in detention.[4]

Many of the provisions in the Convention have some bearing on imprisonment. Article 11 includes the right to join trades unions, Article 12 the right to marry and found a family, and so on. However, the principal provisions relating to inmates' complaints are Articles 3 (prohibiting torture, inhuman or degrading treatment or punishment), 4 (prohibiting the holding of persons in slavery or servitude), 5 (dealing with the conditions under which detention is lawful and allowing review of the legality of detention), 6 (presumption of innocence and the right to a fair trial), and 8 (respect for private and family life, home and correspondence).

Although allegations of breaches of the Convention may be made by any party to it against any other party, all but a handful of complaints are made by individuals against their own governments. These applications are considered first by the Commission, which determines whether they are admissible—that is, whether there is a prima-facie case of a breach of the Convention. If the case is admissible, the Commission then makes an investigation and produces a provisional opinion on the complaint, and tries to seek a settlement between the complainant and the relevant government. If such a settlement proves impossible, the Commission then makes a formal report to the Committee of Ministers on whether or not the Convention has been breached, and the latter makes a determination as to the existence of breach. Only at this point can cases be referred to the Court, and the Court's role is essentially to adjudicate cases where the Commission is unable to produce a settlement. Cases can, however, only be referred to the Court by a government or by the Commission.

These procedures make resort to the Commission and the Court a tortuous matter. In order for a case to be regarded as admissible, all domestic remedies must be exhausted, and the Commission must accept that the case discloses prima-facie evidence of a breach of one or more of the Convention's provisions. Most cases thus

---

[4] Fawcett (1985) and Reynaud (1986) are both useful descriptions of the working of the European Convention, Commission, and Court of Human Rights.

arrive at the offices of the Commission several years after the event, and some 90 per cent of all cases submitted are determined inadmissible on the grounds that no prima-facie breach of the Convention is evidenced. For the 10 per cent that jump this hurdle, there remains the possibility that the Commission or, ultimately, the Court, will determine that no breach occurred. Moreover, the length of the process is daunting. Many, if not most, cases are settled between five and eight years after they were initiated; and in a large proportion of cases the final decision is made some years after the complainant has been released from prison.

To pursue this negative line a little longer, it is also important to appreciate the kinds of prison treatment and conditions that have been determined *not* to be breaches of the Convention. Article 10, dealing *inter alia* with the right to receive and impart information, has not been seen as giving rise to a right for inmates to see instructions and regulations for prison staff (Application 1860/63). Nor does it oblige prison authorities to divulge the names of the officials who sit on internal committees, such as security classification review committees (Application 8575/79). Article 3 proscribes inhuman and degrading treatment; however, certain complaints made by terrorists who were thought dangerous and held in maximum security units were considered inadmissible by the Commission. The circumstances complained of, and apparently regarded as acceptable in these cases, included segregation from all other inmates for periods of up to two months (though they apparently had limited possibilities to associate with other members of their gang in the unit), and intensive staff supervision (Applications 7572/76, 7586/96, 7587/76). A further case of the same nature was ultimately declared admissible, though no breach of Article 3 was found (Application 8463/78). In that case the inmates had been held for up to ten and a half months in conditions that included, at various times, permanent artificial lighting in the cell (later replaced with continuous television and infra-red surveillance in the cell), insulation from noise made by other inmates (by leaving empty the cells above and below the ones they occupied), and restrictions on access to newspapers and radio. Further, the use of a bread-and-beverage diet and a hard bed, for a period of seven days while in disciplinary segregation, while considered out of step with modern standards, was not inhuman or degrading (Application 7408/76). Reynaud (1986: 78–81) also sounds the cautionary note that while complaints alleging physical ill-treatment—for example, the use of excessive force—are treated seriously and investigated at length, most are ultimately declared inadmissible, or the use of force is deemed not to constitute inhuman or degrading treatment.

The ECHR bodies, on this evidence, clearly take a fairly robust line as to what kinds of treatment, conditions, or force may be regarded as compatible with fundamental human rights. And not without good reason. Many of those who bring complaints of ill-treatment before the Commission are serving lengthy sentences for serious offences, and the levels of security and control applied to them can be seen as prudent. Reynaud also describes a number of complaints about treatment which was precipitated by threatened or actual violence on the part of the inmate.

None the less, the generally cautious approach to allegations of inhuman and degrading treatment has not prevented the Court from making forceful criticisms of prison authorities in some instances. In particular, the treatment given to suspects in military barracks in Northern Ireland—wall-standing, hooding, sleep deprivation, and so forth—was considered inhumane (*Ireland* v. *UK*, 1978).[5]

Inmates have, however, been able to obtain a number of very important advances in other areas. The European Court has taken a firm stance on the issue of the length of remand, with periods of more than about three years deemed excessive.[6] The general right of inmates to correspond freely and to express complaints about prison to persons outside was dealt with in two cases, *Golder* (1975) and *Silver* (1983); in a more recent case, *S.* v. *Switzerland* (1992), the monitoring of 'almost all' a remand inmate's correspondence with his lawyer was held to be in breach of Article 8.[7] Rights in relation to fair decision-making procedures, due process, and the separation of judicial and administrative decision-making have also been clearly established. In a decision of 28 June 1984 (the case of *Campbell and Fell*), the Court determined that lengthy

---

[5] *Ireland* v. *UK*, application 5310/71, decision of 18 January 1978. See also the discussion of this case in Fawcett (1985).

[6] See *Clooth* v. *Belgium* (1992), ECHR Reports, Vol. 14, case A/225; 3 years, 2 months and 4 days deemed excessive. In *Birou* v. *France* (1992), Vol. 14, case A/232–B, in which 5 years and 3 months' remand was judged excessive. At least 6 similar cases were decided by the European Court in 1992.

[7] *S.* v. *Switzerland* (1992), ECHR Reports, Vol. 14, case A/220.

periods of loss of remission (in this case amounting to some 570 days), ordered as a punishment by the boards of visitors, did not meet the requirements of Article 6. The key point was that the period of remission at stake, and the charge involved—mutiny—made the case one that should properly have been dealt with by a court or court-like body. The Boards of Visitors, while meeting several criteria of a 'court-like body', did not hold a public hearing, nor allow access to legal representation. While some lesser charges and lesser sentences have been regarded as conforming with Article 6, this case established a point at which judicial rather than administrative procedures should have been used.

However, where decisions have been made in favour of the complainants, governments have not always remedied the matters complained of. The case of *Brogan* (Application 11209/84), though not directly concerned with imprisonment, provides an illustration. This complaint alleged that the period of police detention experienced under the provisions of the UK Prevention of Terrorism (Temporary Provisions) Act 1984 was contrary to Article 5(3) of the European Convention, which requires suspects to be brought 'promptly' before a judge, or released. The Act permitted a maximum detention of 7 days (48 hours by the police, extended by 5 days by the Secretary of State). The four complainants were detained for periods of between 4 days 6 hours and 6 days 16.5 hours. The Commission, in its opinion, found that the two longest periods of detention (5 days 11 hours and 6 days 16.5 hours) breached Article 5(3). The Court subsequently held that all four periods of detention breached this provision. However, the British government did not revise its time limits for detention. Nor did it move control over the detention from the executive (i.e. the Secretary of State) to the judiciary. It effectively re-confirmed the time limits in new anti-terrorist legislation, and used Article 15(1) to derogate from its obligations under the Convention. Article 15(1) provides that, with some exceptions, High Contracting Parties may withdraw from the Convention's requirements 'in time of war or other public emergency threatening the life of the nation'.[8]

---

[8] For a fuller discussion of the Brogan case and its implications, see Tanca (1990).

The European Convention for the Prevention of Torture

This Convention came into force on 1 February 1989. Like the European Prison Rules and the European Convention on Human Rights, it is to a large extent derivative of a United Nations initiative, the 1984 Convention Against Torture. Its main objective is to promote adherence to the provisions of Article 3 of the Human Rights Convention, namely, that 'no one shall be subjected to torture or to inhuman or degrading treatment or punishment'. This it does by the formation of a European Committee for the Prevention of Torture and Inhuman or Degrading Treatment or Punishment (or ECPT), which has a remit to conduct inspections of custodial facilities in member states and to report on them.

The rationale for the establishment of such a committee is that while individuals may apply to the European Committee of Human Rights on matters relating to Article 3 of the Convention on Human Rights, and while the Commission may make fact-finding visits to institutions in their investigation of such complaints, a broader and more regular programme of inspections may lead to correspondingly broader recommendations which are not based on specific facts raised by an individual complainant. The ECPT's first annual report states that 'whereas the Commission's and Court's activities aim at "conflict solution" on the legal level, the CPT's activities aim at "conflict avoidance" on the practical level'. It is also worth mentioning that while the ECHR bodies take a fairly robust line as to what they consider inhuman or degrading, and while the use of the word 'torture' can and does raise hackles among some officials, there is no express implication that physical torture is widespread in any of the member states (though this comment will be somewhat qualified below). The ECPT would probably wish to cast its work in the more positive light of a mission to promote standards of humanity and human dignity in custodial arrangements of whatever kind. In its second Annual Report it states that its role is 'not to publicly criticise States, but rather to assist them in finding ways of enhancing the protection of persons deprived of their liberty from ill-treatment'.

The membership of the ECPT and the nature of its work thus far has been described in Evans and Morgan (1992), and also in the Committee's first two annual reports (Council of Europe 1991,

1992). As at the end of 1991, twenty-three states were parties to the Convention.[9] The membership of the ECPT is, in principle, composed of one member from each party state. The members are elected by the Committee of Ministers of the Council of Europe, from a list of names ultimately constituted through nominations from member states. Members are elected for a period of four years and may serve no more than two terms. Thus the ECPT's membership is comprised of persons who have professional experience or competence in matters germane to human rights. The membership thus includes, for example, Prof. Dr. Günther Kaiser, one of the most prominent German criminologists of the last quarter-century. Although vacancies have occurred on the ECPT, they have primarily arisen through the time taken to replace members who have resigned, or to identify suitable names for inclusion on the list of candidates. Members sit as individuals and not as representatives of their governments; this arrangement is intended to strike a number of compromises between the impartiality of members and the retention of some level of influence by States Parties (that is, the states which are parties to the convention) over the process as a whole. In its work, however, the ECPT is served by a secretariat, and in addition is entitled to call on experts to accompany it on inspections. Since the members of the ECPT themselves tend to be persons without an operational, inspectoral, or empirical research background, it is likely that both secretariat and experts exercise considerable influence at a practical level within the inspection process.

The main activities of the ECPT comprise visits to custodial facilities. It is important to remember in this connection that the brief of the ECPT extends not only to prisons, but to all forms of detention. Thus institutional visits have been made to police stations, detention centres for asylum seekers, and psychiatric clinics. However, the discussion below primarily reflects the situation so far as prisons are concerned. Inspection visits may be programmed some time in advance ('periodic' visits), or be made *ad hoc* on the basis of particular issues arising, for example, within a single insti-

[9] The States Parties were: Austria, Belgium, Cyprus, Denmark, Finland, France, Germany, Greece, Iceland, Ireland, Italy, Liechtenstein, Luxembourg, Malta, the Netherlands, Norway, Portugal, San Marino, Spain, Sweden, Switzerland, Turkey, and the United Kingdom. Czechoslovakia (which has since become two separate republics), Hungary, and Poland, while member states of the Council of Europe as at the end of 1991, had not at that stage become parties to the Convention.

tution. Decisions as to what kinds of visits are made, when, and where, are made on the basis of a number of factors. First, there is an ultimate intention to have made 'periodic' visits to all countries by 1993 and thereafter to visit each country every two years. Second, the countries selected for visits have thus far been chosen by lot, thus avoiding arguments about the reasons for the selection of particular states. However, as of the end of 1992, two *ad hoc* visits had been carried out, and both were to Turkey. It seems that these visits were—and future *ad hoc* visits may also be—based on information received from international human rights organizations.

The procedure for a visit is a compromise between formal, planned visits giving sufficient warning for states to paper over inconvenient problems, and surprise visits with the practical logistical difficulties that they can present. States are informed some time in advance that a visit will take place, but are only told the dates and duration about two weeks in advance, and the provisional list of institutions to be visited is only presented to the state a couple of days prior to the inspection. Each inspection begins and ends with a series of meetings between ECPT delegates, together with the accompanying members of the secretariat, experts and interpreters, and the appropriate national ministers and criminal justice officials. The substance of the inspection is a number of visits to institutions—the team may split up into sub-groups—during which members speak to staff members and inmates, and observe conditions within the establishments. At the second meeting with national authorities, the team may discuss or seek clarification of issues arising from the inspection, and give some brief indications of the nature of their report. The team then reviews and discusses its inspection before individual team members are delegated to draft sections of the report.

The draft reports are currently considered and adopted by the whole ECPT at a single meeting, and then sent to the government concerned. However, the report is considered confidential until and unless the government itself decides either to publish it or to allow the ECPT to do so. Publication may be accompanied by a response from the state.

What of the visits so far conducted and the reports published? In 1990, visits were made to Austria, England, Denmark, and Malta. Reports from the first three countries were published, in each case between seven and nine months after the inspection. The paper on Malta was published in October 1992, after a delay of thirty-three months. In 1991, visits were made to France, Germany, Spain, Switzerland, and Sweden. As of late 1993, only the reports on the German and Swedish visits had been released.[10] No formal reports on the *ad hoc* visits to Turkey have been published; however, in December 1992 the ECPT issued a public statement covering both these visits and a third, periodic, visit which took place in November/December 1992. This is discussed in more detail below.[11]

The reports themselves are highly informative documents, not least because each one summarizes, at a minimum, the legal arrangements for police interrogation and custody, pre-trial detention, and imprisonment in the relevant country. The substance of the reports is also worthy of note. In relation to Austria, the delegation found prison conditions to be largely satisfactory, with the exception of washing arrangements for female prisoners. It did, however, note the poor regimes and conditions in 'police jails'— that is, institutions operated by the police for pre-trial detainees. And it made a number of criticisms of arrangements for police custody, including two specific incidents. In one case an inmate alleged that he had been slapped while under police interrogation; in another, that he had been held for four hours in handcuffs while awaiting interrogation. The report states that while it was itself unable to determine the facts of the cases, it felt that these and similar complaints should not be dismissed out of hand. Some twenty-nine recommendations then followed, some simply being requests for information, but others comprising a number of detailed and linked proposals. The majority of these dealt with police custody, and were directed at improving physical conditions for suspects, for example through overnight provision of mattresses; the prevention of abuse, through increased documentation

[10] However, 2 reports on visits made in 1992 were released in the first half of 1993. A report on Finland appeared 11 months after the visit, and one on the Netherlands only 9 months after the visit.

[11] Full refs. may be found in the ref. section at the end of this book. The report on Austria and the comments of the Austrian Government are Council of Europe (1991b, 1991c) respectively; on England and the response of the English Government, Council of Europe (1991d, 1991e) respectively; on Denmark, Council of Europe (1991f); on Malta, Council of Europe (1992b); on Sweden and the Swedish Government response, Council of Europe (1992c, 1992d) respectively; and for the public statement on Turkey, Council of Europe (1992e).

of detained suspects; and the requirement of police to inform suspects of their right to inform third parties and have legal advice. The response of the Austrian government basically amounted to three observations. First, a law on police duties and powers was being drafted and would address both interrogation procedures and record-keeping. Second, many of the comments made by the ECPT corresponded to long-term aims of the government. But third, action would have to be taken step by step over a long term.

The second report, on England, dealt primarily with Wandsworth, Brixton, Leeds and Holloway prisons. It introduced a division of the ECPT's observations into three categories: comments, requests for information, and recommendations. These three sets of observations amount to eleven pages of single-spaced typescript in the report, and are critical of regime factors including overcrowding, lack of work, and hygiene. Most crucially, the report states at para. 57:

Overcrowding, lack of integral sanitation and inadequate regime activities would each alone be a matter of serious concern: combined they form a potent mixture. The three elements interact, the deleterious effects of each of them being multiplied by those of the two others . . . In the CPT's view, the cumulative effects of overcrowding, lack of integral sanitation, and inadequate regimes amounts to inhuman and degrading treatment. This is a matter that needs to be addressed with the utmost urgency.

This view reflects the ECPT's contention that inhuman and degrading conditions can occur as the result of the 'totality of conditions'. While assessments can be made of specific aspects of prison regimes, attention should also be paid to the way in which they interact. In this regard, the ECPT seems to be following the principle first established in American litigation (*Rhodes* v. *Chapman*, 1981) that a combination of factors, none of them individually unconstitutional, may combine to produce a situation of 'cruel and unusual punishment' contrary to the US Eighth Amendment.

In many of the specifics, the UK government's reply tackles individual criticisms head on. However, on this point—namely that the ECPT found what it considered to be inhuman and degrading treatment—it argued that while conditions at the three male prisons, Wandsworth, Leeds and Brixton, were poor, they were not so poor as to warrant such a description.

England is thus far the only country, or at least the only one on which a report is published, to have been told outright that conditions in several of its major prisons amounted to a breach of Article 3 of the European Convention on Human Rights. Yet neither Denmark nor Sweden, both often considered models of progressive and enlightened corrections, have escaped unscathed. The report on Denmark refers to two cases in which foreign inmates alleged they had been beaten. In relation to the Police Headquarters Prison, one wing of which is used for persons in police detention, the report rather dryly comments (para. 22):

As regards the rubber truncheons about 50 cm long which the delegation observed in a cupboard in the reception office of the Police Headquarters Prison, the CPT recommends the Danish authorities to ensure that their use is strictly limited to the cases mentioned [in an internal circular] on the use of force on prisoners.

The Swedish report makes several comments on different establishments, including criticisms of the space available for exercise in the main remand prison. But, more importantly, it echoes comments made by prison governors on the restrictions on visits and other contacts which prosecutors may impose on remand inmates. These, the ECPT suggests, frequently amount to solitary confinement, and are used in far too liberal a fashion.

The degree of confidentiality that surrounds the work of the ECPT means that it is in no sense a form of public accountability. All discussions take place behind closed doors. The structure of the committee reflects a number of political compromises. The published reports have been, with one main exception, reserved rather than outspoken. Yet the knowledge that assessments of prison conditions are being made by individuals who are relatively influential in their own spheres, and the blunt fact that a refusal to allow publication of a report will be seen by outsiders as an implicit admission of the poor state of custodial facilities, both add to the pressure for higher standards of inmate treatment. It is true that states must volunteer for this kind of scrutiny, and that in so doing they are presumably reasonably confident that they will not create major problems for themselves (though it is also true that refusal to engage in this exercise would also be regarded negatively by outsiders). The kinds of pressures states are subject to from this source are resistible, perhaps over periods of many years. But this is not a short-term exercise. The experience of most Council of Europe initiatives is that results can only be looked for over periods measured in decades.

Given its sensitive role and the sovereignty of nation-states, the ECPT has no sanctions it can use to force compliance with the Convention it polices. The most serious step it can take is to issue a statement on a particular matter. This can only be done on a two-thirds majority of the Committee; and the report or reports which gave rise to the problem remain unpublished. It seems initially that the issue of a public statement was envisaged in situations where a State Party failed or refused to co-operate with the ECPT or to address the issues raised in recommendations. Thus far, the only public statement to have been issued concerns Turkey.

The statement begins by outlining the reasons for the three visits to Turkey, in 1990, 1991, and 1992. There had been a 'considerable number of reports received by the Committee, from a variety of sources, containing allegations of torture or other forms of ill-treatment of persons deprived of their liberty in Turkey. The reports related in particular to persons held in police custody'. The statement then summarizes evidence of torture gathered in the visits, including medical evidence, and the discovery in certain police stations of equipment which could be used for torture and for which no other credible explanation was forthcoming. It observes that no action had been taken following the two *ad hoc* visits and that the situation which gave rise to concern had thus continued without remedy for at least two years. And it reiterates the point that its concern lies solely with police detention rather than prisons or psychiatric hospitals. Finally, the statement indicates the corrective action necessary. It recommends professional training for law enforcement officials and consideration of conditions of service, but also vigorous action by public prosecutors who receive complaints of torture, and the establishment of an effective independent monitoring mechanism. In a discussion of a new (1992) law on terrorism, while welcoming the recognition of fundamental human rights in the body of the law, it criticizes the length of incommunicado detention, other police custody, and an apparent lapse in the provisions for certain suspects to contact a lawyer. Finally, the statement ends with the hope that it will be taken in the constructive spirit in which it was issued.

In the report on England, the conclusion of the ECPT that Article 3 of the European Convention on Human Rights was not observed was simply and roundly denied by the government. This is, then, one of the limits of the ECPT system. The Turkish case demonstrates another limit. Undoubtedly the issue of such a document, against their wishes, is highly embarrassing for the Turkish government. But it is by no means clear that embarrassment in itself is an effective force for change. Not only does there have to be a political will; it must be carried through into decisive action. Even though the government is committed to an eradication of torture, the ECPT statement seems to imply that police conditions of service are too poor to give the government the means fully to control its police force. Attempts to eradicate torture, like attempts to improve prison conditions or inmate rights, assume implicitly that governments possess the means to control the situation. Where this becomes problematic, such attempts are futile until and unless the basic problem of the authority of the state over its servants is remedied.

Consequences of the Enlargement of the Council of Europe

By the early 1990s, political developments in Eastern Europe had had major implications for the Council of Europe, not least in relation to prisons. The reunification of Germany meant that the old East German prison system, with its typically poor levels of provision, had come within the ambit of West German law and administrative procedures and thus was also covered by Council of Europe conventions. Czechoslovakia was admitted to the Council of Europe, but its two constituent republics subsequently separated and have both re-applied for membership. Hungary and Poland were also admitted to the Council of Europe, bringing its membership to 26 countries (excluding, for the moment, the Czech and Slovak Republics). Czechoslovakia, Hungary, and Poland all engaged in extensive discussion with the Council, prior to joining, on matters such as prison rules. Czechoslovakia signed the European Torture Convention in December 1992 (though both new republics must now begin the signing process again), and Hungary in February 1993. By the end of 1992, however, Poland had not become a signatory to the Convention.

In the long term, the impact of Council of Europe membership on the Eastern-European penal systems which now fall within the Council's purview is likely to be extensive. On the one hand, the

Czech, Slovak, Hungarian, and Polish authorities have all signalled their intention to replace or improve their criminal-justice arrangements, and new laws have enshrined or entrenched principles such as respect for the rule of law and inmates' rights. In Poland at least, a large proportion of members of parliament had an immediate interest in penal affairs, since they had had experience of custody as political prisoners.

On the other hand, however, all three countries faced serious problems in implementing these desiderata. Large numbers of legal and criminal justice personnel resigned or were fired; many of those who remained in post none the less had been officials under the old systems, and their capacity to operate within the new frameworks, if not their political sympathies, were in doubt. Plans to require such officials to submit to vetting procedures, in Poland and Germany at least, became problematic (as did the firing of officials for previous political affiliations), bearing in mind the European Convention on Human Rights and its guarantees of freedom of belief.[12] Funding for the criminal-justice system was also jeopardized by the economic situation in Poland, Hungary, and Czechoslovakia, both in terms of government expenditure restraints and the rate of inflation, which meant in Poland, for example, that funds initially voted for 1991 were exhausted within about five months.[13] However, the breakup of Czechoslovakia had less effect on the prison system than might have been expected. The Czech and Slovak Republics had been administered as separate entities with distinct prison systems. While those systems had acted together, for example in publishing joint ministerial circulars, the separation of the republics did not in itself pose any serious practical problems.

Meanwhile, in all these Eastern European countries, the combination of policing problems, amnesties for certain categories of

[12] In addition, the UN International Covenant on Civil and Political Rights states, *inter alia*, 'Everyone shall have the right to freedom of thought, conscience and religion' (Article 18(1)). Article 19 follows this by stating: '(1) Everyone shall have the right to hold opinions without interference', though subsection (3) outlines circumstances under which this right may be limited. Article 26 states: 'All persons are equal before the law and are entitled without any discrimination to the equal protection of the law. In this respect, the law shall prohibit any discrimination and guarantee to all persons equal and effective protection against discrimination on any ground such as race, colour, sex, language, religion, political or other opinion, national or social origin, property, birth or other status'.

[13] Information provided by the director of the Polish prison service to a conference on pre-trial imprisonment, Kazimierz, Poland, 1991.

inmates, and economic turmoil resulted in high crime rates and thus large prison populations; while prisoners' expectations, both of improvements in conditions and of further amnesties, resulted in disturbances. Staffing problems also meant that several prisons in Poland, at least, were effectively taken over by inmate gangs.[14] The old East Germany faces some similar problems, with much of its prison estate in poor condition and certain forms of prison labour no longer being regarded as appropriate in the new situation.[15] Many inmates, too, have complained that social-welfare provisions under the socialist system are no longer available to them. The same applies also to inmate wages, which were higher as a percentage of the average worker's income in the old East Germany than they are in the new.

There is a willingness to change in all these countries, though initial liberal views are being challenged by more politically conservative doctrines as these problems continue. Some of the economic problems may be ameliorated as financial aid becomes available and as the economies grow and stabilize. But the difficulties these countries now face are unlikely to be solved easily or quickly.

Conclusions

There are ways in which the measures discussed above—the European Prison Rules, Commission and Court of Human Rights, and Torture Convention and Committee—have made prison systems more accountable for their treatment of inmates. With some exceptions, these forms of accountability are very largely symbolic; they are based on documents which are aspirational, and are cast in the framework of a partnership and progress towards good practice. Results cannot be expected in the short term. However, these features of the European situation have led, however slowly, in the direction of greater accountability.

There is no overt pressure on member states to participate in these arrangements, but failure to do so sharply raises the question of why there is a reluctance, and implies that state arrangements do not meet what amounts to a European consensus as to minimum standards of provision and practice. In so far as members must account for short-

---

[14] This claim was made in a presentation by Zbigniew Ho da and Andrzej Rzepliński at a conference on pre-trial imprisonment, Kazimierz, Poland, 1991.
[15] See Dünkel (forthcoming).

comings under these measures, they must account ultimately to the Council of Ministers (or the European Court of Human Rights) rather than any national body. The Court is not soft on inmates. The Council is likely to be rather forgiving, in so far as member states know the difficulties of compliance and themselves fall short of the agreed standards in various ways. Much of the discussion at this supra-national level undoubtedly revolves around legal rather than practical changes, redefinitions of situations, and proposals for long-term change that governments know, realistically, will not have to be implemented in the course of their own term of office.

As the English experience shows, too, governments can respond to criticism by insisting on their right to derogate from certain provisions of the European Convention on Human Rights for specific reasons, or by refusing to accept the view of the Torture Committee on the definition of 'inhuman and degrading treatment'. Moreover, and particularly with regard to the Torture Convention, much of the discussion can, in the short term at least, be kept behind closed doors.

Having said all this, persistent criticism, even if behind closed doors, has a gradual attritional effect, and in the long term we can hope that it is effective in promoting improved standards. Thus far, the most significant steps forward have been in the area of inmate rights, for example in relation to correspondence and complaint, as in the *Golder* and *Silver* ECHR cases. Whether significant steps can be made in other areas, such as the physical conditions of prison environments, remains to be seen. The development of the European Committee for the Prevention of Torture is at least capable of putting not only physical abuse, but also observations about the totality of conditions, on the table. Yet despite programmes of prison building and renovation in England, France, and the Netherlands, it seems that practical improvements remain a long-term aspiration. In Turkey, and in Eastern Europe, discussions about such improvements must be linked to wider questions about the authority of the state over its own employees, while the issue of expenditure priorities, already a major block to reform in Western Europe, is being felt much more keenly in the East. It is as yet unclear that inmates will be better treated in the first half of the twenty-first century than they are in the closing years of the twentieth.

13

Accountability and the Private Sector

THE use of private companies to manage prisons is probably the most controversial prison issue of the late twentieth century, and a great deal has been written about it. I do not propose to treat the topic comprehensively. Donahue (1989) summarizes American developments and experience. Matthews (1989) and Ryan and Ward (1989) set out the terms of the English debate, though their discussions pre-date the actual introduction of private prisons. The flavour of the French debate is given in Boulin (1987), while an account of the actual scope and implementation of private-sector involvement in France is to be found in Delteil (1990). My sole concern in this chapter is the question of how private-sector involvement in prison management can or might affect the ways in which prisons are made accountable for what they do. However, some brief scene-setting may be useful.

During the 1980s, right-wing political parties and doctrines experienced a resurgence throughout Europe. On the one hand, such doctrines implied, if they did not overtly call for, increased prison sentence lengths, especially for serious offenders.[1] It was recognized that additional expenditure was needed on the prison system, *inter alia* to house the projected additional prison population. Even though prisons are generally considered to be an unpopular and low-priority item of government expenditure, the decade brought about an expansion of prison building unprecedented in this century, in England and the Netherlands. There were also large increases in the numbers of staff and the cost of payrolls.[2]

---

[1] Through e.g. the changes in the UK parole system (see Maguire *et al.* 1984), and, in France, through the 'Security and Liberty' law of 1981 and its subsequent revisions.

[2] For a description of the English situation, see Morgan (1983). A similar expan-

At the same time, however, right-wing doctrines suggested that if the private sector could be more efficient than government in any area of activity, it should be encouraged to take over provision on a contract basis. More extreme positions (not always accepted within governments) held that the rolling-back of the state should mean that any state function which was capable of being privatized should be handed to the private sector on principle, whether or not the result was increased cost-effectiveness.

Such arguments were being launched in a general situation of economic recession, and the need for anti-inflationary policies resulted in a search for ways to reduce state expenditures. Prisons, under pressure brought about by rising prison populations— themselves partly the product of 'get tough' policies, and more and longer prison sentences—were not immune from these trends. There were, therefore, good and sufficient reasons, doctrinal, economic, and managerial, to seek strategies aimed at cost-effectiveness and, ultimately, at reducing both capital and recurrent expenditure. In England, which is something of a special case because of the power of the Prison Officers' Association, ways of reducing the union grip on the prison system were also being mooted.

The combined effect of the political and fiscal considerations, in England and France, was an argument that the private sector should be given the opportunity to run prisons. Models for private prisons might already be found in the USA. In addition, in England and Scotland, it was proposed that prisons might be floated off from core government expenditure by being given agency status.

What privatization means in practice can be rather variable. It may mean contracting out the whole of the running of a prison. Such contracts can be based on a fee per prisoner per day, or simply an agreed annual sum paid by government to the contractors. However, smaller elements of privatization can be introduced into what is essentially a government operation. For example, Dutch prisons have for some years employed private contractors to operate prison kitchens without prison labour. Outside contractors may operate workshops with prison labour supervised by their own managerial staff, as happens in France. Prison buildings may be

sion in the numbers of staff, though not the prison capacity (which was already sufficient) occurred in Germany. In e.g. Baden-Württemberg, the number of inmates grew by 17% between 1976 and 1985; the number of prison staff grew by 43% in the same period; yet overtime was still considered to be a major problem.

leased, or a contract signed covering maintenance as well as initial building; plans for this were drawn up in France. Prisoner escorts may be by private coach, with a civilian driver and prison staff simply to provide security, as is sometimes the case in England. The major debate, however, concerns the privatization of entire establishments. Within this, most of the concern has been about issues which are germane to any discussion of accountability. They include the questions of how prison conditions and regimes will be monitored, the powers granted to private contractors, the extent to which it would be possible to overcrowd private prisons (and, by implication, levels of overcrowding in the remainder of the prison system), and the responsibilities of private contractors for security, both generally and in the event of major disturbances.

The first section of this chapter deals with the ideological arguments for and against private, or quasi-private, prisons. The second reviews the American experience, which provided the models that were further considered in both England and France. Three further sections review the English arrangements for privatization; the English and Scottish discussions of the no less important concept of 'agency'; and the ultimately less-radical French version of private-sector involvement. Although comments on the problems and prospects of private-prison accountability are made along the way, they are presented in a rather more formal way in the final part of the chapter.

Ideological Arguments

The notion that prisons should invariably be the responsibility of a centralized state administration is historically no more than 150 years old, and in several countries rather less than that. The arguments for the assumption of state responsibility were essentially that private, local, municipal jails were poorly funded and badly run, and exploited and mistreated their inmates. In the 1980s, however, the reverse argument came to be applied. State-prison systems, first in the US but later in England and France, were seen as inefficient, poorly-managed, and with impoverished physical conditions and regimes. One solution—increasingly held to be *the* solution—was to allow private companies or other non-governmental organizations to operate institutions on a contractual basis.

This was held to have a number of practical, economic advantages. First, bureaucracy could be streamlined and many financial decisions delegated to institutional level. Private-sector financial controls were thought to be more flexible and cheaper to operate than centralized governmental procedures. Nor would private companies be constrained by at least some kinds of public policy considerations. If they wished they could provide all inmates with television in cells; this has been claimed to be cost-effective in preventing boredom and thus disciplinary problems. Government agencies wishing to make such improvements would first have to overcome political and treasury opposition. Second, and more importantly, private organizations would not be tied to government pay scales or, indeed, to existing agreements with unions. They could operate more cheaply by adjusting pay scales, while also seeking new agreements as to productivity and manning levels. Third, the resources freed through these increased efficiencies could be absorbed by the state—in other words, prisons would be cheaper to operate—or ploughed back into regime improvements. Fourth, from the government's point of view, prison-system costs could be more easily controlled through fixed contracts and/or agreed charges per inmate per day, while many of the unforeseen financial risks would be shouldered elsewhere.

At the same time, arguments based on political doctrine were being deployed. The weakest such argument, because it is patently untrue, has been that the delegation of prison-management decisions to private companies will take the political heat out of the prison issue. Notwithstanding the political debate generated by privatization proposals, this argument ignores the point that in so far as governments contract companies to run prisons, the ultimate responsibility for regimes and good governance remains with the contract issuer, since poor performance should result in the termination of a contract. Similarly, the perceived need for a private-prisons solution can be treated as a symptom of the persistent overcrowding of prisons. It thus feeds the reductionist debate about how the size of the prison population can be justified.

The stronger and more important argument, however, concerns the 'rolling back of the state'. This doctrine adopts the view that while governments should ensure that certain kinds of facilities are provided, they are under no obligation to provide those services directly. Any area of government responsibility that can be performed equally or more efficiently by a private organization should be contracted out, and state expenditure—and thus taxes—can be reduced. No area is sacrosanct. Health, education, social welfare, criminal justice, and even tax assessment and collection and certain military defence facilities can be operated in this way. The force of this view has not been diminished by the blunt fact that the record of the pro-privatization English Conservative government was, since 1979, one of increasing centralization and a reduction in the powers of local authorities.[3] And any brief inspection of the degree to which criminal justice was in state hands would in any case conclude that a large number of facilities, from juvenile homes and probation hostels to immigration detention facilities, were actually operated by private companies or voluntary organizations. What was being proposed was a change in scale, with contracting-out becoming a core rather than a peripheral feature of criminal justice.

The argument on the level of principle, in England at least, was effectively spiked by the pre-existence of private, or at least non-government, institutions of various kinds for offenders. (In France, there was more argument about the constitutionality of private prisons, and this will be discussed later.) Opposition to privatization therefore rested on practical arguments, most of which were related to a concern with accountability. How, for example, would the Prison Department ensure that the terms of a contract were being met? Would the combined costs of the contract, plus the monitoring arrangements, really turn out to be cheaper than the direct operation of prisons? Would a 'private prisons lobby' emerge with a vested interest in a continuing large prison population? What powers, in relation to the use of force, for example, would be provided to private prisons staff? What complaints and disciplinary mechanisms would be installed in private establishments? What would happen if the private companies, faced perhaps with a riot, decided to withdraw on short notice? What would be the impact of private prisons on the remainder of the prison system, for instance in terms of overcrowding, if private establishments had agreed capacity limits which could not be exceeded? And, most

---

[3] During the period in question a number of state-owned enterprises, including energy utilities and British Airways, were sold off, though it was frequently argued that the end result was that the shares were quickly acquired by foreign companies. It was only in 1992/3, with widespread opposition to the privatization of the rail network, that the privatization initiative appeared to encounter serious opposition.

importantly, would the terms of contracts be regarded as commercially privileged information and thus kept confidential?

The American Experience

Three factors together explain the popularity of privatization in America. First, the number of prisoners and the incarceration rate increased dramatically from the 1970s to the mid 1980s. Blumstein (1988: 232) states that:

In 1970, there were 196,429 prisoners in state and federal prisons, and the incarceration rate was 96 per 100,000 population . . . By the end of 1986, the prison population had nearly trebled, and the incarceration rate had more than doubled to 227 per 100,000 population.

This rise occurred despite a long-term trend towards increased use of community corrections, so that the proportion of offenders being sent into custody had dropped.[4] Second, litigation against state-prison systems in the 1970s often used the constitutional provision against 'cruel and unusual punishment', arguing that the totality of prison conditions, including overcrowding, were unconstitutional. The courts frequently ordered prison systems to reduce their populations, using a variety of mechanisms including automatic parole of certain inmates when numbers increased, and refusing to admit new inmates from from the county jails into state penitentiaries (thus increasing overcrowding in the county jails). But the only long-term solution, other than drastic restrictions on the courts' use of prison, was to build new establishments. Third, however, American states have traditionally raised finance for prison construction, and indeed for most major projects, from the issue of public bonds. The issue of bonds must usually be preceded by a referendum, and the issue of bonds for prison building were not infrequently voted down.[5]

---

[4] Scull (1984) probably remains the best discussion of this broader context. His argument is that government budgetary problems from the 1950s onwards made decarceration, for both mental patients and offenders, an acceptable policy option which was then espoused by liberal medical and criminological opinion on the utterly different grounds of the psychologically debilitating effects of total institutions. The end result may have been a shambles, with mental patients living on social welfare in cheap lodging houses, and offenders released to incoherent and possibly inappropriate community programmes and ghetto neighbourhoods. But it was at least a low-cost shambles.

[5] For one e.g. of this (in New York) and a discussion of the issues, see Jacobs and Berkowitz (1983).

Many privately operated halfway houses, drug rehabilitation centres, and other peripheral institutions existed in the US by the early 1970s. In 1975, a subsidiary of the RCA Corporation opened a unit for 'hard-core' juvenile delinquents, with a capacity of about twenty, in Pennsylvania (Borna 1986). But faced with the triple pressure of budgets, court orders, and population, it is not surprising that some jurisdictions decided to solve all three problems by contracting prisons to private companies in the 1980s. It is, perhaps, surprising that only a small number of jurisdictions have in fact taken this step.

Donahue (1989: 151) states, 'In the late 1980s, there were around twenty or thirty private correctional facilities, with the exact number depending upon the precise definitions of *private* and of *correctional facility*'; he also acknowledges, but does not include in his total, almost 2,000 juvenile custodial facilities run by a variety of non-governmental organizations and private companies. The institutions involved were primarily county jails, low-security prisons, remand and juvenile facilities, and detention centres for illegal immigrants.[6] Stephan (1992: 20) notes that in 1990, some 67 facilities were contracted out by states, of which only 21 were primarily institutions for confinement (the others were community-based institutions with work release programmes or similar). All but two of these institutions held fewer that 500 inmates. As Delteil (1990: 194) points out, fewer than 1 per cent of all incarcerated offenders in the US are in private as opposed to state or federal institutions. And despite the rapid increase in the number of privately-operated facilities, Borna's (1986) judgement is still accurate: 'the prisons-for-profit industry is still in its embryonic stage. Many of the companies in it are barely past the start-up phase, but they have a tempting market'. This makes the scale of privatization in England and France, discussed further below, all the more remarkable.

The American experience of these institutions has been mixed. First, private prisons have not been unopposed. The ACLU, the National Institute of Corrections, the National Sheriffs' Association, the American Bar Association and various unions have all been critical of privatization on both ethical and practical grounds. Second, claims about financial economy are problematic, because private

---

[6] There are 4 key players in the US, with the lion's share of the market: Corrections Corporation of America, Behavioral Systems Southwest, Pricor, and more recently a Wackenhut/Bechtel joint venture.

establishments are usually smaller and have less troublesome populations than do state or federal institutions. Donahue (1989: 160–61)
summarizes available data in the following way. For juvenile institutions, private institutions appear to have running costs only 3 per
cent lower than public ones, and are overall slightly more expensive
because they must budget for future investment from current income.
Privately-operated centres for illegal immigrants were 20 per cent
more expensive than the INS's own institutions, though much of this
can be attributed to variations in land and labour costs rather than
management efficiency. In so far as direct comparisons are possible,
it appears that private centres are more efficient because they pay
about 15 per cent less to their staff, representing a reduction of about
9 per cent in total operating costs. However, they are typically hiring
from the private-security market, and employing a less well-educated
workforce with a larger proportion of part-time staff. Finally, costs
to the county or state may not drop in line with this if the contract
specifies a fee per prisoner per day. Donahue (1989: 166–7) cites the
experience of Hamilton County, Tennessee, which negotiated a rate
of $24 per person per day, but then introduced tougher sentences for
drink-driving. The contract did not separate fixed and variable costs,
but instead made a direct link between fees and inmate numbers.
Thus the fees payable to the company were much greater than the
cost of running the jail directly.[7] It should be mentioned, however,
that many subsequent contracts have been rather more sophisticated.

One further contentious issue is that of whether inmates in private institutions experience a better regime. In principle this could
be settled in research terms by comparing prison regimes with the
American Correctional Association standards. In practice, a
requirement to comply with materially relevant standards could be

---

[7] This discussion relates to Silverdale work farm. However, another institution in
the same country, the Hamilton County Penal Farm, was also operated by the CCA
and was estimated to have saved approx. US$1 per day per prisoner, with total cost
savings of 3.8% in fiscal year 1985/6, 3.0% in 1986/7 and 8.1% in 1987/8 (Logan
and McGriff 1989). Logan and McGriff's figures were generated by a fairly complex
calculation. The estimate of what direct operation would have cost was based on
agency costs plus indirect costs, including depreciation, interest charges, and prison-
related expenses paid from the budgets of other state agencies. The estimate of private operating costs included not only the direct contract cost, but the costs to the
state of monitoring the contract, and other residual state-prison-related functions
depreciation and interest charges; and the 'rebate' aspects of local taxes paid by the
CCA. These calculations reveal the complexity of trying to make private/public
comparisons in the prisons sector.

written into the contract (not all standards would be relevant, since some deal with matters still handled by the state, such as parole decision-making). But detailed and comprehensive data on these matters is absent. The information we have has mainly been impressionistic, based, for example, on visits to private institutions by delegations of English members of parliament, prison officials, and members of the Prison Officers' Association.[8] Their reports suggest that they have generally seen what they wanted to see. On the one hand they have reported better physical conditions with more facilities available to inmates. On the other there have been adverse comments on the competence of staff, and Weiss (1989, citing Cannon 1986) notes that a proposal to license a private prison company as a private security agency in Bay City, Florida, would have resulted in their staff requiring less than 50 rather than a state norm of 320 hours of training.

Privatization in England

The impetus towards private prisons in England had only one thing in common with America: the increasing prison population. Budgetary problems involved matters of expenditure rather than the raising of capital, and existed largely between government departments, the Prison Department, the Home Office, and the Treasury. Long-standing arguments about staffing levels and efficiency had been, if not settled, then put onto a new footing by the introduction of the Fresh Start package in 1987. Capital had been allocated for the building of new prisons in the early 1980s, and several became operational towards the end of the decade.

The primary pressure for privatization was not practical but political. The Conservative government that came to power in 1979 started by forcing local authorities to put out for tender some local services, such as rubbish collection. It followed by requiring the National Health Service to contract out ancillary functions such as

---

[8] Though some investigative journalism has taken place. See Mangold's (1985) generally complimentary description of Silverdale, a prison run by the CCA, which also comments unfavourably on some of the more revolutionary aspects of private prisons, such as Behavioral Systems Southwest's proposals for small prefabricated 'modules' rather than cells. And it raises some questions of accountability, esp. to do with the private companies' abilities to 'buy up' experienced prison wardens so that the pool of expertise for 'monitors' (in English parlance, the 'controllers') would be correspondingly narrowed.

cleaning, and reorganizing budgets so that health authorities were forced to act more like private businesses. Several nationalized industries, including gas, British Airways, and telecommunications, were later floated off with the issue of shares. The proposal to privatize some prisons is thus also a continuation of an established policy, and, as Ryan and Ward (1989) put it, a signal that 'nothing is sacred'. It would also have a spinoff benefit, namely that the power of the Prison Officers' Association would be further weakened. The final imprimatur was put on this proposal by Lord Windlesham, a former minister and Chairman of the Parole Board, who in 1987 advocated the creation of private-sector remand establishments. His view—which was, and is, widely shared—was that conditions for remand prisoners were disgraceful, and that a private-sector establishment under Home Office management would be a way to ease the problem.

The Criminal Justice Act 1991 duly included among its many provisions a specific authority (in s. 84) for the Home Secretary to contract out the management of new remand prisons.[9] In addition, it allowed the Home Secretary to make an order by statutory instrument which would allow him virtually unfettered freedom to contract out almost any other type of prison or prison function, including court escorts. The first private remand prison, The Wolds, opened on 6 April 1992. This was one of the establishments built as part of the 1980s prison expansion programme, with a design capacity of 300 in normal accommodation, and intended for remands. It was contracted to Group 4 Remand Services Ltd., who were able to assume control over it from its opening.

The document which formed the basis for competitive tendering was published in May 1991. It set out what amounted to contractual standards in a wide variety of areas, including security and incidents, health and hygiene, food, reception, discharge and record-keeping, and regime activities (Group 4 has committed itself to providing fourteen hours per day out-of-cell activities). The text of the contract finally signed by Group 4 has not been made public, and neither has any information as to costs been released, apparently because of concern about commercial confidentiality. In this respect the British government ought to have a great deal to

---

[9] Many of the issues involved in contracting out for remand establishments had previously been identified in a report commissioned by the Home Office from a management consultancy (Deloitte Haskins-Sells 1989).

answer for. There can be no reason for such secrecy except to prevent the cost basis of the contract from being discovered by possible competitors. Yet not only should this information be made public as a matter of legitimate public interest in government expenditure, the very idea that the terms of successful government contracts should not be publicly available seems to run counter to the idea of a commercial 'level playing field' on which different firms can properly compete. It is notable that at least two other countries which now have private prisons, the USA and Australia, are compelled to provide financial and contractual information on the privatization contracts under the terms of their Freedom of Information legislation.

The British attitude to secrecy notwithstanding, the agreement with Group 4 appears to be along the following lines. The government retains the power to terminate the contract either if it is breached or if there is a change of government policy (answer to House of Commons question, 3 February 1992). The management of the prison is conducted by a Director, appointed by the contractor. Apart from day-to-day management, the Director's duties include responsibilities in relation to security arrangements, incident and emergency planning and management, liaison with various local and national government services, co-operation with bodies such as the boards of visitors and HM Prison Inspectorate, and the making of proper arrangements for prison visitors, religious observances, and the like. The Director's work is monitored by a Controller (a governor from the Prison Department) who has, in addition, specific responsibilities for conducting inmate adjudications, authorizing removal from association, segregation, and the use of special restraints, and for investigating allegations against prison custody officers.

Key decisions affecting inmates—transfers, discipline and so on—thus remain outside the powers of the private company. The company may not refuse to accept prisoners sent to it, 'subject to the limits on the number and category of prisoners to be held there' (answer to House of Commons question, 3 February 1992), a phrase which indicates that this prison, at least, will not be overcrowded. The complaints mechanisms are as for state-run prisons; there is a Board of Visitors; and HM Prison Inspectorate can inspect it. In the event of a riot, Prison Department officers and/or the police may be called (there was one riot shortly after it opened). However, the contractor is only liable for the costs of disorder which come about as a result of its failure to carry out the contract properly (answer to House of Commons question, 3 February 1992).

The points above meet some of the criticisms that have been made of private prisons, in particular in relation to complaints and discipline. It is also clear that private prisons will not only be monitored by the pre-existing machinery, but also by an on-site Controller. In a perhaps ironic twist, given the government's longstanding reluctance to develop prison regime standards, private prisons will be evaluated in terms of performance indicators which are specified in the tender document.[10] The issue of the use of force by the private contractor's staff was partly resolved through s. 86 of the 1991 Criminal Justice Act, which sets out the powers and duties of the contractor's staff (designated as 'prison custody officers' to distinguish them from prison officers). These staff have the power to search any prisoner confined in a prison, any other person in or seeking to enter a prison, and any articles in the possession of such persons, subject to provisions in the Prison Rules.[11]

[10] e.g. would be as follows. For security, the 5 criteria include no escapes, no key compromises, and completion of agreed searching programmes. For health, safety, and hygiene, the 12 criteria include proper investigation and reporting of all accidents, publication of a fire precautions manual, an absence of proven cases of breaches of existing legislation or codes of practice, accurate maintenance of specified temperatures, accurate records of daily outdoor exercise, and an absence of complaints about clothing and laundry. For reception, registration, and discharge, the 10 criteria include accurate entry of information onto Prison Department computer systems, medical screening of receptions, identification of inmate physical and welfare needs, and compliance with legal requirements as to warrants and release on bail. For regime activities, 9 criteria include delivery and dispatch of inmate correspondence, and the functioning of the grievance procedure. For inmate services, 10 criteria include arrangements for medical services, welfare services, and the provision of meals. Under the regime activities head, 8 criteria include prisoners being free to leave their cells outwith routine lock-ups and emergencies, library visits, a minimum 6 hours per week of supervised physical education, access to the prison shop twice a week, and accurate recording of disciplinary charges. For incidents, the 5 performance criteria include arrangements for resolving minor incidents, for isolating major ones, testing exercises, and provision of debriefing sessions and facilities. While these may not be a comprehensive set of minimum standards, they none the less do set clear expectations as to normal levels of staff and management performance, both explicitly and in terms of what is necessarily implied in accomplishing such tasks. Interestingly, the tender brief for Strangeways, released subsequently, contains no explicit performance criteria even though it places a large number of specific requirements on contractors.

[11] However, s. 86(2) states that such powers extend only to requiring a person to remove his outer coat, jacket or gloves; it does not permit strip searches.

They have specific duties to prevent escapes, prevent or detect other unlawful acts, and ensure good order and discipline (s. 86(3)); and in executing these duties they may use 'reasonable force' (s. 86(4)). Presumably the definition of reasonable force in this context will be clarified through litigation, as it was in respect of nurses in mental institutions following the 1983 Mental Health Act. Finally, the fears that the contracts would be regarded as commercially confidential, and thus remain unpublished, were well-founded.

It is, as yet, too early to pass judgement on some of the wider questions about privatization. It is not yet clear whether any cost benefits have emerged from the contracting-out system. It is clear that a 'private prisons lobby' exists, but its influence is not easy to gauge. And it is not yet clear how any breach of contract would be resolved if discovered. None the less, since the private prisons initiative has a primarily political motive, there appears to be no intention to perform a proper evaluation of The Wolds. Even before it opened, the government had laid plans to contract out a 649-place local prison being built at Blakenhurst. On 5 December 1991, the Home Secretary announced his ultimate intention to privatize half the prison estate. And this was followed up in mid-1992 by a further announcement that Strangeways, on completion of repairs following the 1990 riot, would also be contracted out, together with at least two further establishments.[12] In the event, an in-house team won the bid to run Strangeways. Meanwhile, the first Prison Inspectorate report on The Wolds, published in 1993, is one of the most damning reports published on any prison to date; its major criticisms were that the staff had lost control of the prison to inmate groups, and that while physical conditions may have been adequate, the regime was lethargic. Whether such failings are unique to The Wolds remains to be seen; but unfortunately, it is likely that about 10 per cent of the prison population will be under private management before any lessons from this first experience of privatization can be properly learned and acted upon.

[12] The tender documents for Strangeways explicitly state that about 20 of the inmates can be expected to be Category A, and about 30% will be sentenced to terms of over 4 years. Thus, in contrast to the American model, in which private prisons are primarily intended for low-security inmates, the English plan now clearly envisages the use of private management in handling comparatively 'heavy' prisoners.

Executive Agencies: England and Scotland

There is an alternative model to that of entrusting prisons to the
private sector, and that is to loosen the ties between government
and the prison department in such a way that the latter can act
rather like a corporation. This is the concept of the 'executive
agency'. In this form of organization, the prison department is
wholly 'owned' by government, and its employees remain civil servants, but its functions are streamlined, it becomes responsible for
providing its own support services, and it acquires greater authority over the way in which its budgets are used. Equally, it is no
longer automatically tied to civil-service master pay scales or other
agreements, and senior management can be brought in from outside the ranks of government officials. It may—and in the English
case, will—have a board of management and a Chief Executive
rather than a Director General. It will be funded through fees
agreed with government and may have to compete against private
corporations for 'new business'. Control over the agency would be
through the creation of a Supervisory Board, chaired by the relevant minister, which would advise the Home Secretary on policy.
At the same time, the Chief Executive would not only have to satisfy his own board of management, but also be accountable to the
Home Secretary advised by the Supervisory Board. The advantage
of this model is that, in theory at least, it enables the prison service
to cut away some of the management restrictions and financial
constraints imposed by the civil service, so that there can be gains
in efficiency and flexibility.

The recommendation to move to agency status was first made in
England by Sir Raymond Lygo in December 1991, and the change
to executive agency status was made on 1 April 1993.[13] An earlier

---

[13] Lygo (1991). Interestingly, the Lygo Report is rather unspecific on several key
issues. It does not e.g. include a clear statement of what 'agency status' might be
taken to mean. This has to be inferred from the description of the arrangements
that he proposes, though it is more fully described in a series of 'Agency Updates'
issued by the prison service. Equally, while Lygo describes his aims as marrying
together 'continuing clear political accountability for the major policy and resource
issues' and 'a greater degree of managerial authority for day-to-day operations'
(para. 16), little is said about the determination of resource levels, budgets, and
expenditure beyond the comment in para. 41 that 'It cannot be right that . . . a governor with up to 1000 staff cannot suspend or indeed dismiss a member of staff on
his own authority. Similarly, it cannot be right that a governor with responsibility
at least in theory for budgets of up to £10 million has, in practice, authority to
affect expenditure of a tiny fraction of that budget'.

decision to move to agency status was made in Scotland, and the Scottish system became an executive agency in 1992. The advantages of agency status have been sufficiently compelling for groups such as the Prison Reform Trust, which has frequently been critical of the prison service and the Home Office, to find in favour of it.[14]

The proposed changes, however, clarify and sharpen some of the problems with the current structures which are likely to be carried forward into the new arrangements, and these problems have a clear bearing on matters of accountability. First, the Chief Executive will be responsible, as the Director General was, for the efficient and effective use of resources. He and his board of management will prepare an annual plan for the agency, as was previously done for the service, on the basis of which resources will be negotiated. The Home Secretary, presumably aided by his Supervisory Board, will determine the extent of resources, and remain ultimately accountable to parliament. However, none of this will affect the role of the Treasury, which is the funder of these resources and which will continue to exercise what Kirkpatrick (1986) has described as its 'corrosive effects'. Equally, although the Chief Executive is explicitly given responsibility for negotiating matters such as pay, his hands are again tied by the Treasury. The overall effect is to perpetuate a system in which those who take resource-allocation decisions remain insulated from the practical effects of their decision-making. Second, one of the intentions of agency status is to allow the agencies to take fuller responsibility for their own use of resources, while allowing ministers to be removed from most day-to-day matters. Lygo argued (para. 15) that:

recent events like the Strangeways riots and the Brixton escapes will inevitably require the Home Secretary to account to the House of Commons at least for the major issues raised. At the same time Ministers should not become involved in day-to-day operational matters. Neither should they be expected to answer detailed questions on such matters.

The framework document for agency status identifies six specific types of events about which the Home Secretary would automatically receive reports, and they clearly encompass the kinds of incidents about which parliamentary questions may be asked.[15] Yet the

[14] See Prison Reform Trust (1992).

[15] Escape of a Category A prisoner, apparent inmate suicide, serious disturbances involving a number of prisoners and damage or injury, any matter (including

result of having a Home Secretary, supported by a Supervisory Board probably chaired by a junior minister, being advised by and issuing instructions to the Chief Executive of the prison agency and his board of management, is effectively to increase the number of people amongst whom responsibility can be shuffled, not for specific events, but for the policies which created the situations in which they happened.

Private-Sector Involvement in France

France has, if anything, a stronger and more recent history of private sector involvement in corrections. For one thing most inmates work in prison workshops operated by private companies, with private-sector management but prison staff providing security and control. For another, the major form of custody for juveniles, education surveillée, comprises a network of institutions which, while under government supervision, are largely owned and operated by non-governmental organizations or associations. Stefani *et al.* (1982: 745 ff. *et seq.*) describe the system as one which deals not only with juvenile offenders, but also those in 'moral danger' or who are 'maladapted'. The Service de l'Education Surveillée, a department of the Ministry of Justice, runs its own establishments, but also exercises a 'technical and administrative' control over more than 300 private institutions. Regimes are controlled via agreements with the Ministry of Justice; juvenile judges (the juges des enfants) and ministry inspectors monitor establishments; and the majority of institutions are members of regional associations with remits including administrative co-ordination and professional training.

So far as its prisons are concerned, France faces many of the same problems as England. But despite its long tradition of resort to the private sector, both in prison labour and for juvenile custody, its prison privatization initiative envisages private-sector involvement on a completely different model to the English approach.

Delteil (1990) dates serious French interest in privatization to the autumn of 1986, when the then minister of justice, Albin

incidents) likely to arouse parliamentary or public concern, national or other serious industrial action or disputes, and major changes in an institution's functions or the closure of an institution.

Chalandon, visited the USA and was 'seduced' by the prospect of a private solution to the French prison overcrowding problem. Faced with a prison population of some 51,000 in 1987, his initial proposal was to develop a private sector of some 25,000 new prison places to complement the then existing capacity of 34,500, so that the private sector would operate roughly 40–45 per cent of the expanded prison estate.[16] This vision was watered down in the following years.

First, there was legal argument about the constitutionality of private prisons. Some of the issues had already been resolved in principle through the arguments about the privatization of a television station, TF1. Others concerned the extent to which the administration and execution of justice could be divided. Boulan (1987: 19) observed that while the administration of justice was incontestably covered by the constitution and could not be delegated, the actual operation of penal establishments was not seen by the Conseil Constitutionnel as having the character of 'judicial administration'. While new laws may need to be passed, constitutional issues need not arise. Moreover, whatever the outcome, constitutional guarantees as to equality of treatment under the law would apply; those guarantees already envisaged some differentiation of prison regimes for different classes of inmates; and no-one was arguing that inmates in private establishments should constitute a different legal category of prisoner (see also Favoreu 1987).

Second, the cost estimates—F325,000 per cell, or F8,000 million for the programme, proved too expensive (Deltcil 1990: 195). The number of new cells to be built was reduced first to 15,000 and then by another 2,000, so that the project became known as 'Programme 13,000'. In terms of establishments, it was to comprise twenty-five prisons, of which four would be maisons centrales. However, twenty-five existing small or decrepit establishments were also slated for closure as the new prisons came on line.

The new establishments were to be opened by 1991. But the management of the institutions was not to be left to private enterprise. The four consortia involved would build the prisons (the final budget for this being F4,200 million), and were also con-

---

[16] Delteil also makes a wry aside about the reception of this plan. Chalandon had previously been involved in the sale of low-cost housing subsequently found to have a high rate of structural defects. His critics rather sarcastically remarked that prisons built under his plan would very likely have similar problems.

tracted to provide a number of services to inmates at a fixed rate per inmate per day over a ten–year period. This fee—initially fixed at F120 per inmate-day—was to cover food, plant maintenance, inmate transport, medical and drug rehabilitation services, and work or vocational training for inmates, with the objective of providing inmate wages at 60 per cent of the national minimum wage within three years. Some of these services had long been provided by the private sector in France and elsewhere, but only in certain establishments. Fleury-Mérogis, the largest French prison, already had a private maintenance contract. But in all cases, prison management, security, and control would remain the task of the prison administration and DAP-employed prison staff would work in them. The only new factor was, therefore, the scale of contracting-out of ancillary services.

Finally, Delteil notes that some of the new prisons are high-tech institutions which employ control measures such as electronic identity cards to control inmate movement. These devices have also proved controversial, and in one prison (Neuvic-sur-Isle, in Dordogne) were the subject of an inmate protest and subsequently a riot in September 1990, some two months after the establishment opened.

Private Prisons: Problems and Prospects

Private prisons clearly offer hard-pressed governments a way out of penal overcrowding. If they are built and operated by private contractors, they constitute an addition to the prison stock which can be paid for out of recurring, rather than capital, appropriations. This is the French model. In the US and England, however, the majority of private prisons are in fact existing, or built but not yet operational, state institutions. They do not constitute an addition to the prison estate beyond that already on-stream or planned, and may potentially exacerbate the overcrowding in the rest of the system, since the contracts under which they are run do not envisage overcrowding.

None the less, for the inmates in them, private prisons have the potential to offer improved prison regimes; and ironically, in England, they were a key factor in persuading the administration that standards were necessary, if only for the purpose of monitoring contractual compliance (though this does not address the issue of how such standards may differ from those envisaged within the minimum-standards debate).

In the short term, in England at least, most problems of day-to-day accountability can be solved through the insistence that private prisons operate under the scrutiny of boards of visitors, inspectorates, and the like; through the installation of a controller on-site; and through the requirement to comply with performance criteria. Other problems—relating to limits on the use of force, and how to deal with non-compliance with performance criteria—will undoubtedly come to be resolved through existing legal and administrative channels as and when they emerge. The courts may find in the coming years, for example, that they have to rule on the interpretation of performance criteria contained in contract clauses.

In the longer term, however, problems of 'top end' accountability remain. First, England has adopted the practice of applying performance criteria to private prisons, and the adoption of agency status for the prison department carries with it the obligation to adopt key performance indicators. There is no longer, therefore, a question of principle about applying the same standards across private and state-run prisons—though these need not be, of course, standards for inmate treatment along the lines proposed by Casale (1985). But the question remains as to how such performance criteria and key performance indicators will be translated into criteria or standards for inmate treatment. Second, is it likely that the reduced state-run prison system, which will have to absorb overcrowding, will become a second-class service? Third, Lilly and Knepper (1991) have claimed to be able to discern a 'correctional-commercial complex' in the US, with flows of information, influence, and personnel between government and commercial operations. They argue that this complex, which defines its activities as being in the public interest, exercises significant influence over correctional policy and yet operates without public scrutiny. It may be more accurate to say that it is only publicly accountable at the point that policies are made, contracts tendered for, and prisons monitored. It is probably not accurate to describe England as having even a nascent 'corrections-commercial complex' (and it may be too grand a description even of the US situation). Yet one can imagine situations arising in which the Home Office is faced with a choice between compromising on prison regimes for prisons out to tender, or continuing to run the institutions directly. One can also imagine a situation in which private companies place pressure on the Home Office to alter the conditions of contracts, or to allow companies greater discretion in the running of prisons, on the grounds that no company could reasonably be expected to meet the conditions imposed. Equally, one might imagine a situation in which private companies offer suffienctly attractive salaries and benefits that they 'poach' senior governors from the state sector, who are then monitored by relatively junior governors who could be intimidated by the former's reputation. And finally, it is not too far-fetched to imagine companies bidding for work such as the administration of the parole system or the calculation of remission. At least one Swiss-based company is currently used by governments worldwide for tasks such as the calculation of taxes and duties on imports and exports. It could be argued that the administration of parole is not, in principle, any different as an administrative process from such activities.

Such problems are likely to become most pressing in England. France has not gone down the path of privatization in the same way, and the US has been far less radical than England intends to be, both in terms of the scale of private imprisonment and the types of inmates held in private prisons.

The bottom line is that private prisons may offer cheaper alternatives to state-operated imprisonment, and may enable some juggling between capital and recurring costs. But they are essentially a solution to a problem of overcrowding, while the funding of prisons, the creation of prison spaces, and the level of regimes, will remain subject to the willingness of the Treasury to allocate appropriate budgets. Private prisons do not resolve any of the pressing problems about the way in which imprisonment is used, and the nature of imprisonment. Government will get only what it pays for; the prison agency will be able to get only what the Treasury is prepared to allow it to buy, and private prisons do not remove from government the responsibility to make adequate provision for prisoners and prison regimes.

Finally, in the US, and to a large extent in France, the appeal of privatization lay in its practical ability to minimize capital expenditure by shuffling costs onto recurrent budgets. In England, the motive for private prisons is primarily a matter of political doctrine. The issue is not primarily whether private contractors do in fact enable costs to be cut, though this is clearly one objective. It is intended as a signal that privatization is possible in any area of government activity, and as an indication of the government's commitment to a pro-corporate culture. This difference in the motivation for private prisons may well affect the way in which the success of private prisons is evaluated. It may also—though it is too early to say how far—affect the extent to which there is political willingness to hold private contractors to account.

14

Conclusions

ANY discussion of accountability in relation to prisons has to deal with a variety of sources of authority and procedures. In addition, it must address a large number of issues, including the role of the courts and of supranational bodies such as the Council of Europe, inmate rights, standards and performance indicators, discipline, and political initiatives such as privatization. Linkages between these topics exist only in a very abstract way. Matters which might in principle be linked together, such as prison conditions and inmate rights, seem to be related only loosely. The tools at our disposal, meanwhile, do not amount to a theory, but to a bundle of perspectives and sensitizing concepts. A few general premisses seem none the less to be fairly clear.

First, accountability is about control. In particular it is about controlling and legitimating the ways that power and authority are exercised. In relation to prisons, those powers and authorities are concerned with policy, resourcing and resource use, organizational performance, and the treatment of prisoners; and they include substantial levels of internal powers which apply only to the prison situation, such as administrative segregation.

Second, the sources of accountability are primarily legal and political. Managerial authority is simply the executive arm of political intention, though managers are also accountable to the courts for the lawfulness of their actions. The political constraints imposed upon them may, however, include independent inspectors or lay committees, intended to bolster the legitimacy of the prison system by making it more 'transparent'.

Third, although accountability is often described in terms of structures and systems, accountability is not inherent in any particular set of formal arrangements. Official procedures, regulations, inspectors, inquiries, and so forth are simply the practical arrangements by which accountability is claimed to be accomplished. On this view, accountability is a process which depends crucially upon a series of formal and informal organizational relationships, and thence on negotiations and agreements: that this set of arrangements can be said to 'work', that it is 'effective', that this or that body is 'independent', that all parties concerned will depict their relationship as one that involves accountability, and so on.

Fourth, however, we cannot be doctrinaire about the precise nature of the relationships involved nor the currencies in which accountability is rendered. Groups entirely unconnected with prisons, such as the media, clearly have some role to play. Some of the relationships involved may be conducted behind closed doors. And at the bottom of the pyramid, insofar as prison officers must (usually) use reason, argument, explanation, and sometimes a blind eye to help them run their prisons, accountability can be as much an issue in the wings and workshops as it is in cabinet meetings or parliamentary debates.

Propositions about accountability are, on the basis of these general premises, likely to be about issues of control, law, politics, management, and legitimacy. They are likely to concentrate on the differences between formal structures and informal relationships within the prison organization, and between it and other interested parties. And they are likely to be propositions about the determinants of change within the prisons and in the way that prisons are run. The only clear link across these propositions is that they deal with ways of representing and evaluating the performance of prison organizations and the treatment of inmates; the uses to which those representations and evaluations are put; the motivations of those who use them; and the results of those uses.

The following sections, drawing on the discussions in previous chapters, are arranged more or less thematically. The first section deals with the relationship between control and discretion, the second with accountability and interaction in prisons, and the third with longer-term and larger-scale changes in prison organization, the impact of law, and legitimacy. The four sections that follow deal with the relationships between accountability for regularity and accountability in major crises, the link between administrative control within prisons and political control over them; the sources of prison reform; and the resourcing and management problems brought about by change. The final section offers some brief comments on the prospects for prison reform.

Control, Discretion, and Rights

Prisons not only accommodate inmates; they are expected to control them. The primary method of control is to make inmates dependent on staff. This dependency has several aspects, but one is to do with privileges—goods and services which inmates desire but which are granted at the discretion of staff. Privileges thus represent a gap between inmate entitlements or rights, and the actual level of provision which can be made.

The exercise of discretion is not incompatible with concepts of administrative fairness and justice. However, control over this discretion is problematic for two main reasons. Grievance procedures are not wholly effective because the matter complained about is usually access to higher levels of provision, or different kinds of provision, than those to which the inmate could claim any entitlement; and a complainant must demonstrate that the decision complained about was prima-facie unreasonable, or, in Germany, that it breached a legal right. In practice, of course, most inmate rights are procedural rather than substantive, so that the extent to which inmates can claim rights in any area of prison life is highly restricted.

Staff, meanwhile, usually protest that they have no formal decision-making power. This is true but disingenuous. In all four countries, though seemingly to a greater extent in the Netherlands and England, staff do have the ability to reward and punish inmates through the way they handle their jobs. They can, for example, answer requests quickly or slowly. Moreover, staff are rarely held accountable for the way in which their behaviour may affect inmates. If they provoke an inmate who subsequently insults or attacks someone, smashes his cell, or creates some disturbance, the responsibility for the inmate's behaviour is the inmate's alone. There would have to be gross and clear provocation before staff could be brought to account for the inmate's acts.

Prisoners are not, however, completely dependent on staff. Leaving aside the issue of contraband, there are ways in which they can and do try to increase the range and level of provision they experience. Their major problem is that decision-making seems often to be arbitrary. Many prison officers are felt to be inconsistent in their approach to discipline; decisions made by governor-grade staff appear not to be based on clear or consistent criteria. For the moment it is not important whether this is in fact the case or not. The point is that this is how inmates experience imprisonment; and the result is that they cannot make clear links between behaviour and outcomes, at least in small matters such as getting extra visits, swapping items with other inmates, and so forth. Inmate challenges to such apparent inconsistencies form the bread and butter of the lower levels of most complaints procedures. Yet the problem of dealing with such matters is permanent and intractable.

One argument in this connection has been that the provision of clear inmate rights to specific levels of services and facilities would enable inmates to know precisely what they are entitled to, grievance procedures to establish the facts of the matter, and decisions to be reached in a rational manner. Although it is fair to say that there have been substantial improvements in inmate rights over the last twenty or so years, the current situation is hardly an ideal one. Moreover, the promise that rights allegedly hold out is unlikely to be realized.

It is commonly accepted that inmates should have all rights which are not removed, expressly or by implication, by the fact of imprisonment. It is also broadly accepted that imprisonment involves the exercise of a number of special administrative powers not applicable to the general population, and that concepts such as 'natural justice' have some relevance, at least in so far as a duty is imposed on officials to exercise their powers fairly. A quick survey of the four countries reveals, however, that acceptance of these broad principles has led only to relatively weak and mainly procedural rights being granted to inmates, thus minimizing the exposure of administrations to challenges based on the substantive conditions experienced by the generality of inmates. Such improvements as have been seen in the last twenty or so years have come about through lengthy legal struggles. The inevitable conclusion is that however powerful is the image of reform through the introduction of rights, prison reform is in practice more a matter of political willingness, and there is a greater willingness to introduce privileges than rights—not least because the former do not entail enforceable claims against the administration. Legal traditions, meanwhile, have resulted in major differences in the opportunities formally given to inmates to call for legal review of administrative decisions that affect them. Broadly, this is easiest in the Netherlands and Germany, and most difficult in France and England. Yet the mere existence of such frameworks is a poor guide to the ability of inmates successfully to challenge administrative decisions, still less to the ways in which prison institutions actually operate. The Dutch system does not prevent administrators from taking their intended action, and provides remedies by way of additional association or small amounts of money at a later date. The German system has no effective remedy if the administration is intransigent.

Moreover, most inmates themselves appear to see arguments over rights as marginal to their own position. Much of the dispute is at the margins of entitlements, not the provision of core services; it is, to put it crudely, at the level of whether the food is hot enough rather than whether there is any. Standing on one's rights can cause more problems than benefits because of the structure of the prison situation, as many French inmates find out when they wear their own clothes but have no laundry facilities. There is always the fear that insisting on rights results in one being labelled as a troublemaker, with informal reprisals possible. Realistically, those who know and insist on their rights are typically those who have already achieved this reputation; who see themselves as in conflict with the administration; and for whom rights may very well be of crucial importance in fighting administrative decisions about them. For such inmates, the question of rights is an immediate and critical one. For the majority, however, there is a clear current of feeling which suggests that given the choice between increased rights and better day-to-day provision—crudely, more videos and more association—inmates would very likely choose the latter. Having said this, the Woolf Report in England has come down firmly in favour of the development of inmate rights; it is careful not to raise hopes too high, and uses the language of 'legitimate expectations' and 'justice in prisons'. The report has clearly effected a major shift in the political willingness to consider and extend inmate rights. How far this will be translated into practice remains to be seen.

The points above are not intended as an argument against rights. They are, however, intended to indicate that rights are not a panacea. The likelihood of producing a comprehensive set of rights covering all situations is remote. Frameworks of rights are unlikely to be accepted, by inmates or staff, as an appropriate way of handling the majority of matters in prison, though they clearly can and should exist in relation to measures and conditions which may be considered punitive. In the wider sphere of prison life, the main virtue of substantive rights has been to exert an upward pressure of privileges so that staff can maintain the margin of discretion which they use for control purposes.

Accountability, Interaction, and Administrative Structure

The comments above have indicated, among other things, that issues of accountability are ingrained into the everyday aspects of prison life. Many of those issues concern informal aspects of relationships. This being the case, it is worth exploring two propositions on the formal/informal relationships of accountability. First, many of those informal aspects are matters of interpersonal style, designed to mitigate the full weight of the formal relationship in the interests of getting a job done. Second, it is possible to manipulate formal structures in order to reach informal ends.

Morgan and Maggs (1985) describe accountability as taking place within three modes: of direction, stewardship, and partnership. Directive modes imply hierarchical and formal relationships of reports and instructions. Stewardship implies a relationship in which the director of a service is allowed a degree of managerial discretion, and is required periodically to account for the way in which it has been used. The partnership model is one in which there is, in principle, a high degree of co-operation between a service and those to whom it accounts; where the expertise of the service providers and the concerns of the scrutineers are thought to be equally important; and where the co-operation is directed towards refining and implementing 'good practice'. One of the interesting thing about these distinctions is that relationships which are formally of one type turn out in practice, much of the time, to be conducted in a style which is more appropriate to another mode of accountability. Depicting oneself as a partner rather than a steward may be, in most normal circumstances, the best way of getting the job done. It also allows, as one member of a Dutch Commissie van

Toezicht suggested, the presentation of the 'overseer' as an informed and knowledgeable insider, and facilitates discussion of informal as well as formal aspects of the situation.

In many large organizations, including prisons, despite the formal hierarchical structure, the giving of direct and explicit orders is relatively rare. There is of course a recognition that suggestions and requests from senior staff are backed by formal authority and can be given as orders; yet the maintenance of good relations and high levels of performance often depend upon co-opting staff into willing co-operation. So far as inmate-staff relationships are concerned, the giving of direct orders is bound up with disciplinary issues. Staff everywhere recognize that giving a direct order will require them, if it is disobeyed and seen by other staff or inmates to be disobeyed, to follow a formal course of action. Indeed, in dealing with recalcitrant inmates, staff would often have to make an explicit and direct order which the inmate disobeys before there are clear grounds for a disciplinary report.

At other levels in the system, parallel situations arise. Boards of Visitors and prison inspectorate staff in England, for example, have extensive powers to enter premises, consult records, and speak to inmates. This does not mean that entry into premises, consultation of records, and communication with inmates is automatically conducted as a matter of right. Clearly the authority exists and its exercise can be insisted on. Yet a great deal of effort goes into smoothing over the situation, building bridges with staff and seeking their confidence, and developing styles of personal presentation that stress co-operation and partnership. There is in consequence some room for negotiation and 'game-playing' in terms of the depiction of the relationships between the various participants. This is not to suggest that accountability is compromised by such games; in fact, rather the reverse. Such negotiations, games, and collusive depictions of relationships appear to be a necessary component in ensuring that scrutiny and oversight take place in a regular, smooth fashion.

On the second proposition, it seems that the difference between the formal modes of accountability and the informal styles of 'getting things done' can create problems for the nature of accountability in prisons. One English example can demonstrate this point. In the post-1985 English discussions of prison disciplinary procedures, it was postulated that governors would vary their response to indiscipline, depending upon the ease of use of, and their degree of accountability for, disciplinary and administrative measures.[1] A significantly greater degree of accountability for disciplinary measures could result in the rise of an informal disciplinary system constructed out of administrative measures. It was also suggested, having regard to the 'dispositional' character of prison discipline, that formal procedures which appeared not to result in speedy and effective punishment would be supplemented or replaced by informal reprisals against inmates, such as filing requests for visiting orders behind the office radiator. While both hypotheses may be true, they remain essentially untested. On the one hand, the differences in levels of accountability for disciplinary and administrative measures turned out to be much less clear than was originally supposed. On the other, the fairly chaotic nature of prisons suggested that systematic use of informal measures would be extremely hard to sustain, while the level of unintended arbitrariness that inmates felt existed in decision-making would mute the effects of any intended discrimination against individuals.

In short, the issue that had to be addressed was whether changes in the formal structures would have any unintended influence on the practices of staff and governors. This became an issue because the regulations governing administrative segregation, although in principle not permitting governors to use them calculatedly as an alternative disciplinary mechanism, could none the less have allowed this to happen. The problem did not arise in the other three countries because, ironically, discipline has never been so forcefully separated from disciplinary concerns, nor are there such marked differences as between discipline and other administrative measures in the channels by which they can be challenged by inmates.

Law, Legitimacy, and Long-Term Trends

The administration of prisons, like the process of government generally, has changed greatly in the last century. Barak-Glantz (1981), Jacobs (1977), and others have described a broad trend, over the latter part of the last century and most of this, away from authoritarian and charismatic forms of administration and towards ratio-

---

[1] These discussions took place after the *Tarrant* decision, which turned out to be the beginning of the end of board of visitor disciplinary hearings; see Chs. 3 and 10.

nal-bureaucratic systems. In the same vein, prisons have moved away from militaristic norms of behaviour, accepted the presence of trades unions, and employed increasing numbers of professional staff, such as social workers and psychologists. The reasons for these changes are connected with shifts in penal discourse, and in particular with the rise of the treatment and rehabilitation ethics. The promise of rehabilitation has never been realized. Prison administrations have thus found themselves, at least since the 1960s, stuck between a formal set of arrangements that were drawn from a rehabilitative ethic, and a practice often more closely based on warehousing. In the 1970s and 1980s, the law came to be seen and used more often in the prison setting, as the issue of inmate rights became increasingly central to debates on imprisonment. From the mid-1980s onwards, 'scientific-management' concepts, performance criteria, strategic plans, and the like have also come to feature in the administration of prison systems. These developments can be seen as layered; that is, each new development comes on top of, but does not displace, previous concerns. The original and fundamental issues of means and processes—regime delivery, and 'getting through the day'—remain ultimately more important in the prisons than the ends of imprisonment.

Yet all these changes have had implications for the nature of accountability. Crudely speaking, the expectations placed on prisons have increased, in terms of what kinds of regimes inmates are subject to, and in terms of the information that must be provided about organizational performance. Both the provision of regimes, and the provision of information about regimes, are matters which have implications for the legitimacy of prison systems.

For example, no-one would now argue that inmates should not have access to social work and psychological services (though the mode of delivery of such services is periodically debated). The acceptance of professionals into the prisons has meant, in practice, the acceptance of persons who are accountable in their professional capacities to authorities other than the prison administration. These persons find themselves marginalized within the prisons, not least because they are usually excluded from discussions in security matters. They are at odds with the authoritarian and quasi-military vestiges of prison life, and yet must guard against being manipulated—or being seen to be manipulable—by staff and inmates alike. They usually work out some *modus vivendi* within the prison, making themselves useful to staff, or taking on some of the administration of social and cultural activities. In short, the price of their acceptance in the prison is to downplay their professional role, while their very presence enables lip-service to be given to one plank of the legitimation of prisons, the rehabilitation ethic.[2]

In some other respects, such as the impact of law on the institutional order, the four countries have developed rather differently. This is probably due to the differences in their legal cultures. The Romano-Germanic view of law has led to the codification of many aspects of administration in the head prison law, the judicialization of decision-making, and the insistence that higher administrators be legally trained. The acceptance of the view that the length of imprisonment is to be judicially determined has, notwithstanding the arguments about decision-making procedures, resulted in institutions such as the French JAP. Common-law systems, such as England, have largely seen prison management as an executive task for which no legal training is necessary. And in policy terms, they have been content to see early release, for example, as an administrative prerogative. However, in Europe (though not in America) it is interesting to note how rarely the courts have been prepared to intervene in matters which are ultimately the result of long-term and cumulative administrative and budgetary problems. Such problems affect prison conditions quite directly, though they affect the generality of inmates through mechanisms such as overcrowding. In the US, such situations lead to litigation based on claims of cruel and unusual punishment, having regard to the totality of prison conditions. In Europe, such situations have never been the subject of successful court cases. Despite the codification of prison law in France and Germany, and the highly judicialized structure of the prison administration, arguments based on concepts such as 'cruel and unusual punishment' have generally had to be argued within the European Court of Human Rights, with little or no success at national level. The general line within Europe has thus been to progress on narrow fronts, addressing relatively specific legal issues.

Finally, one can add that systems of accountability have devel-

---

[2] This discussion assumes that 'treatment staff' have relatively few powers. In the US, as Jacobs (1977) indicates, they were sometimes the dominant staff group in the prison, which led to another set of problems, e.g. claims that security and order were being compromised.

oped more or less haphazardly as responses to practical problems of providing some legitimation of the prison systems. In the Netherlands, the creation of a new complaints system in 1977 was the result of more than a decade of unease about arbitrariness in handling decisions and, more broadly, discussion about the legal status of prisoners. The new procedure was thus a way of re-grounding the legitimacy of the administration. In England, the creation of an independent prisons inspectorate was in an immediate sense a response to a recommendation of the May Committee, but the Committee itself was reflecting a long-standing unease that inspection needed to become a more forceful and self-standing operation. An in-house inspection team was perceived as too closely linked to operational concerns, and incapable, therefore, of giving the prison system a clean bill of health. If this can be seen as a practical solution to a problem of how the prison system should be made accountable, the changes in, and subsequently termination of, boards of visitors' disciplinary powers can be seen as addressing the issue of confidence in a mechanism of accountability, since it was ultimately intended to give the boards greater legitimacy as overseers of the prison system.

Regularity, Crises, and Responses

Formal arrangements for accountability revolve around procedures such as inspections, reviews, grievance procedures, and regular reporting to superiors of both routine matters and 'incidents'. These arrangements, which might be characterized as providing routine accountability for the regular operation of prisons, are, by and large, built around a domain assumption that the general scheme of things is acceptable. Where there are major and systemic problems ingrained into the organization, or where there are upheavals—internally, such as major riots, or externally, such as political or fiscal crises—the arrangements for dealing with regularity are neither intended nor able to cope. The ways of addressing such problems must be regarded as a distinct form of accountability. Tiryakian (1970) argues than certain kinds of social events create a heightened sense of time; events appear to move quickly, to be pregnant with possibilities, and to be about to give rise to some new and significant social phenomenon. While Tiryakian's argument is directed at broad social movements, much the same, though writ small, could be said of prison organizations in particular periods. There is a qualitative and phenomenological difference, then, between questions of regularity, and processes of accountability which arise out of, or indeed create, events which can be seen as convulsions, shocks, or upheavals.

The best recent example of such an extraordinary situation is the 1990 wave of prison disturbances in England, followed by the Woolf report. Prison riots are not in themselves new; and it is widely believed that riots in one institution can lead to copycat disturbances elsewhere. However, the image of riots unfolding in a rapid sequence across one third of a prison system raised the spectre of unknown but violent forces suddenly unleashed and out of control. There were, it was widely supposed, major problems in the prison system which resulted in what some organizational theorists describe as a catastrophic—that is, sudden and overwhelming—failure.[3] The Woolf report conducted its inquiries in a very public way, and was the subject of constant media attention. High expectations were created as to its ability to offer accurate diagnoses and solutions. When the report was published it generated intense debate, and its central conclusion—that prisoners should legitimately expect to be treated with a sense of justice and that the administration should take steps to meet this expectation—was held to be a major indictment of the prison service (and certain senior management arrangements) and a call for a fundamental change of direction. The results are still working their way through the prison administration, although they have been to some extent overtaken by subsequent events such as privatization.

Such 'extraordinary' events, and the special forms of accountability that evaluate them, though not common, cannot be seen as rare. They have occurred, certainly in England, France, and the Netherlands, roughly every decade. The Groningen riots of the 1970s in the Netherlands, and the changes resulting from the McKinsey report on the Dutch Ministry of Justice in the 1980s, may be seen in a similar light. The latter presents, however, the interesting case of a ministry being called to account for its expenditure, and the results of that process bringing in train a series of changes so quick and dramatic that an administrative breakdown

---

[3] This is sometimes described as a 'catastrophic failure'; see Bignell and Fortune (1984) for a fuller discussion of such failures in nuclear power stations, the construction industry, transport, etc.

begins to occur. In this case Governors no longer saw the need to file routine reports because it was no longer clear who would need them or why, and the non-production of reports was not challenged. Equally, they found it more efficient to use informal means of processing requests and proposals through the administration, since the high level of flux made it increasingly uncertain that formal procedures would give results. The odd country out is Germany, which appears to have had no such major convulsions for at least forty years; it is by no means clear that the passing of the StVollzG can be counted in this light. The explanation for the lack of such incidents may lie in a long-term commitment to maintain prison capacity even where it was under-utilized; questions of efficiency appear to have been less important than in the other countries.

Accountability, Control, and Political Willingness

Accountability is in large part about control. More particularly, it is about ensuring that authority is used for its intended purposes, and about control over the way in which powers vested in authorities are exercised. Governments, in modern, democratic states at least, must commit themselves to the maintenance of such controls in order to maintain confidence in, and the legitimacy of, the organizations and decision-makers whose actions are under scrutiny.

For this reason, if no others, accountability is an inherently political process. On the one hand, governments can reasonably be expected to create frameworks of oversight intended to ensure that their policies are being carried out. Accountability is, in this view, no more than one aspect of the cycle of government policy, given shape and form through laws, and implemented through agency regulations and policies which are, finally, intended as the determinants of day-to-day managerial and administrative decisions. This might be described as a managerial and administrative view of accountability. On the other hand, the maintenance of legitimacy implies that governments must also create machinery capable of ensuring that the functions of government are being carried out efficiently, effectively, economically, lawfully, and with propriety and humanity—though precisely how these different concerns are blended together is again a political issue which is central to the concept of legitimacy.

Two broad points follow on from this observation. First, mechanisms of accountability, even those intended to deal with matters of efficiency and effectiveness, become part of the machinery of legitimation. Even if the policy issues are settled, the mechanisms intended to ensure that those policies are being executed also have a legitimating function. They must command the confidence, not necessarily of the public, but of informed, concerned, and politically influential 'outsiders'; they must command the compliance, if not the respect, of staff; and they must be seen to be working effectively. In this sense, they must appeal not only to a particular policy or doctrine, but to broader concerns of good government. Grievance procedures, inspections, and the like, must indicate in their practice that equity of treatment and acceptable standards are officially promulgated intentions, and that derogation from these expectations can be remedied.

Second, if the legitimacy of prisons depends on effective oversight, the other side of this coin is that the mechanisms of accountability must command the support and attention of governments. On this score we may reasonably have doubts about the capacity as well as the willingness of governments to support the relevant bodies, and to act decisively on the information these mechanisms provide. On issues such as prison conditions, one must reckon with the unwillingness of governments to create rods for their own backs, either through committing themselves to improved conditions, creating inmate rights, or adopting regime standards. There are limits to what the administration and the government are prepared to be accountable for and the terms in which they will be made accountable, and one must also reckon with the corrosive influence of treasuries in this regard. Such commitments are sometimes made, but usually after a great deal of foot-dragging. In England, the prison administration consistently opposed the general adoption of standards until it because clear, in the privatization debate, that they would be needed as a way of monitoring the contractors—and even then, the talk was of performance measures and benchmarks, rather than standards as such.

So far as listening to messages and acting on them is concerned, England again provides a useful example. Many reports of the independent prison inspectorate have been highly critical of prison conditions. The response of the administration and the Home Office, in the main, has been to make no more than vague commitments to change as and when resources permit. The areas in which the inspectorate has had most success have been those where, for whatever reason, a prior commitment to reform already existed within the administration. The inspectorate reports provided clarity, direction, and urgency for such matters. However, the English example is comparatively mild; the state apparatus is, broadly speaking, in command of its public servants. Questions of the capacity of the state to effect change have arisen even more sharply in relation to the findings of the European Torture Committee in Turkey, where the basic issue is not the willingness, but the ability, of the Turkish government to control the actions of its own police force.

Reform and Riot

Fundamental changes in prison systems appear to spring from two sources. The primary direction of change is top-down; some changes, however, occur from the bottom up. These types of change have somewhat different implications for accountability and need to be discussed separately.

Top-down change occurs where governments seek to resolve major systemic, and often financial, problems (such as dependence on overtime). Alternatively, as with privatization in England, they may wish to bring the system into line with a particular political doctrine, and perhaps use the prisons as a means of signalling the seriousness of government intent. If major changes can be brought about in a service which is the undivided responsibility of central government, and which is structurally difficult to change because of the security issues involved, where can change not be made?

In this kind of situation, processes of accountability can become enmeshed with reform in three main ways. They may be the midwife of reforms, that is, the key participant in showing the need for them. An English example would be the role of the prison inspectorate. They may be the tool of reform. 'Increased accountability' is often cited as one of the arguments to legitimate change, while new forms of accountability may be imposed in order to monitor the implementation of the new arrangements. This was arguably the case for Fresh Start, which followed on from, and subsumed, a number of managerial controls intended to measure value for money. Or they may be the object of reform, where the processes alleged to legitimate the authority of the agency themselves become problematic. The best example here would be the changes in prison disciplinary procedures in England from the mid-1980s to 1992.

However, wholesale change is relatively rare. On the whole, reform is a gradual and piecemeal business, often involving small advances made by inmates in the courts. Governments do not usually engage in reforms that are not forced upon them, or which appear to be giving something for nothing. They are not usually prepared to make commitments (other than election promises) that they may be unlikely to fulfil. And such undertakings are even less likely in the case of prisons, which not only suffer from less-eligibility arguments as compared with education, health, or social welfare expenditure, but which are in effect the dustbins of the criminal justice system. The amount of government money spent on prisons is and continues to be substantial, especially in periods of tough law and order policies. Yet the kinds of provisions made (mainly new prison spaces), and the primary concerns of government (mainly that the money means improved efficiency), have not given priority to matters such as improved prison conditions unless or until those conditions have been repeatedly and credibly condemned as scandalous. Serious discussions of minimum-regime standards have been watered down into proposals for benchmarks and performance indicators.

Prison privatization, in England, France, and the US, illustrates some of these issues. In the US, privatization was a practical solution to a problem of capital expenditure. In England, it has been undertaken partly to reduce costs, partly to reduce the power of the prison officers' union, and partly for reasons of political symbolism. It is, interestingly, a case of reform being carried through not because of problems within the prisons (though these clearly exist), but because it fits the government's wider political aims. It carries with it one irony, which is that privatization entails operating contracts spelling out, in detail, criteria for adequate performance. These can only loosely and indirectly be considered as providing 'regime standards'. Yet if the private sector is expected to meet them, there is no matter of principle preventing their application to non-privatized institutions. In effect, we may see standards being created by a back door, even if they are not enforceable by inmates. In addition, privatization, along with agency status, and increasing emphasis on performance targets and the like, has created a change in the language of evaluation. How far this linguistic reform will be translated into changes in the prisons themselves remains, as with the extent to which performance criteria can or will be enforced, a matter for the future.

There is also a bottom-up dimension to change, though this is less common and less certain in its effects. This is fundamentally a matter of inmates taking matters into their own hands and onto the prison roof. Governments, for obvious reasons, have difficulties in treating such protests seriously. They cannot, for reasons of face, be seen to give in to the protesters. To do so would be to surrender legitimacy. For this reason the French reforms in the 1970s were a kind of brinkmanship, since they were formulated immediately prior to, but implemented after, a series of riots.

Two brief points may be made about riots and prison accountability. The first is that the fact that a major disturbance erupts is often indicative of some administrative breakdown. In this sense riots are indications that processes of accountability have been inadequate, or, if adequate, were not heeded. A major riot thus requires a special form of accountability—usually an inquiry—that will, among other things, ask why the riot was not foreseen. Second, in the immediate aftermath of a major incident, conditions frequently worsen and individual ringleaders are punished. Moreover, they cannot force governments to improve conditions that they are set against improving. However, over the long term— which may mean periods of five to ten years—such events do come to be seen as caused by poor conditions, and attempts are made to improve conditions. Riots can be seen as sporadically effective, over a long period, in forcing governments to address fundamental systemic problems. They underline the need for, and seem to give added urgency to, proposals for improvements that would otherwise be seen as low-priority.

Resourcing, Management, and Organizational Change

Two issues seem paramount under this head. First, formal and general statements about managerial accountability rarely apply to the management of prisons. For example, Garrett (1980: 130), in a discussion of civil service accountability, suggests that:

An accountable manager is one to whom specific authority over part of an organization's resources has been delegated and who is required to answer for the results he has obtained from the deployment of those resources.

Accountability implies the delegation to managers of authority over money and manpower; a form of organization in which managers can be made responsible for the activities of sub-units; a strategic planning framework in which the objectives of those managers can be related to corporate objectives; an arrangement of control information so that progress towards the attainment of those objectives can be monitored and a procedural system for securing managerial commitment to unit objectives and for reviewing results.

Yet prison governors, even those with their own budgets, have relatively little authority over crucial areas such as staffing levels and staff disciplinary procedures, and may have to seek higher authority for even relatively minor types of capital expenditure. There have been attempts in England and the Netherlands to make governors more responsible for their prisons in matters such as the use of resources, performance measures, and linkage of operational matters to organizational objectives. This has been done through the creation of institutional plans and contract arrangements. Yet higher officials often find themselves juggling resources so that planned and contracted arrangements cannot take place. It is not so much that governors—and, for that matter, higher officials—are not accountable, as that the planned allocation of resources and hence the planned performance levels change rapidly, while the control information systems and the strategic planning framework are revised so frequently that questions can legitimately be asked about their effectiveness.

Beyond this, it also seems clear that treasuries are in a position to question proposed budgets closely, to demand justifications not just for line items but for the policies which lie behind them, and to demand or impose cost-cutting measures. They have a 'corrosive effect'. At the same time, they are relatively insulated from the realities of day-to-day prison life, and are not held responsible for failures in the prison system that result from underfunding.

Second, England and the Netherlands in particular have experienced rapid organizational change. While change is probably the only constant feature of large organizations, there is qualitative evidence to suggest that the extent and speed of change, in the late-1980s Netherlands at least, outstripped the capacity of the prison administration to function effectively. The result was that prison governors and administrators had to rely heavily on informal contacts and personal favours to get their jobs done; and that some prison governors tried to influence departmental policy, for example by using the media.

Lessons for Reform

For prison reformers, the message appears to be that any improvements in prison conditions are achievable only over the long term, and opportunities for reform are constrained by several factors.

First, a political willingness to countenance reform is paramount. If senior administrators and ministers have no clear and pressing reason to accept that a problem exists, nothing will be done about it. The lesson of the English prison inspectorate is that it is most able to affect prison developments when a problem becomes politically explosive, and the inspectorate can offer solutions. Outside such situations, its observations—often sternly critical of the whole fabric of the organization—are met with procrastination. One must strike while the iron is hot, and follow agendas set not by the processes of accountability themselves, but by wider political considerations.

Second, the nature of reform that can be contemplated is influenced by wider social and legal considerations. Standards provide one example. It seems that pressure for the introduction of minimum regime standards has been greatest in the common-law countries (England and America) rather than countries influenced by the Romano-Germanic legal tradition. The most likely explanation is that the concept of a penal code in the Romano-Germanic structure includes the specification of relationships between inmates and the administration which bears some semblance to prison standards. None the less, as I have argued, it is the weakest rather than the strongest version of standards, desiderata which may in principle form the basis of inmate litigation but which in practice may turn out not to be explicit rights. The situation in England, meanwhile, has been one of growing pressure for standards which could be effectively monitored and enforced, against a history of a foot-dragging prison department. How quickly standards will be developed now that Woolf has come down in their favour, and privatization has given urgency to the idea of performance criteria, has yet to be seen.

Third, the utility of reform has to be seen against a backdrop of unintended consequences and inadequate provision. There is no point in allowing inmates to wear their own clothes unless laundry facilities are also provided. The provision of rights must be accompanied by a means for enforcing them. The acceptance of lawyers into prison disciplinary hearings is robbed of its point if the lawyers cannot handle cases promptly.

Fourth, and finally, we need to counsel against having too high expectations. This is not to say that we should be cynical; only that we should take a realistic view of the kinds of reforms likely to be achievable. Prisons are essentially about control, and accountability is about the legitimation of that control. In the last instance, processes of accountability play second fiddle to wider fiscal and policy concerns of governments. It is tempting, but overly idealistic, to hope that prisons will be significantly better places in the twenty-first century than they have been in the latter half of the twentieth.

Appendix A

Data Collection

THIS book emerged out of several related data-gathering exercises. The idea of a project on accountability and prisons arose out of the issues discussed, in relation to boards of visitors in English prisons, in Maguire and Vagg (1984). The completion of the research project on the boards led to the idea of a more general book on accountability and prisons in England, which I was involved in editing (Maguire *et al.* 1985). The opportunity to extend the study of complaints and disciplinary procedures to other European countries was given a push by the offer, in late 1985, of a Council of Europe Criminological Fellowship, which took me to France and the Federal Republic of Germany.

The 1987–1988 Fieldwork

The initial results of the Council of Europe Fellowship formed part of the proposal for a larger piece of fieldwork which was funded by Economic and Social Research Council. This project, which made comparisons between England, France, the Netherlands, and Germany, took place in 1986–8 (grant No. E00232186). Its aims were:

- to determine the conditions under which the use of agencies formally outside the prison system to conduct disciplinary hearings may lead to a growth in the use of administrative rather than formal disciplinary measures;
- to determine the extent to which the formal recognition of prisoners' rights and prison regime standards leads in practice to compliance with those rights and standards;
- to investigate the practical consequences for the prison system and for prison regimes of the existence of independent agencies mandated to conduct inspections of prisons; and
- to examine the extent to which these agencies are necessarily interdependent, and to which the failure of any one agency to meet its objectives has a deleterious effect upon others.

This project involved a number of elements: meetings with senior prison officials and academic researchers in the four countries; the collection of published and unpublished reports, articles, policy documents and statistics; and visits to several institutions in each of the countries which included interviews with inmates, staff, and governors and, where possible, the collection of information from files in establishments.

The fieldwork took place primarily in 1987 and early 1988. The following establishments were visited in the course of the research:

England   A small training prison and a medium-sized local prison, both in the then Prison Department South-West Region.

France   A large maison d'arrêt, a 'centre pénitentiaire' (this comprised a very large adult male remand/short-sentence unit, a smaller young-offender institution, and a women's prison; only the adult male and youth sections were visited), a 'maison centrale' (high-security long-term prisoners), and a 'centre de détention' (medium-security long-term prisoners).

Germany   A young-adult-offender establishment in Hessen, and a prison for adult males of all sentence lengths in Baden-Württemberg.

The Netherlands   A 'half-open' prison, a women's remand centre and prison and a men's remand centre (both part of the same penal complex), and a high-security prison for long-term prisoners (technically, each wing of this prison was regarded by the prison administration as a separate institution).

Formal, structured interviews were conducted during these visits with the types and numbers of respondents shown in Table A.1.

Three points are worthy of mention in relation to these interview data. First, the English interview schedules were translated into Dutch, French, and German with the help of colleagues in those countries. Interviews in France were conducted in French by JV; in Germany, in German by two colleagues at the Max Planck Institute and in a few cases in English by JV; and in the Netherlands, in Dutch and in one case in German by a Dutch colleague, and also in some cases in English by JV. Second, neither Dutch, French, nor German has a direct translation for the English term 'accountability'. In consequence, interviews were invariably prefaced by an explanation of the concept and of the reasons for the study. These explanations, and the discussions that they provoked, were themselves of interest as data. Third, in the Dutch section of the fieldwork, it was suggested by the Ministry of Justice that interviews with inmates should concentrate on persons known to be 'litigious', many of whom had extensive prison histories in addition to their experience of grievance mechanisms.

TABLE A.1.

	England	France	Germany	Netherlands
Governors	2	3	2	5
Discipline staff (all grades)	7	6	7	10
Other staff (e.g. social workers, psychologists)	1	1	4	7
Inmates	10	6	8	14

Since the purpose of the interviews was to obtain qualitative rather than quantitative data, and it was never our intention to generate a 'representative sample' of inmates, we followed this suggestion, and obtained data of such a quality that we followed this pattern in France and West Germany. By this time, unfortunately, the English interviews had already been completed with a group of inmates who by and large proved not to be so litigious. However, we have been able to rely to some extent on data from our own and others' previous research (Maguire and Vagg 1984; Austin and Ditchfield 1985) to remedy this deficiency.

In addition to these interviews, a number of informal interviews were carried out with the following:

England:	Regional psychologist, clerical officers dealing with inmate petitions, governor grades in the regional manpower office, an assistant regional director, members of the prison inspectorate; and staff of the National Council for the Welfare of Prisoners Abroad.
France:	Members of the prison administration research and legal departments; the inspector; members of CESDIP (a CNRS research institute). During the Council of Europe-sponsored preliminary project, informal interviews were conducted with several prison staff, two governors, a juge de l'application des peines, a lawyer practising in the area of prison litigation, and members of CRIV (a CNRS research institute).
Germany:	Members of prison administration legal and policy departments, including several with responsibilities for inspection; a judge from the prisoners' complaints court; academic researchers.
The Netherlands:	Members of the prison administration legal department, regimes department, and research unit; a

regional director, the inspector; three members of Commissies van Toezicht (one also an academic lawyer); senior staff of the national ombudsman; one academic researcher.

All these interviews were those conducted on the basis of the researcher's own agenda, but without a formal and previously duplicated interview schedule.

The Process of Writing Up

The mass of data, interviews, and other items were sifted gradually and analysed according to themes—rights, inspection, complaints, discipline, and so on. However, a number of events intervened. I moved from England to Hong Kong, and became involved in other research. Writing up the project became increasingly delayed. At the same time, new developments in prisons had to be incorporated into the analysis. In England, for example, these included the 1990 Strangeways riot and the Woolf Report, changes to boards of visitors' disciplinary powers, a restructuring from regional to area offices, the introduction of new procedures for inmate grievances, several significant court cases, operational changes within the prison inspectorate, the contracting out of The Wolds, and the planning of agency status for the prison department. All these developments had to be studied and incorporated into the analysis, and the data available on these developments was primarily documentary (though frequent communication with English and European academics at a variety of venues also helped me assess the material I was reading).

The finished product is, then, compiled from two main sources; my own fieldwork, conducted mainly in 1987–8, and the use of documentary sources and discussions with ex-colleagues in the four countries to update and expand on the material collected during the fieldwork.

Appendix B

Chronology of Main Developments in Prisons in France, England and Wales, the Netherlands and Germany, 1971–1992

1971

England and Wales. —

France. February: at MA Aix-en-Provence, two inmates wound a guard and take a hospital orderly and social assistant hostage. They seize weapons but are killed before they can escape. July: at Lyon, a prisoner is sent a gun in a parcel and kills a guard. September: at Clairvaux, prisoners Buffet and Bontemps take a hospital orderly and a guard hostage. The hostages' throats are cut. October: at Baumettes (Marseilles) a prisoner attempts to escape, takes a hospital orderly hostage, and is killed by a guard. November: René Pleven, Minister of Justice, prohibits Christmas parcels for prisoners. Protests occur in several prisons in the following days. December: insurrection at MC Toul. Ended by force, and 75% of inmates transferred to other prisons. Commission of inquiry set up under Robert Schmelck (avocat général de la cour de cassation and ex-director of the prison administration).

West Germany. A national Prison Commission presents a draft bill (the Commission was set up to address shortcomings of former attempts to draft a bill with a joint set of regulations for the prison services of all the German constituent states. It was also to propose reforms). A court decision determines that the constitutional rights of prisoners can only be curtailed by act of parliament on or the basis of such an act. The doctrine of a 'special relationship of dominance' is not sufficient ground for restrictions on rights.

The Netherlands. Prisoners in Groningen jail revolt. No deaths or serious injuries. Official inquiry fails to identify cause. Coornhert Liga presents counter-report emphasizing lack of attention to prisoners' rights, work situation, bad pay, etc. (The Coornhert Liga has contributed substantially to the emergence of lawyers interested in the underprivileged and in penitentiary reform.)

1972

England and Wales. August: prisoners at Albany locked in cells to facilitate searches following an escape. They smash furniture etc., and create small fires. November: violent demonstrations at Gartree prison following a failed escape attempt.

France. January: publication of the Schmelck report on the riot at MC Toul. It calls into question the severe discipline imposed in the prison. Insurrection at Charles III prison, Nancy. René Pleven takes a prison reform project before the council of ministers (the reforms are promulgated by decree on 12 Sept. and subsequently by the law of 29 Dec.). In summary the reform allows prison leave and remission and gives to the JAP powers to grant parole where the sentence is not more than 3 years. Prisoners also obtain the right to read journals. November: Buffet and Bontemps executed.

West Germany. —

The Netherlands. Ministerie van Justitie publishes 'Rapport Beek', which makes recommendations for a new organizational model changing the tasks of officers and instituting team work-meetings.

1973

England and Wales. —

France. May: mutiny at MA St-Paul, Lyon. June: murder of Dr Fully, medical inspector of prisons.

West Germany. —

The Netherlands. —

1974

England and Wales. Circular Instruction 10/74 provides for temporary transfer of 'difficult' prisoners from dispersal prisons to other prisons. Subsequently becomes known as the 'ghosting' procedure due to the speed with which inmates can be moved (sometimes repeatedly) between establishments. Instigation of two 'control units' for difficult prisoners.

France. July: a law of amnesty releases 1,437 prisoners sentenced to 3 months or less. July/August: 89 disturbances in prisons, including 9 mutinies and in the course of which 6 prisoners die. Eleven prisons completely or partially destroyed. August: the council of ministers approves reforms proposed by Lecanuet and Dorlhac. Giscard d'Estaing visits prisons in Lyon. October: announcement of a 'grâce' of 6 months for all those who have 'observed good behaviour during the month of July 1974', and 1,266 prisoners released.

West Germany. —
The Netherlands. —

1975

England and Wales. Abandonment of the two 'control units'. Judgement in *Golder* (ECHR case) results in 'internal ventilation rule'—no communication of grievance to lawyer etc. until internal channels of complaint are exhausted.

France. May and July: decree and law bringing into effect certain reforms. The 'progressive system' for long-term prisons is abandoned. Prison regimes are 'diversified'. New articles of the CPP institute suspension or part-suspension of sentence under various circumstances, special reductions for educational achievement or for 'exceptional progress' in rehabilitation and enlarged possibilities for parole. Regimes of CDs are lightened, with the possibility of prison leave for 5 days and once per year of 10 days (only 3 days in other prisons). Prisoners in CDs may make telephone calls in certain important personal or family circumstances, and may write to any person (in other prisons they may only write to restricted categories, e.g. those who also hold permits to visit them). In MCs, security is reinforced notably with the creation of 11 Quartiers de Sécurité Renforcée.

West Germany. —
The Netherlands. —

1976

England and Wales. August/September: riot at Hull during which buildings are taken over by prisoners and extensive damage caused.

France. —

West Germany. New penal reform bill brought before parliament.

The Netherlands. —

1977

England and Wales. —

France. Hunger strikes against the creation of the QSRs.

West Germany. New prison law ('Strafvollzugsgesetz') implemented. It gives a number of specific rights to prisoners, e.g. on visits, home leave (up to 21 days per year), productive work, medical care etc. Most can however be restricted by administrative discretion. Prisoners may also take grievances to a special prison court ('Strafvollstreckungskammer'). The general principles of the bill are rehabilitative, stressing reintegration of the prisoner into society, counteracting damage caused by imprisonment, attachment of prison life to life outside, though still also supporting the idea of incarceration serving to protect the public against crime. Suicide of three prisoners (RAF members) in Stuttgart Stammheim.

The Netherlands. Change in prison law ('Beginselenwet Gevangeniswezen') which also implements changes in prison rules ('Gevangenismaatregel') and house rules of individual establishments. The major change is the introduction of a new prisoner complaint system by which prisoners' rights could be effected. Extension of the right to uncensored correspondence; prisoners may also now receive and send letters as often as they want, and make telephone calls at their own expense (maximum 5 minutes, once per week).

1978

England and Wales. October: riot at Gartree following rumours that staff had beaten up a prisoner. Serious damage to buildings.

France. January: assistant governor and two guards held hostage by two inmates at Clairveaux. The inmates are killed by a special firearms unit. May: escape of Mesrine and Besse from the QHS at La Santé. A third inmate killed by a guard. The director of the prison administration is sacked and the governor of La Santé transferred to the central administration. Hunger strikes against QSRs continue.

West Germany. —

The Netherlands. —

1979

England and Wales. March: extensive damage caused by rooftop protest at Parkhurst. April: Disturbance at Hull, C wing. August: prisoners on D wing, Wormwood Scrubs, refuse to work in protest against lack of facilities. Allegations of excessive force in restoring order (two years later, police and prosecuting authorities concluded there was insufficient evidence to prosecute officers). 'May Committee' set up *inter alia* to review regimes in long term prisons—concludes the 'dispersal' system should continue. Also recommends the establishment of an independent Prison Inspectorate. Case of St Germain; BoV adjudications are judicially reviewable for correctness of procedure, must follow rules of natural justice.

France. Reorganization of DAP following a December 1978 decree. The effects are broadly that more decisions concerning more prisoners are taken centrally; and more individualization of sentences should be possible (educational opportunities etc.). A special 'security brigade' is also put in place.

West Germany. —
The Netherlands. —

1980

England and Wales. Prison officer industrial action starts, continues to 1981. Two army barracks brought into use as emergency prison camps, staffed by troops. From this time on, police cells are in use more or less continuously to hold some prisoners.

France. July: a 'grâce' of one month releases 1,410 prisoners having one month or less left to serve.

West Germany. —

The Netherlands. Visiting hours in remand prisons extended from half-hour to 1 hour per week or 2 hours every 2 weeks.

1981

England and Wales. January: independent Prison Inspectorate comes into being. Significant legal cases decided during the year include *Williams*; which determined that the 1974–5 'control units' breached prison rules but no legal redress possible; *Mealy*, which decided that prevention of access to material witnesses in disciplinary hearings is unfair; and *Payne*, which decided that the rules of natural justice are not applicable to parole procedure.

France. February: 'security and liberty' law passed. June: commission set up to examine the retention of the QSRs. July: 4,775 prisoners benefit from a 'grâce' of 3–6 months, depending on the length of their sentence. August: amnesty releases 1,437 prisoners with sentences of 6 months or less.

West Germany. Legislation on drugs amended, to enable deferral of sentence if convicted person agrees to detoxification treatment.

The Netherlands. The Commissie van Hijlkema publishes a report on capacity problems stressing structural long-term provision necessary to maintain the quality of the buildings and a desperate need for extension of prison capacity. Ministerie van Justitie publishes 'GIS-nota', a report on standardized establishment structure and the functions of guards. Restrictions introduced on the freedom of inmates to furnish cells as they please. New regulations on the placing of TV in cells.

1982

England and Wales. Several important legal cases decided. In *Raymond*, it was determined that prison authorities may not prevent communication with a court. In *Fox-Taylor*, it was determined that in BoV disciplinary hearings, prison must draw attention of prisoner to existence of any material witnesses who may support his case. In *McConkey*; it was determined that in BoV adjudications, evidence of association with an offence is not in itself a sufficient basis for the imposition of punishment.

France. February: QSRs abandoned. New arrangements concerning the use of isolation are introduced. December: new reforms announced, in particular the introduction of 'open' visits (i.e. the abandonment of 'hygiaphone'-style separation of visitors and prisoners).

West Germany. Act amending the criminal courts. They may now order that the execution of a life sentence shall be suspended on probation once 15 years have been served in prison.

The Netherlands. Introduction of AVR, a general leave arrangement which enables long-term prisoners to visit their homes under certain circumstances.

1983

England and Wales. May: Criminal Justice Act 1982 comes into force, making substantial changes to the young offender population; principally, determinate-length 'Youth Custody' replaces the indeterminate 'Borstal' and the Detention Centre orders for more minor offences are halved in length. In addition, time spent on remand now counts towards sentence for YC. The recalculations of sentence lengths enable approximately 1,000 young offenders to be released immediately. Strikes by prisoners at Albany, followed by damage to wings and a rooftop protest. June: prisoners damage fittings etc. in D wing Wormwood Scrubs. November/December: change of parole rules—minimum qualifying period for parole reduced from 12 to 6 (effectively 10.5) months (or one-third of sentence— no change) but some categories of inmates serving long sentences may no longer expect early release on parole. Legal cases decided this year include: *Silver* (ECHR), which prompts 'simultaneous ventilation rule' (i.e. prisoners may complain to a lawyer at same time as they initiate complaint procedure within prison) and *Tarrant*, in which it was determined that prisoners should be legally represented at BoV adjudications under certain circumstances.

France. January: disturbances in prisons, particularly Fleury-Mérogis and Baumettes, in the period waiting for the reforms announced in December 1982 to take effect. Reforms of December 1982 come into effect, allowing open visits; lightening daily regimes; facilitating links with families etc.; and giving powers to inspect public health in prisons to IGAS. The reforms applied in 1975 to prisoners in CDs are now extended to all sentenced prisoners. Prisoners in CDs may now make one telephone call per month. Prison uniforms no longer obligatory except under special (security-related) circumstances. June: the 'security and liberty' law is revoked.

West Germany. Campaign of Bar Association results in abolition of the practice of using pre-trial detention for young offenders for 'educative' or 'deterrent' purposes.

The Netherlands. Change in prison regulations as to the conditions of a systematic right of search of inmates' body and clothes.

1984

England and Wales. March: following the *Tarrant* decision, the Home Secretary sets up the 'Prior Committee' to review prison disciplinary procedures. July: approx 2,000 prisoners released following implementation of November/December 1983 parole rules. Legal judgements in the year include *Campbell and Fell* (ECHR) which determined that BoVs are suitable bodies to conduct disciplinary hearings. in *King*, judicial review of governors' adjudications was denied. The correct procedure is to petition the Home Secretary, whose decision may be reviewed judicially. In *Anderson*, communications with solicitors or court proceedings were determined not to be capable of giving rise to internal disciplinary proceedings for a 'false and malicious allegation'. In *Findlay*, the Home Secretary was held to have discretion to alter the parole rules though he must consider prisoners in the light of those rules. In *Gunnell*, a prisoner was held not entitled to an oral hearing before parole board. In *Freeman*, the nature of imprisonment *per se* was held not to invalidate the possibility of consent to (experimental) medical treatment. In *McAvoy*, transfer of prisoners was deemed not reviewable by courts. In *McGrath*, it was determined that prison officers' orders to prisoners must be grounded in lawful authority and not arbitrary.

France. January: case of *Caillol* decided: the use of the QPGS, being a measure of internal order within prisons, was not susceptible of recourse to an administrative judge for review on the grounds of excess of power. Decree places prisons under the supervision of the public health ministry. March: new rules on prison discipline, entry of paperback books into prison etc. July: new law on provisional (remand) detention.

West Germany. January/February: RAF members in prison go on hunger strike. The Prison Administration Act of 20 December abolishes social therapy as a separate measure of rehabilitation and prevention, though the Act is not yet brought into force. (Committal to a social-therapeutic institution is currently a particular type of sentence).

The Netherlands. Introduction of permission to have TV in cells in all local and closed prisons. Censorship of all letters in local and closed prisons no longer obligatory; only random checks, and checks where governor thinks it necessary.

1985

England and Wales. Prison Department begins to publish Standing Orders (previously not freely available to prisoners or the general public). Four of the 16 SOs are published—3c (calculation of sentence), 4 (privileges), 5 (communications) and 12 (civil prisoners). An undertaking is made to publish further SOs as and when they are revised. October: report of Prior Committee on prison disciplinary procedures recommends the replacement of BoVs by Prison Disciplinary Tribunals with legally qualified chairmen (the proposal is not adopted).

France. January: new legal provisions concerning the treatment of prisoners in hospitals or special hospitals for detained persons. New circular on the enforcement of prisoners' rights on remand (i.e. enforcement of law of 9 July 1984). April: a guard is killed by inmates at Lyon-Montluc prison. May: disturbances in about 40 prisons. Fleury-Mérogis building D4 and D1 infirmary are destroyed. Damage at MA Montpellier. At Fresnes, a prisoner falls from the roof. July: a 'grâce' permits the release of 2,763 prisoners one or two months before the end of their sentence. August: modification of the CPP to facilitate individualization of short sentences, improve relations with social workers and create an interministerial co-ordination committee on prison hygiene. December: authorization for prisoners to have TV in cells. Remission granted for all prisoners (previously only for those serving 3 months or more).

West Germany. The Prison Administration Amendment Act of 27 February removes the obligation imposed on prison authorities to take coercive measures in cases of acute risk of death of an inmate. This obligation is now only imposed where the prisoner is no longer in a situation which prevents him freely determining his own will.

The Netherlands. Introduction of additional rule in the House Rules for remand prisons, laying down the right of all inmates whether convicted or remand to consult their own general medical practitioner or specialist (previously only a privilege for those held on remand). Changes in rules concerning: hunger strikers, use of teargas in prisons, making of contracts between press and inmates. Change in Election Act, enabling prisoners to vote by proxy. Budgetary cuts result in alterations to regimes. Prisoners now only work half-days, spending other half in recreation, and spend more time in cells at weekends.

1986

England and Wales. April/May: POA industrial action, banning over-time. Almost immediately there are disturbances and riots in over 40 establishments. One wing of Bristol destroyed; Northeye cat. C prison burned out.

France. March: decree reforms organization and functioning of proba-tion committees. September: introduction of laws relating to penal sanc-tions and the 'fight against crime and delinquency'.

West Germany. 23rd Criminal Law Amendment Act of 13 April extends the possibilities of early release from prison.

The Netherlands. Introduction of rules concerning unexpected illness during home leave. Extra days on home leave claimed by convicts due to this will be served in prison, postponing the release date by the corre-sponding number of days (there is consequently a reduction in the number of days of illness claimed by prisoners). CRvA decides disciplinary punish-ments may only be imposed after prisoner has been heard by the governor and a place for the punishment to take place has been found. CRvA decides that in any decision made in prison, a lack of facts concerning the inmate must be interpreted in the inmate's favour. August: escape of pris-oners from high-security wing at Den Haag prison.

1987

England and Wales. July: HMCIP's report on the 1986 disturbances crit-icizes prison staff who refused to carry out orders, but also the prepared-ness of the Department and the general prison conditions at that time. Rules changing remission announced. Existing and future sentences of 12 months or less will now attract half remission (previously one-third—and this proportion remains unchanged for all those with longer sentences). When implemented in August, these new rules result in the release of about 3,500 prisoners. September: 'Fresh Start', a new scheme of prison officer attendance and functions, designed to reduce dependence on over-time, begins to be implemented in the first establishments.

France. January: new circular on the imprisonment of women with their children. June: law enabling the construction of 15,000 new prison places in the private sector.

West Germany. —

The Netherlands. New regulations concerning conditional sentences and conditional release. Conditional sentences now possible for up to 3 years (previously 1 year); for those between 1 and 3 years only one-third may be conditional. Full or partial revocation may be ordered by the trial court for a new offence. The former 'conditional release' is transformed into unconditional release. It is executed after two-thirds of sentence, provided a minimum 6 months (previously 9 months) is served. State Secretary for Justice announces a number of new policy measures to repress drugs in prison, including the establishment of 'drug-free sections' in a number of prisons. At this time a drug-free unit is operational in Overamstel complex, Amsterdam.

Council of Europe. Publication of the European Prison Rules, replacing the previous Standard Minimum Rules for the Treatment of Prisoners.

1988

England and Wales. February: judgements in *Leech* and *Prevot* determine that prison governor's disciplinary hearings are susceptible of judicial review. April: new regulations restricting the abilities of remand prisoners to receive food etc. from visitors come into force. Home leave extended to 2 days every 6 months for category C prisoners after their parole eligibility date. Reduction in censorship for category C and certain YC and female prisons; only a random 5% of mail in or out will be read and prisoners may send and receive as many letters as they wish. The installation of payphones for prisoners in selected establishments is announced. June: riot at Haverigg low-security closed prison. Many buildings destroyed, 26 prisoners break out—most recaptured. Over 50% of prisoners transferred elsewhere. Riot blamed on (a) petty restrictions, e.g. that prisoners must put photos etc. only on cell notice boards (this in turn was due to the renovation of the buildings) and (b) the attempt to use low-security prison capacity effectively, leading to 'unsuitable' inmates being allocated there.

France. April: mutiny at MC Ensisheim; buildings severely damaged.

West Germany. —

The Netherlands. —

1989

England and Wales. —

France. —

West Germany. —

The Netherlands. —

Council of Europe. February: the European Convention for the Prevention of Torture and Inhuman or Degrading Treatment of Punishment (ECPT) comes into force. It provides *inter alia* for a Committee to undertake inspection visits to member countries.

1990

England. Riot followed by inmate occupation of Strangeways prison, Manchester, 1–25 April. Riots and major disturbances also took place in many other prisons during this period. September: reorganization of four regional offices into fifteen 'area' offices.

France. September: inmates in one of the new 'semi-private' prisons, Neuvic-sur-Isle, Dordogne, protest and subsequently riot against the use of electronic identity cards in the establishment.

Germany. Reunification of East and West Germany. West German laws, including prison laws, become applicable to the old East Germany. Some disturbances are claimed to be 'copycat' incidents following the screening of the Strangeways, England, riot on television.

The Netherlands. —

1991

England. Publication of the Woolf report on the 1990 prison riots. It makes extensive criticism of many aspects of prison conditions and management, and argues for a new approach to inmates based on the concept of 'justice in prisons'. 1991 Criminal Justice Act empowers the Home Secretary, *inter alia*, to make contracts for privately run prisons. December: publication of the Lygo Report, arguing that the prison service should be given 'executive agency' status. The Home Secretary announces long-range plans to privatize half the prison estate. March: the ECPT Committee's report on a 1990 visit to the UK is published along with the government's response. The report considers conditions in three prisons, Brixton, Wandsworth, and Leeds, to constitute inhuman and degrading conditions. The government denies this.

France. —

Germany. It becomes clear that conditions in the old East German prisons are far below acceptable standards. However, the imposition of West German law led to a reduction in inmate pay for East German prisoners.

The Netherlands. —

1992

England. Boards of Visitors' disciplinary powers are removed with effect from 1 April. The Wolds, a privately run remand facility, is opened. Home Secretary announces plans to privatize Strangeways prison and several others. An inquiry is announced into a series of suicides by young prisoners at Feltham young offenders' institution.

France. August: a prison officer is killed by an inmate at maison d'arrêt Bonne-Nouvelle de Rouen. Industrial action ensues and the government seeks talks with staff unions on security matters 'without delay'.

Germany. —

The Netherlands. —

The information above was compiled from the following sources:

England and Wales: Home Office (1984b; 1991), HM Chief Inspector of Prisons (1987), Maguire *et al.* (1985), Richardson (1985), Treverton-Jones (1989), NACRO Briefings (various dates), Prison Information Bulletin (various dates), press cuttings.

France. Dablanc (1980), Delteil (1990), Dorlhac de Borne (1984), Faugeron (1991), Favard (1987), Fize (1984), Prison Information Bulletin (various dates), press cuttings.

The Netherlands. Kelk (1983; 1991, forthcoming), Ministerie van Justitie (1984), Prison Information Bulletin (various dates).

Germany. Dünkel and Rössner (1991), Feest (1982; 1988), Kaiser *et al.* (1983), Prison Information Bulletin (various dates).

Appendix C

Glossary

amnestie (Fr.), a means of reducing the prison population by releasing selected categories of inmates. An amnestie was announced in 1981. See also 'grâce collective'.

Antrag (Ger.), a formal request for information directed from members of parliament to a government body. Normally an Antrag can only be allowed if it is signed by a specified number of parliamentary members.

area (Eng.), since 1990 the English prison system has been administratively divided into 15 areas. See also 'region'.

atelier (Fr.), workshop.

Beirat (pl. *Beiräte*) (Ger.), voluntary bodies attached to each penal institution with some powers of oversight, but primarily involved in the rehabilitation and resocialization of inmates.

Beklagcommissie (Neth.), the subcommittee of a CvT (q.v.) which actually hears inmate complaints.

Board of Visitors (abbr. BoV) (Eng.), attached to every prison, comprised of volunteers drawn from a local community, and appointed by the Home Secretary, whose brief is essentially to oversee the prison and hear inmate complaints.

Borstal (Eng.), a type of training institution for young offenders; Borstal sentences were indeterminate. The system was superseded by Youth Custody Centres (q.v.) in 1982.

brigadier (Neth.), a team leader with 8–12 prison officers under his command. The designation officially no longer exists, having been changed in staff reorganizations, but is still more widely used than the new official title.

bunker (Neth.), slang term for a small high-security cell block at the high-security prison in Den Haag.

cahier d'observation (Fr.), an 'observation book' issued to each member of staff in which he is expected to record any observations he may have on individual inmates that pertain to security. The book is then inspected by senior officers and governors on a daily basis.

canteen (Eng.), the prison 'shop' at which inmates may purchase items such as cigarettes, coffee, tea, sweets, batteries, groceries, and so forth. This is sometimes physically in the form of a shop to which each inmate has regular access; in other institutions, inmates order goods by filling in forms and the items are subequently delivered to their cells.

Centrale Raad van Advies (abbr. CRvA) (Neth.), the central advisory board which advises the Dutch prison service on policy proposals and also adjudicates on appeals of decisions of the CvTs (q.v.).

centre de détention (abbr. CD) (Fr.), prison for long-term inmates, in principle offering a less restrictive regime than a maison centrale (q.v.).

Centre National d'Observation (abbr. CNO) (Fr.), a unit at Fresnes prison, near Paris, which assesses and makes recommendations on allocation for certain categories of prisoner, including all inmates whose sentence will run for 10 years or more from the point at which it is confirmed.

certiorari, writ of (Eng.), a writ to the Divisional Court requesting that it review an administrative decision on the grounds that it was not properly made.

chef de détention (Fr.), the Surveillant Chef (q.v.) with particular responsibility for security in an establishment.

Chief Officer (abbr. CO) (Eng.), the highest uniformed staff grade. Since Fresh Start (q.v.), officially re-designated as 'Grade 4' or 'Grade 5' (depending on seniority) along with the erstwhile assistant governor grades.

Circulaires de l'Administration de la Justice (Fr.), administrative circulars issued by the DAP (q.v.).

Circular Instruction (abbr. CI) (Eng.), a form of administrative instruction issued by the Prison Department.

Code de Procédure Pénale (abbr. CPP) (Fr.), the French code of criminal prodecure, which also includes legal provisions concerning imprisonment, prison management and organization, and so on. In addition to the Articles of the Code, it also includes administrative regulations (règlements d'administration publique), decisions of the conseil d'état, and decrees (décrets).

Commissie van Toezicht (abbr. CvT) (Neth.), a board of volunteers appointed to each prison with a remit including oversight of prison conditions and the hearing of inmate complaints.

commission de l'application des peines (abbr. CAP) (Fr.), a committee in each prison comprising the JAP (q.v.), the prison governor, and various other prison staff. It advises the JAP on various issues concerning the treatment of inmates.

Confédération Française du Travail (abbr. CFDT) (Fr.), one of the trades unions to which many prison staff are affiliated.

Confédération Générale du Travail (abbr. CGT) (Fr.), one of the trades unions to which many prison staff are affiliated.

Conseil d'État (Fr.), the highest administrative court dealing with reviews of administrative decisions and policies.

Detention Centre (abbr. DC) (Eng.), institutions for young offenders sentenced to Detention Centre Orders. There are two types of establishment; 'junior' (aged 14–17), and 'senior' (aged 17–21). The sentences are comparatively short—2 to 4 months—though the regime is more 'brisk' than in YCC (q.v.). DCs tend to be used more often for young first offenders.

Direction de l'Administration Pénitentiaire (abbr. DAP) (Fr.), official name of the French prison service.

director (trans.), the direct English equivalent of the French, German, and Dutch term for a prison governor (q.v.).

dispersal prisons (Eng.), designated closed training prisons with high levels of security, which are used to hold high-security-risk inmates. However such inmates usually constitute only a minority of their penal population.

Fleury-Mérogis (Fr.), the largest prison complex in Europe, to the south of Paris. It comprises a maison d'arrêt, youth, and female prisons, holding in total over 6,000 inmates. The national prison training centre is also located there.

Force Ouvrière (abbr. FO) (Fr.), one of the trades unions to which many prison staff are affiliated.

Fresh Start (Eng.), a management initiative begun in England in 1987 which attempted to 'buy out' prison staff overtime and introduce new working practices.

gevangenis (pl. *gevangenissen*) (Neth.), prison.

governor (Eng.), the head of a prison (though some governor-grade staff are employed in area, regional, or headquarters offices). The equivalent term most often used in France, Germany and the Netherlands is usually translated as 'director'.

Governors' Orders (Eng.), memoranda of instruction issued by a prison governor for an individual prison; often specify 'house rules' (q.v.).

grâce collective (Fr.), a large-scale reduction of sentence lengths for serving inmates, designed to reduce the prison population. Grâces were announced in France in 1980, 1981, and 1985. See also 'amnestie'.

Handakte (Ger.), a file on an individual inmate kept in the prison, and available to staff.

Her Majesty's Inspector of Prisons (abbr. HMCIP) (Eng.), the head of the English Prisons Inspectorate, the body with primary responsibility for inspecting prison establishments. He reports directly to the Home Secretary rather than to the Prison Department.

Home Office (Eng.), the government department with ultimate responsibility for prison affairs.

house rules (slang), rules made by the governor of a prison, usually relating to the local arrangements for visits, exercise, association, inmate clothing, etc.

Hoofd begeleiding (Neth.), a member of a prison management team with specific responsibility for regime and programme co-ordination.

Huis van Bewaring (abbr. HvB) (Neth.), equivalent to the English local prison (q.v.).

Inspection Générale des Affaires Sociales (abbr. IGAS) (Fr.), a government inspectorate whose remit can in some circumstances extend to prisons.

Inspection Générale des Services Judiciaires (Fr.), a govenment inspectorate whose remit includes, for example, the inspection of arrangements for released prisoners.

Inspection des Services Pénitentiaires (abbr. ISP) (Fr.), the department within the DAP (q.v.) with primary responsibility for inspecting prisons.

Juge de l'Application des Peines (abbr. JAP) (Fr.), a judge appointed to each prison, with a remit (a) to advise to the governor on the treatment of prisoners, and (b) to make decisions affecting inmates' length of stay in the establihsment, eg. home leave and parole. However, the JAP is restricted in several ways; on some issues he may only make recommendations to the Ministry of Justice, and on others he must accept the advice of the CAP (q.v.).

jugendarrestanstalten (abbr. JVA) (Ger.), youth custody centres.

Land (pl. *Länder*) (Ger.), constituent state of the Federal Republic of Germany. Each Land has its own prison service. There is no federal prison service.

landing (Eng.), prison wings, especially in older Victorian-designed prisons, are divided into 'landings', i.e., floors. The ground floor is often called 'the 1's', the first floor 'the 2's', and so on.

Lockerungsmassnahmen (Ger.), 'relaxation measures'—a collective name for facilities such as visits, exercise, association, home leave, and so on.

local prison (Eng.), multi-purpose establishment holding pre-trial detainees, usually remanded from specific courts; short-term prisoners; and long-term inmates at the beginning of their sentence and awaiting allocation elsewhere.

maison centrale (abbr. MC) (Fr.), prison for long-term inmates, in principle offering a higher level of security than a Centre de Détention (q.v.).

maison d'arrêt (abbr. MA) Fr.), French equivalent of a local prison (q.v.).

missions de contrôle générale (Fr.), general inspections of prisons conducted by ISP (q.v.).

missions d'enquête (Fr.), inquiries conducted by ISP (q.v.) following escapes or incidents.

Oberlandsgericht (abbr. OLG) (Ger.), the supreme court in each Land which hears, *inter alia*, appeals of prisoner complaints from the Strafvollstreckungskammern (q.v.).

Over-Amstel (Neth.), a large establishment in Amsterdam with some six separate prisons, in high-rise blocks, on a single site, surrounded by a common wall and having a number of other common facilities including kitchen and visits area.

Parliamentary Commissioner for Administration (abbr. PCA) (Eng.), official name for the ombudsman.

pavilion (trans. Neth. *paviljoen*), a unit of accommodation for inmates, usually designed for about 20–25 inmates.

Premier Surveillant (Fr.), the grade between surveillant (i.e. basic grade officer) and Surveillant Chef (q.v.).

prétoire (Fr.), inmate disciplinary hearing.

Principal Officer (abbr. PO) (Eng.), the staff grade below Chief Officer grade. Since Fresh Start (q.v.), redesignated officially as 'Grade 6'.

Prison Officers' Association (abbr. POA) (Eng.), the main trades union for prison discipline staff.

Projet 15,000 (Fr.), a plan initially proposed under minister Albin Chalandon in 1986 to build 15,000 new prison places. Various services for inmates, from medical care to food, would be provided in these new prisons by private companies. The project was subsequently scaled down and re-named 'Projet 13,000'.

region, the layer of administration between individual institutions and headquarters. The French prison system is divided into nine regions. Only one German Land prison system is large enough to warrant administrative division into two regions. The Dutch prison system has experimented with various regional divisions. The English system was divided into four regions until 1990. They were superseded by 15 'areas'.

relaxation measures (trans. Ger.), see Lockerungsmassnahmen.

remand centre, an establishment primarily used for holding pre-trial detainees.

Rule 43 (Eng.), the Prison Rule specifying the conditions under which inmates may be segregated from the main inmate population for other than disciplinary reasons. Inmates so segregated are usually referred to as 'Rule 43s'.

satellite, a small prison managed as an integral part of a larger nearby prison; a common feature of German prison systems.

'Sécurité et Liberté' (Fr.), the 'security and liberty' law of February 1981, enabling those convicted of a fairly wide range of violent offences or certain serious property offences to be given longer sentences which would be served under less permissive conditions as regards day-release, home leave, and parole. The law was partly revoked in 1983 by the newly elected left-wing government.

security period (Fr.), a period during which day-release, home leave or parole cannot be granted—the first two-thirds of a sentence of 3 years or more, and half of any sentence of 10 years or more. Instituted in November 1978.

Senior Officer (abbr. SO) (Eng), the staff grade immediately above basic grade and below Principal Officer (q.v.). Since Fresh Start (q.v.), officially redesignated as 'Grade 7'.

sous-directeur (Fr.), an assistant governor of a prison. Literally, 'under-director'.

sozialtherapeutische Anstalt (Ger.), establishment or part of establishment with a psychological and social programmes for 'resocializing' prisoners.

Standing Order (abbr. SO) (Eng.), a form of administrative instruction issued by the Prison Department.

Strafvollstreckungskammer (pl. -*kammern*) (Ger.), special courts with a special responsibility to consider prisoner complaints and certain other matters such as parole.

Strafvollzugsgesetz (abbr. StVollzG) (Ger.), German national prison law.

surveillant (Fr.), a basic-grade prison officer. Occasionally referred to in slang as a 'maton'.

Surveillant Chef (Fr.), the most senior of the uniformed staff grades, equivalent to the English Chief Officer (q.v.).

training prison (Eng.), establishments for medium and long-term inmates, usually offering a less crowded regime than local prisons and opportunities for training, including vocationally-oriented workshops.

Tribunal Administratif (Fr.), the lower, regional, administrative court dealing with complaints relating to administrative decisions.

walking sentence, a person who is sentenced to imprisonment but for whom there is no place in prison. In the Netherlands such persons are released and when a place is available for them they are told to report to the prison. If no place becomes available within a specified time period the prison sentence is set aside.

Vollzugsdienstleiter (Ger.), literally 'prison service leader'; uniformed staff grade equivalent to the English Chief Officer (q.v.).

vulnerable-prisoner unit (abbr. VPU) (Eng.), a separate unit within a prison used to hold inmates thought to be at risk of victimization by other prisoners, for example because of the nature of their crimes, their having acted as informers, or their being 'easily led' by others.

Vollzug (Ger.), prison. One sometimes sees the terms geschlossener Vollzug (closed prison) and offener Vollzug (open prison).

Youth Custody Centre (abbr. YCC) (Eng.), establishment for offenders up to the age of 21 sentenced to a Youth Custody Order.

References

Actes (1982). 'Dix variations sur le thème du changement' (Ten Variations on the Theme of Change), anonymous article, *Actes*, 37 (June).

ADVISORY COUNCIL ON THE PENAL SYSTEM (1968). *The Regime for Long-Term Prisoners in Conditions of Maximum Security*. London: HMSO.

AGER, T. J. (1986). *Evaluation of the Accountable Regimes Project at Shepton Mallet, 1982–84*. Prison Dept., Directorate of Psychological Services Report, Series II, No. 143. London: Prison Dept. Directorate of Psychological Services.

ALEXANDER, R. (1992). 'Cruel and Unusual Punishment: A Slowly Metamorphosing Concept', paper presented to the American Society of Criminology 44th Annual Meeting, New Orleans, Nov.

AMERICAN CORRECTIONAL ASSOCIATION (1990). *Standards for Adult Correctional Institutions*, 3rd edn. Laurel, Md.: American Correctional Association.

AUSTIN, C., and DITCHFIELD, J. (1985). 'Internal Ventilation of Grievances: Applications and Petitions', in Maguire *et al.* (1985).

BADEN-WÜRTTEMBERG, MINISTER OF JUSTICE (1983). *Strafvollzug in Baden-Württemberg* (Prisons in Baden-Württemberg). Stuttgart: Justiz-ministerium Baden-Württemberg.

BARAK-GLANTZ, I. L. (1981). 'Towards a Conceptual Schema of Prison Management', *Prison Journal*, 61/2.

BARRE, M.-D. (1987). *Résistible progression des effectifs de la population carcerale en France?* (A Resistible Increase in the French Penal Population?) Paris: CESDIP.

BERGHUIS, B. (1987). 'De Sprang: An Evaluation of a Special Regime in a Remand Centre', in M. J. M. Brand-Koolen (ed.), *Studies on the Dutch Prison System*. Amstelveen: Kugler.

BIBAL, D., KENSEY, A., LAGRANGE, M. and MEURS, D. (1983). *Le travail en prison* (Prison Labour). Paris: Ministère de la Justice, Service des Études, de la Documentation et des Statistiques/Centre National d'Études et de Recherches Pénitentiaires.

BIGNALL, V., and FORTUNE, J. (1984). *Understanding Systems Failure*. Manchester: Manchester Univ. Press.

BIRKINSHAW, P. (1985a). 'An Ombudsman for Prisoners', in Maguire *et al.* (1985).

—— 1985b). *Grievances, Remedies and the State*. London: Sweet and Maxwell.

BLUMSTEIN, A. (1988). 'Prison Populations: A System Out of Control?', in

M. Tonry and N. Morris (eds.), *Crime and Justice: A Review of Research*, (Chicago: Univ. of Chicago Press).

BORNA, S. (1986). 'Free Enterprise Goes to Prison', *British Journal of Criminology*, 26/4 (Oct.).

BOULAN, F. (ed.) (1987). *Les Prisons dites 'privées'* ('Private' Prisons). Paris: Economica/Presses Universitaires d'Aix-Marseille.

BRAND-KOOLEN, M. J. M. (1987). 'The Dutch Penal System and its Prisons: an Introductory Note', in M. J. M. Brand-Koolen (ed.), *Studies on the Dutch Prison System*. Amstelveen: Kugler.

BRYDENSHOLT, H. H., VAN DER GOOBERGH, B., ALMEIDA, M., and SHAPLAND, P. (1983). *Prison Management*. Strasbourg: Council of Europe.

BURNEY, E. (1985). *Sentencing Young People: What Went Wrong With the Criminal Justice Act 1982*. Aldershot: Gower.

CALAVITA, K. (1983). 'The Demise of the Occupational Safety and Health Administration: a Case Study in Symbolic Action', *Social Problems*, 30/4 (Apr.).

CANNON, A. W. (1986). 'Private Sector Management Contracting in Prisons', unpublished paper, Committee on Corrections, Probation, and Parole, Senate, State of Florida.

CASALE, S. (1984). *Minimum Standards for Prison Establishments*. London: National Association for the Care and Resettlement of Offenders.

—— (1985). 'A Practical Design for Standards', in Maguire *et al.* (1985).

CHAPLIN, B. (1985). 'Accountable Regimes: Concept and Significance'. Unpublished MS.

CHATTERTON, M. (1987). 'Assessing Police Effectiveness: Future Prospects', *British Journal of Criminology*, 27/1.

CLEMMER, D. (1940). *The Prison Community*. New York: Holt, Rinehart, and Winston. 2nd edn., 1958.

COLLINS, W. P. (1980). 'Public Participation in Bureaucratic Decision-making: A Reappraisal', *Public Administration* 67: 465–77.

COUNCIL OF EUROPE (1986). *Prison Regimes: Study Prepared by the Select Committee of Experts on Prison Regimes and Prison Leave*. Strasbourg: Council of Europe.

—— (1987). *European Prison Rules: Recommendation No. R(87)3 Adopted by the Committee of Ministers of the Council of Europe on 12 February 1987 and Explanatory Memorandum*. Strasbourg: Council of Europe.

—— (1991a). *European Committee for the Prevention of Torture and Inhuman or Degrading Treatment of Punishment, 1st General Report on the CPT's Activities Covering the Period November 1989 to December 1990*. Strasbourg: Council of Europe.

—— (1991b). *Report to the Austrian Government on the Visit to Austria Carried Out by the European Committee for the Prevention of Torture and Inhuman or Degrading Treatment or Punishment from 20 May 1990 to 27 May 1990*. Strasbourg: Council of Europe.

—— (1991c). *Comments of the Austrian Government on the Report by the European Committee for the Prevention of Torture and Inhuman or Degrading Treatment on its Visit to Austria from 20 May 1990 to 27 May 1990*. Strasbourg: Council of Europe.

—— (1991d). *Report to the United Kingdom Government on the Visit to the United Kingdom Carried Out by the European Committee for the Prevention of Torture and Inhuman or Degrading Treatment or Punishment from 29 July 1990 to 10 August 1990*. Strasbourg: Council of Europe.

—— (1991e). *Response of the United Kingdom Government to the Report by the European Committee for the Prevention of Torture and Inhuman or Degrading Treatment or Punishment on its Visit to the United Kingdom from 29 July 1990 to 10 August 1990*. Strasbourg: Council of Europe.

—— (1991f). *Report to the Danish Government on the Visit to Denmark Carried Out by the European Committee for the Prevention of Torture and Inhuman or Degrading Treatment or Punishment from 2 to 8 December 1990*. Strasbourg: Council of Europe.

—— (1992a). *European Committee for the Prevention of Torture and Inhuman or Degrading Treatment of Punishment, 2nd General Report on the CPT's Activities Covering the Period 1 January to 31 December 1991*. Strasbourg: Council of Europe.

—— (1992b). *Report to the Maltese Government on the Visit to Malta Carried Out by the European Committee for the Prevention of Torture and Inhuman or Degrading Treatment or Punishment from 1 to 9 July 1990*. Strasbourg: Council of Europe.

—— (1992c). *Report to the Swedish Government on the Visit to Sweden Carried Out by the European Committee for the Prevention of Torture and Inhuman or Degrading Treatment or Punishment from 5 to 14 May 1991*. Strasbourg: Council of Europe.

—— (1992d). *Response of the Swedish Government to the Report by the European Committee for the Prevention of Torture and Inhuman or Degrading Treatment or Punishment on its Visit to Sweden from 5 to 14 May 1991*. Strasbourg: Council of Europe.

—— (1992e). *European Committee for the Prevention of Torture and Inhuman or Degrading Treatment or Punishment: Public Statement on Turkey*. Strasbourg: Council of Europe.

COUVRAT, P. (1985). 'Les Recours contre les décisions du Juge de l'Application des Peines' (Recourse Against Decisions of the JAP), *Revue de science criminelle et de droit pénale comparé*, 1.

CRESSEY, D. R. (1959). 'Contradictory Directives in Complex Organizations: The Case of the Prison', *Administrative Science Quarterly*, 4.

—— (ed.) (1961). *The Prison: Studies in Institutional Organization and Change*. New York: Holt, Rinehart, and Winston.

CRESSEY, D. R. (1965). 'Prison Organization', in J. March (ed.), *Handbook of Organizations*. Chicago: McNally.

DAVID, R. and BRIERLY, J. (1985). *Major Legal Systems in the World Today*. London: Stevens.

DAY, P. and KLEIN, R. (1987). *Accountabilities: Five Public Services*. London: Tavistock.

DELOITTE HASKINS-SELLS (1989). *A Report to the Home Office on the Practicality of Private Sector Involvement in the Remand System*. Unpublished report.

DELTEIL, G. (1990). *Prisons: la marmite infernale* (Prisons: The Infernal Stewpot). Paris: Syros Alternatives.

DIIULIO, J. J. (1987). *Governing Prisons: A Comparative Study of Correctional Management*. New York: Free Press.

DITCHFIELD, J. (1990). *Control in Prisons: A Review of the Literature*. Home Office Research Study 118. London: HMSO.

—— and AUSTIN, C. (1986). *Grievance Procedures in Prisons*. Home Office Research Study 91. London: HMSO.

DONAHUE, J. D. (1989). *The Privatization Decision: Public Ends, Private Means*. New York: Basic Books.

DORLHAC DE BORNE, H. (1984). *Changer la prison* (To Change the Prison). Paris: Plon.

DUNBAR, I. (1985). *A Sense of Direction*. London: Home Office.

DÜNKEL, F. (1987). *Die Herausforderung der geburtenschwachen Jahrgänge: Aspecte der Kosten-Nutzen-Analyse in der Kriminalpolitik*. Freiburg im Br.: Max Planck Institute.

—— (1993). 'Imprisonment in Transition: The Situation in the New States of the Federal Republic of Germany', paper presented to the Eleventh International Congress of Criminology, Budapest, Aug.

—— (forthcoming). 'Germany: Recent Developments', in J. Vagg, (ed.), *Prevention and Punishment: Dangerousness, Long-Term Prisoners and Life Imprisonment, an International Perspective*. Chicago: Univ. of Illinois Office of International Criminal Justice.

—— and ROSNER, A. (1982). *Die Entwicklung des Strafvollzugs in der Bundesrepublik Deutschland seit 1970* (The Development of Prisons in the Federal Republic of Germany since 1970). Freiburg im Br.: Max Planck Institute.

—— and RÖSSNER, D. (1991). 'Federal Republic of Germany', in D. van Zyl Smit and F. Dünkel (eds.), *Imprisonment Today and Tomorrow: International Perspectives on Prisoners' Rights and Prison Conditions*. Deventer: Kluwer.

EICHENTHAL, D. and JACOBS, J. B. (1991). 'Enforcing the Criminal Law in State Prisons', *Justice Quarterly*, 8/3.

ERATT, S. and NEUDEK, K. (forthcoming). 'The Life Sentence Prisoner', in J. Vagg, J. (ed.), *Prevention and Punishment: Dangerousness, Long-Term*

Prisoners and Life Imprisonment, an International Perspective. Chicago: Univ. of Illinois Office of International Criminal Justice.

ERKELENS, L. H. (1984). 'Main Characteristics and Some Special Topics of the Drug Policy of the Dutch Prison Administration', paper presented to the Eighth World Conference of Therapeutic Communities, Rome, Sept.

—— and VAN DER WORP, R. A. (1983). *Regelovertreders in Bankenbosch: Een Dossier-onderzoek de Samenhang Tussen het Vóórkomen van Incidenten, Strafduur en Detentiefase.* The Hague: Ministerie van Justitie.

ETZIONI, A. (1967). *A Comparative Analysis of Complex Organizations.* New York: Free Press.

—— (ed.) (1969). *Readings on Modern Organizations.* Englewood Cliffs, NJ: Prentice-Hall.

EVANS, M., and MORGAN, R. (1992). 'The European Convention of the Prevention of Torture: Operational Practice', *International and Comparative Law Quarterly*, 41.

FAGART, T. (1982). 'Une situation bloquée' (A Blocked Situation), *Déviance et société*, 5/2.

FAUGERON, C. (1991). 'France', in D. van Zyl Smit and F. Dünkel (eds.), *Imprisonment Today and Tomorrow: International Perspectives on Prisoners' Rights and Prison Conditions.* Deventer: Kluwer.

FAVARD, J. (1981). *Le Labyrinthe pénitentiaire* (The Penitentiary Labyrinth). Paris: Centurion.

—— (1987). *Des prisons* (Prisons). Paris: Gallimard.

FAVOUREU, L. (1987). 'Approche constitutionelle du principe de la participation du secteur privé au fonctionnement du service publique pénitentiaire' (A Consitutional Approach of Principle to Private-Sector Participation in the Functioning of the Public Prison Service), in F. Boulin (ed.), *Les Prisons dites 'privées'* ('Private' Prisons). Paris: Economica/Presses Universitaires d'Aix-Marseille.

FAWCETT, J. (1985). 'Applications of the European Convention on Human Rights', in Maguire *et al.* (1985).

FEEST, J. (1982). *Imprisonment and the Criminal Justice System in the Federal Republic of Germany.* Arbeitspapiere des Forschungsschwerpunktes soziale Probleme: Kontrolle und Kompensation No. 8. Bremen: Universität Bremen.

—— (1988). 'Prisoners' Protest in Germany: Legislation, Implementation and Evaluation', paper presented to conference on Aspects of European Criminal and Penal Policy, Bristol, Jan.

FITZGERALD, E. (1985). 'Prison Discipline and the Courts', in Maguire *et al.* (1985).

FIZE, M. (1984). *L'Isolement cellulaire dans les établissements pénitentiaires français* (Cellular Isolation in French Penal Establishments). Service des Études, de la Documentation et des Statistiques, Travaux et Documents

29, 2 vols. Paris: Ministère de la Justice/Direction de l'Administration Pénitentiaire.

FLANAGAN, T. J. (1982). 'Discretion in the Prison Justice System: A Study of Sentencing in Institutional Disciplinary Proceedings', *Journal of Research in Crime and Delinquency* (July), 216–37.

—— (1983). 'Correlates of Institutional Misconduct Among State Prisoners', *Criminology*, 21/1, 29–40.

FOUCAULT, M. (1979). *Discipline and Punish: The Birth of the Prison*. Harmondsworth: Penguin.

FRANKE, H. (1990). *Twee Eeuwen Gevangen* (Two Centuries of Imprisonment). Utrecht: Spectrum.

GALTUNG, J. (1961). 'Prison: The Organization of Dilemma', in D. R. Cressey, (ed.), *The Prison*. New York: Holt, Rinehart, and Winston.

GARFINKEL, H. (1984). *Studies in Ethnomethodology*. Cambridge: Polity.

GARLAND, D. (1985). *Punishment and Welfare: A History of Penal Strategies*. Aldershot: Gower.

GAROFALO, J., and CLARK, R. D. (1985). 'The Inmate Subculture in Jails', *Criminal Justice and Behavior*, 12/4, 415–34.

GARRETT, J. (1980). *Managing the Civil Service*. London: Heinemann.

GENDERS, E., and PLAYER, E. (1989). *Race Relations in Prisons*. Oxford: Clarendon Press.

GIDDENS, A. (1985). *The Nation-State and Violence*. Cambridge: Polity Press.

—— (1986). *The Constitution of Society*. Cambridge: Polity Press.

GOSTIN, L. and STAUNTON, M. (1985). 'The Case for Prison Standards: Conditions of Confinement, Segregation, and Medical Treatment', in Maguire *et al.* (1985).

GRAHAM, J. (1987). 'The Declining Prison Population in the Federal Republic of Germany', *HORPU Research Bulletin* 24. London: Home Office Research and Planning Unit.

GRAPENDAAL, M. (1990). 'The Inmate Subculture in Dutch Prisons', *British Journal of Criminology*, 30/3, 341–57.

GRAMSCI, A. (1971). *Selections from Prison Notebooks*. London: Lawrence and Wishart.

HALL, S., CRITCHER, C., JEFFERSON, T., CLARKE, J., and ROBERTS, H. (1978). *Policing the Crisis: Mugging, the State, and Law and Order*. London: Macmillan.

HALL-WILLIAMS, J. E. (1975). *Changing Prisons*. London: Peter Owen.

—— (1984). 'The Need for a Prison Ombudsman', *Criminal Law Review*, 87.

HEINZ, W. (1988). 'The Problems of Imprisonment, Including Strategies that Might be Employed to Minimize the Use of Custody', paper presented to the European Colloquium on Research on Crime and Criminal Policy in Europe, Oxford, July.

HIRST, J. (1990). 'Little to Cheer in New Complaints System', *Prison Report* 3. London: Prison Reform Trust.

HM CHIEF INSPECTOR OF PRISONS (1987a). *Report of an Inquiry by Her Majesty's Chief Inspector of Prisons for England and Wales into the Disturbances in Prison Service Establishments in England Between 29 April and 2 May 1986.* London: HMSO.

—— (1987b). *A Review of Prisoners' Complaints.* London: Home Office.

—— (1989a). *Prison Sanitation: Proposals for the End of Slopping-out.* London: Home Office.

—— (1989b). *Annual Report 1988.* London: HMSO.

—— (1989c). *HM Remand Centre Risley.* London: Home Office.

—— (1990a). *HM Prison Armley (Leeds).* London: Home Office.

—— (1990b). *HM Prison Brixton.* London: Home Office.

—— (1990c). *HM Prison Haslar.* London: Home Office.

—— (1991). *HM Young Offender Institution Eastwood Park.* London: Home Office.

—— (1992a). *HM Prison Canterbury.* London: Home Office.

—— (1992b). *HM Prison Parkhurst.* London: Home Office.

HM PRISON SERVICE (1984). *Current Recommended Standards for the Design of New Prison Establishments.* London: HM Prison Service.

—— (1989). *An Improved System of Grievance Procedures for Prisoners' Complaints and Requests: A Report by a Working Group.* London: Home Office.

—— (1992). *An Independent Complaints Adjudicator for Prisons: A Consultation Paper.* London: HM Prison Service.

HOFFMAN, H. (1990). Isolation in Normalvollzug (Isolation in normal prisons). Pfaffenweiler: Centaurus.

HOME OFFICE (1966). *Report of the Inquiry Into Prison Escapes and Security.* Cmnd 3175. London: HMSO.

—— (1977). *Report of an Inquiry by the Chief Inspector of the Prison Service into the Cause and Circumstance of the Events at HM Prison Hull During the Period 31 August to 3 September 1976* (the Fowler Report). London: HMSO.

—— (1979). *Committee of Inquiry into the UK Prison Services* (the May Report). Cmnd 7673. London: HMSO.

—— (1984a). *Criminal Justice: A Working Paper.* London: HMSO.

—— (1984b). *Managing the Long-Term Prison System: The Report of the Control Review Committee.* London: HMSO.

—— (1985). *Report of the Departmental Committee on the Prison Disciplinary System* (2 vols.; the Prior Report). London: HMSO.

—— (1986). *The Prison Disciplinary System in England and Wales.* Cmnd 9920. London: HMSO.

—— (1991). *Prison Disturbances, April 1990: Report of an Inquiry by The Rt. Hon. Lord Justice Woolf (Parts I and II) and His Honour Judge*

Stephen Tumim, pt. 2 (the Woolf Report). Cmnd 1456. London: HMSO.

—— /PA MANAGEMENT (1986). *Study of Prison Officers' Complementing and Shift Systems, vol. i: Report.* London: HM Prison Service.

HOWARD, J. (1929, first published 1777). *The State of the Prisons.* 3rd edn. (abr.). London: Dent.

HUBER, B. (1978). 'Safeguarding of Prisoners' Rights Under the New West German Prison Act', *South African Journal of Criminal Law and Criminology*, 2/3, 229–38.

IRWIN, J. (1970). *The Felon.* Englewood Cliffs, NJ: Prentice-Hall.

—— (1985). *The Jail: Managing the Underclass in American Society.* Berkeley, Calif.: Univ. of California Press.

JACOBS, J. B. (1977). *Stateville: The Penitentiary in Mass Society.* Chicago: Univ. of Chicago Press.

—— (1980). 'The Prisoners' Rights Movement and its Impacts, 1960–1980', in N. Morris and M. Tonry (eds.), *Crime and Justice: An Annual Review of Research, ii.* Chicago: Univ. of Chicago Press.

—— and BERKOWITZ, L. (1983). 'Reflection on the Defeat of New York State's Prison Bond', in J. B. Jacobs, *New Perspectives on Prisons and Imprisonment.* Ithaca, NY: Cornell Univ. Press.

JANKOWSKI, L. J. (1992). *Correctional Populations in the United States, 1990.* Washington, DC: US Dept. of Justice, Bureau of Justice Statistics.

JEFFERSON, T. and GRIMSHAW, R. (1984). *Controlling the Constable: Police Accountability in England and Wales.* London: Frederick Muller/Cobden Trust.

JONES, C. A. G., (1993). 'Auditing Criminal Justice', *British Journal of Criminology*, 33.

Justice (1981). 'Le Juge de l'Application des Peines: peau de chagrin' (The JAP: Skin of Sorrow), anonymous article, *Justice*, 84.

—— (1982). 'Mouvement des prisonniers: une nouvelle donnée' (The Prisoners' Movement: New Fact), anonymous article, *Justice*, 92.

KAISER, G., KERNER, H-J., and SCHÖCH, H. (1983). *Strafvollzug* (Prison). Heidelberg: C. F. Müller.

KALINICH, D. B., and STOJKOVIC, S. (1985). 'Contraband: The Basis for Legitimate Power in a Prison Social System', *Criminal Justice and Behavior*, 12/4, 435–51.

KELK, C. (1983). 'The Humanity of the Dutch Prison System and the Prisoners' Consciousness of Their Legal Rights', *Contemporary Crises*, 7, 155–70.

—— (1991). 'The Netherlands', in D. van Zyl Smit and F. Dünkel (eds.), *Imprisonment Today and Tomorrow.* Deventer: Kluwer.

—— (1992). 'Dangerous Offenders and Dutch Criminal Law', paper presented to an international seminar on dangerous and long-term offenders, Prague, April.

—— (forthcoming). 'The Netherlands', in F. Dünkel and J. Vagg, (eds.), *Waiting For Trial: International Perspectives on the Use of Pre-Trial Detention and the Rights and Living Conditions of Prisoners Waiting for Trial*. Freiburg im Br: Max Planck Institute.

—— and DE JONGE, G. (1982). 'Le Droit de plainte au Pays Bas' ('The Right of Complaint in the Netherlands), *Déviance et Société*, 6/4, 391–6.

KING, R. D. and ELLIOTT, K. W. (1977). *Albany: Birth of a Prison—End of an Era*. London: Routledge and Kegan Paul.

—— and McDERMOTT, K. (1989). 'British Prisons 1970–1987: The Ever-Deepening Crisis', *British Journal of Criminology*, 29/2.

—— and McDERMOTT, K. (1990). '"My Geranium is Subversive": Some Notes on the Management of Trouble in Prisons', *British Journal of Sociology*, 41/4.

KIRKPATRICK, J. (1986). 'Service Objectives and the Corrupting Influence of the Treasury', *Public Money* (Sept.).

LAMBERT, J. L. (1986). *Police Powers and Accountability*. London: Croom Helm.

LANDFRIED, C. (1982). 'Legal Policy and Internal Security', in K. von Beyme and M. G. Schmidt (eds.), *Policy and Politics in the Federal Republic of Germany*. Aldershot: Gower.

LILLY, J. R., and KNEPPER, P. (1991). 'Towards an International Perspective on Privatization in Corrections', paper presented to the British Criminology Conference, York, July.

LINDBLOM, C. E. (1969). 'The Science of "Muddling through"', in A. Etzioni (ed.), *Readings on Modern Organizations*. Englewood Cliffs, NJ: Prentice-Hall.

LOGAN, C. H. and McGRIFF, B. W. (1989). 'Comparing Costs of Public and Private Prisons: A Case Study', *NIJ Reports*, 216.

LOMBARDO, L. X. (1981). *Guards Imprisoned: Correctional Officers at Work*. New York: Elsevier.

LOS, M., and ANDERSON, P. (1976). 'The "Second Life": A Cross-cultural View of Peer Subcultures in Correctional Institutions in Poland and the United States', *Polish Sociological Bulletin*, 4.

LYGO, R. (1991). *Management of the Prison Service*. London: HM Prison Service.

LYNXWILER, J., SHOVER, N., and CLELLAND, D. (1983). 'The Organization and Impact of Inspector Discretion in a Regulatory Bureaucracy', *Social Problems*, 30/4.

McCLEERY, R. H. (1961). 'The Governmental Process and Informal Social Control', in Cressey (1961).

McCONVILLE, S. (1981). *A History of English Prison Administration, 1750–1877*. London: Routledge and Kegan Paul.

McDERMOTT, K., and KING, R. D. (1988). 'Mind Games: Where the Action is in Prisons', *British Journal of Criminology*, 28/3.

McDermott, K., and King, R. D. (1989). 'A Fresh Start: The Enhancement of Prison Regimes', *Howard Journal of Criminal Justice*, 28/3.

Maguire, M. (1985). 'Prisoners' Grievances: The Role of the Boards of Visitors', in Maguire *et al.* (1985).

—— Pinter, F., and Collis, C. (1984). 'Dangerousness and the Tariff', *British Journal of Criminology*, 24/3, 250–68.

—— and Vagg, J. (1984). *The Watchdog Role of Boards of Visitors.* London: Home Office.

—— —— and Morgan, R. (eds.) (1985). *Accountability and Prisons: Opening Up a Closed World.* London: Tavistock.

Mangold, T. (1985). 'Profiting From Crime', *Listener*, 19 Sept.

March, J. G., and Simon, H. A. (1984). 'The Dysfunctions of Bureaucracy', in D. S. Pugh (ed.), *Organization Theory.* Harmondsworth: Penguin.

Marshall, G. (1984). *Constitutional Conventions: The Rules and Forms of Political Accountability.* Oxford: Clarendon Press.

Martin, J. P. (1975). *Boards of Visitors of Penal Institutions* (the Jellicoe Report). London: Barry Rose.

Matthews, R. (ed.) (1989). *Privatizing Criminal Justice.* London: Sage.

Meijboom, L. (1987). 'Drug Users in Remand Centers', in M. J. M. Brand-Koolen (ed.), *Studies on the Dutch Prison System.* Amstelveen: Kugler.

Melossi, D., and Pavarini, M. (1981). *The Prison and the Factory: Origins of the Penitentiary System.* London: Macmillan.

Menard, M., and Meurs, D. (1984). *Les Aggressions commises par les détenus contre les membres du personnel dans les établissements pénitentiaires de la metropole* (Assaults Committed by Inmates Against Staff in Penal Establishments in France). Service des Études, de la Documentation et des Statistiques, Travaux et Documents 26. Paris: Ministère de la Justice/Direction de l'Administration Pénitentiaire.

Mestre, J.-L. (1987). 'Historique du recours a "l'initiative privée" en matière pénitentiaire' (A History of Recourse to the 'Private Initiative' in Prison Matters), in Boulan (1987).

Metcalfe, L., and Richards, S. (1987). 'The Efficiency Strategy in Central Government: An Impoverished Concept of Management', *Public Money* (June).

Ministère de la Justice (1986). *Rapport général sur l'exercice 1986* (Annual Report of the Prison Administration 1986). Paris: Ministère de la Justice/Direction de l'Administration Pénitentiaire.

—— (1987). *Rapport Général sur l'Exercice 1987* (Annual Report of the Prison Administration 1987). Paris: Ministère de la Justice/Direction de l'Administration Pénitentiaire.

Ministerie van Justitie (1980). *Rights, Restrictions and Opportunities for Prisoners in a House of Detention.* Pamphlet. The Hague: Ministerie van Justitie.

MORGAN, R. (1983). 'How Resources Are Used in the Prison System', in NACRO (ed.), *A Prison System for the 80s and Beyond: The Noel Buxton Lectures 1983–1983*. London: NACRO.

—— (1985). 'Her Majesty's Inspectorate of Prisons', in Maguire *et al.* (1985).

—— (1991). 'Being Part of an Inquiry: Research and Policy-making', *Socio-Legal Newsletter* (spring).

—— and BRONSTEIN, A. (1985). 'Prisoners and the Courts: The US Experience', in Maguire *et al.* (1985).

—— and MACFARLANE, A. (1985). *After Tarrant: Advice for Boards of Visitors Regarding Assistance for Prisoners at Adjudications*, 2nd edn. London: Association of Members of Boards of Visitors.

—— and MAGGS, C. (1985). *Setting the PACE: Police Community Consultation Arrangements in England and Wales*. Bath Social Policy Papers 4. Bath: Univ. of Bath Centre for the Analysis of Social Policy.

—— MAGUIRE, M., and VAGG, J. (1985). 'Overhauling the Prison Disciplinary System: Notes for Readers of the Prior Committee's Report', in Maguire *et al.* (1985).

—— and RICHARDSON, G. (1987). 'Civil Liberties, the Law and the Long-term Prisoner', in A. E. Bottoms and R. Light (eds.), *Problems of Long-Term Imprisonment*. Aldershot: Gower.

NEALE, K. (1991). 'The European Prison Rules: Contextual, Philosophical and Practical Aspects', in J. Muncie and R. Sparks (eds.), *Imprisonment: European Perspectives*. London: Harvester Wheatsheaf/Open Univ.

NEVILLE BROWN, L., and GARNER, J. (1983). *French Administrative Law*, 3rd edn. London: Butterworths.

NIJBOER, J. A., and PLOEG, G. J. (1985). 'Grievance Procedures in The Netherlands', in Maguire *et al.* (1985).

NORDRHEIN-WESTFALEN, MINISTER OF JUSTICE (1984). *Strafvollzug in Nordrhein-Westfalen* (Prisons in Nordrhein-Westphalia). Dusseldorf: Justizminister des Landes Nordrhein-Westfalen.

NORMANTON, E. L. (1966). *The Accountability and Audit of Governments: A Comparative Study*. Manchester: Manchester Univ. Press.

PARSONS, T. (1952). *The Social System*. Glencoe, Ill.: Free Press.

—— and SHILS, E. (eds.) (1962). *Toward a General Theory of Action*. New York: Harper and Row.

PAUCHET, C. (1982). *Les Prisons de l'insécurité* (The Prisons of Insecurity). Paris: Editions Ouvrières.

PIERRE, M. (1991). 'Les Prisons de la IIIᵉ République (1875–1938)' (The Prisons of the 3rd Republic, 1875–1938), in J.-G. Petit, N. Castan, C. Faugeron, M. Pierre, and A. Zysberg, *Histoire des Galères, Bagnes et Prisons* (A History of Galleys, Labour Camps and Prisons). Paris: Privat.

PRISON OFFICERS' ASSOCIATION (1984). *The Prison Disciplinary System: Submission to the Home Office Departmental Committee on the Prison Disciplinary System*. London: POA.

PRISON REFORM TRUST (1992). *Comments on the Report by Admiral Sir Raymond Lygo: Management of the Prison Service*. London: PRT.

PUGH, D. S. (ed.) (1984). *Organization Theory*. Harmondsworth: Penguin.

QUINN, P. (1985). 'Prison Management and Prison Discipline: A Case Study of Change', in Maguire *et al*. (1985).

REINER, R. (1985). *The Politics of the Police*. Brighton: Wheatsheaf.

REYNAUD, A. (1986). *Human Rights in Prisons*. Strasbourg: Council of Europe.

RHODES, G. (1982). *Inspectorates in British Government: Law Enforcement and Standards of Efficiency*. London: Royal Institute of Public Administration.

RICHARDSON, G. (1984). 'Time to Take Prisoners' Rights Seriously', *Journal of Law and Society*, 11.

—— (1985). 'The Case for Prisoners' Rights', in Maguire *et al*. (1985).

RYAN, M., and WARD, T. (1989). *Privatization and the Penal System: The American Experience and the Debate in Britain*. Milton Keynes: Open Univ. Press.

SALAMAN, G., and THOMPSON, K. (eds.) (1973). *People and Organizations*. London: Longman.

SCHICK, H. (1986). 'Everyday Life of Male Prisoners', in B. Rolston and M. Tomlinson, (eds.), *The Expansion of European Prison Systems*. Working Papers in European Criminology 7. Belfast: European Group for the Study of Deviance and Social Control.

SCULL, A. (1984). *Decarceration: Community Treatment and the Deviant—a Radical View*, 2nd edn. Cambridge: Polity.

SEYLER, M. (n.d.). *L'Application de la réforme de 1975* (The Application of the 1975 Reform). Paris: Ministre de la Justice/Centre National d'Études et de Recherches Pénitentiaires.

SHÄFER, K. H. (1987). *Anstaltsbeiräte: die institutionalisierte Öffentlichkeit? Eine empirische Untersuchung über die Tätigkeit der Anstaltsbeiräte an den hessischen Vollzugsanstalten* (Institutional Boards: Institutionalized Openness? An Empirical Study of the Work of the Institutional Boards in Prisons in Hessen). Heidelberg: C. H Müller.

SHAW, S. (1990). 'Two Cheers for New Grievance Procedure', *Prison Report* 3. London: Prison Reform Trust.

STEFANI, G., LEVASSEUR, G., and JAMBU-MERLIN, R. (1982). *Criminologie et science pénitentiaire* (Criminology and Penal Science). Paris: Dalloz.

STEPHAN, J. (1992). *Census of State and Federal Correctional Facilities, 1990*. Washington, DC: US Dept. of Justice, Bureau of Justice Statistics.

STERN, V. (1987). *Bricks of Shame: Britain's Prisons*. Harmondsworth: Penguin.

SYKES, G. (1958). *The Society of Captives: A Study of a Maximum Security Prison*. Princeton, NJ: Princeton Univ. Press.

TANCA, A. (1990). 'Human Rights, Terrorism and Police Custody: The Brogan Case', *European Journal of International Law*, 1(1/2).

TAYLOR, L. (1980). 'Bringing Power to Particular Account: Peter Rajah and the Hull Board of Visitors', in P. Carlen and M. Collison (eds.), *Radical Criminology*. Oxford: Martin Robertson.

THOMAS, J. E. (1972). *The English Prison Officer Since 1850*. London: Routledge and Kegan Paul.

TIRYAKIAN, E. (1970). 'Structural Sociology', in J. McKinney and E. Tiryakian (eds.), *Theoretical Sociology*. New York: Appleton Century Crofts.

TOURAINE, A., WIEVIORKA, M., and DUBET, F. (1987). *The Workers' Movement*. Cambridge: Cambridge Univ. Press.

TREVERTON-JONES, G. D. (1989). *Imprisonment: The Legal Status and Rights of Prisoners*. London: Sweet and Maxwell.

UNITED NATIONS (1984). *Standard Minimum Rules for the Treatment of Prisoners and Procedures for the Effective Implementation of the Rules*. New York: UN Dept. of Public Information.

—— (1988). *Code of Conduct for Law Enforcement Officials*, 3rd edn. New York: UN Dept. of Public Information.

US DEPARTMENT OF JUSTICE (1980). *Federal Standards for Prisons and Jails*. Washington, DC: US Dept. of Justice.

USEEM, B., and KIMBALL, P. (1989). *States of Siege: US Prison Riots, 1971–1986*. New York: Oxford Univ. Press.

VAGG, J. (1985). 'Independent Inspection: the Role of the Boards of Visitors', in Maguire *et al.* (1985).

—— (1991). 'Correcting Manifest Wrongs? Prison Grievance and Inspection Procedures in England and Wales, France, Germany and the Netherlands', in J. Muncie and R. Sparks (eds.), *Imprisonment: European Perspectives*. Brighton: Harvester Wheatsheaf.

VAN DER LINDEN, B. (1987). 'The Impact of Medium-Term Incarceration: A Comparative Exploration, in M. J. M. Brand-Koolen (ed.), *Studies on the Dutch Prison System*. Amstelveen: Kugler.

VINSON, T., BROUWERS, M. and SAMPIEMON, M. (1985). *Impressions of the Dutch Prison System*. The Hague: Ministry of Justice Research and Documentation Centre.

—— —— —— (1987). 'Management Strategies and Tactics', in M. J. M. Brand-Koolen (ed.), *Studies on the Dutch Prison System*. Amstelveen: Kugler.

WEBER, M. (1970). 'Bureaucracy', in H. H. Gerth and C. W. Mills, *From Max Weber*. London: Routledge and Kegan Paul.

WEIS, C. (1991). 'German Democratic Republic', in D. van Zyl Smit and F. Dünkel (eds.), *Imprisonment Today and Tomorrow: International Perspectives on Prisoners' Rights and Prison Conditions*. Deventer: Kluwer.

WEISS, R. P. (1989). 'Private Prisons and the State', in Matthews (1989).

WRIGHT, M. (1985). 'Mediation in Prisons', *Prison Service Journal* (Jan.), 17–20.

WU, H. H. D. (1992). *Laogai—The Chinese Gulag*. Boulder, Colo.: Westview Press.

YBEMA, S. B., and WESSEL, J. (1978). 'Redress of Grievances Against Administrative Action', in D. C. Fokkema, J. M. J. Chorus, E. H. Hondius, and E. Ch. Lisser. *Introduction to Dutch Law for Foreign Lawyers*. Deventer: Kluwer.

ZELLICK, G. (1981). 'The Prison Rules and the Courts', *Criminal Law Review*, 602.

Cases Cited

English cases

European Commission and Court of Human Rights

French cases

US cases

Name Index

Subject Index

Juge d'instruction (France) 167 n.,
    177
Juge de l'Application des Peines
    (France) 34, 37–8, 103, 129, 142,
    166–9 and n., 177, 222 n., 227–8,
    234, 242, 261, 320
'justice' approach to prisons 112–13

King 232 n.
Kiss (ECHR case) 233
Knechtl (PCA case) 39, 187

La Santé prison (France) 122 n., 169,
    264
language, and accountability 11, 139,
    141, 142 n., 266–7
'lay-downs', *see* transfers
League of Nations 205
Leech 171 n.
Leeds prison (England) 201, 203, 283
legislation, primary, of prisons 27,
    29–31, 33, 56–7, 170, 216 n.; *see
    also* Code de Procedure Pénale,
    Strafvollzugsgesetz
legitimacy, concept of 318–21, 324
'less eligibility' 143–4, 161–2
library, prison 173
life sentences, discretionary 172 n.
local prisons (England), functions of 23
Lockerungsmassnahmen (Germany), *see*
    relaxation measures
Lygo Report 303–5

Maastricht prison (the Netherlands) 25
McKinsey, management consultants 43,
    124, 322
McKonkey 232 n.
Maisons centrales (France), functions of
    24
Maisons d'arrêt (France), functions of
    24
management, objectives 110, 224–5; and
    administration 115; models of, 2, 9
    and n., 111–13, 129–30, 136, 138,
    141–3, 146–7, 149–51, 316;
    structures of 21–3
Manterola 168–9
Marseille prison (France) 180
May Committee (England) 43, 50, 321
Maze prison (N. Ireland) 250
media 249, 265–7
mediation 181, 183–4
medical records, inmate 49 n.

medicine, in prison, and inspection
    200–1
Melun prison (France) 68 and n., 249
Ministers, roles of 22, 169, 177; *see also*
    Home Secretary

Nahar 144 n.
Nancy prison (France) 246, 260, 261,
    262
Napoleonic Code (France) 6, 30 n.,
    194–5, 213–14, 320
National Association for the Care and
    Rehabilitation of Offenders
    (NACRO) (England) 143 n., 187 n.
National Council for Civil Liberties
    (England) 156
National Health Service (England) 299
National Observation Centre (Fresnes
    prison, France) 24
National Sheriff's Association (United
    States) 296
'natural justice', concept of 137, 171,
    217
negotiation, between prisoners and staff
    76; *see also* riots, 'shared powers'
    management model
Neuvic-sur-Isle prison (France) 307
New York City Board of Correction
    209
Nîmes prison (France) 262
Nordrhein-Westfalen (Germany) 22, 37
Northeye prison (England) 246

Oberlandsgericht (Germany) 41, 165
Ombudsman 39, 175 n., 177, 187–8
Over-Amstel institution (the
    Netherlands) 25, 73 n., 127
overcrowding 16–17, 23 n., 28, 160,
    170, 186 and n., 295; and prison
    regimes 29, 44; and remand
    prisoners 18
oversight, of prisons, arrangements for
    36–41; *see also* Boards of Visitors,
    Commissie van Toezicht,
    Inspection, Juge de l'Application
    des Peines

PA Management Consultants 52
'pains of imprisonment' 63
Parkhurst prison (England) 201
Parliamentary Commissioner for
    Administration (England), *see*
    Ombudsman